AF804214

Engineering Electromagnetics

Applications

Engineering Electromagnetics
Applications

edited by
Rajeev Bansal

Taylor & Francis
Taylor & Francis Group
Boca Raton London New York

CRC is an imprint of the Taylor & Francis Group,
an informa business

The material was previously published in *The Handbook of Engineering Electromagnetics* © Taylor & Francis 2004.

Published in 2006 by
CRC Press
Taylor & Francis Group
6000 Broken Sound Parkway NW, Suite 300
Boca Raton, FL 33487-2742

© 2006 by Taylor & Francis Group, LLC
CRC Press is an imprint of Taylor & Francis Group

No claim to original U.S. Government works
Printed in the United States of America on acid-free paper
10 9 8 7 6 5 4 3 2 1

International Standard Book Number-10: 0-8493-7363-8 (Hardcover)
International Standard Book Number-13: 978-0-8493-7363-3 (Hardcover)
Library of Congress Card Number 2006040452

This book contains information obtained from authentic and highly regarded sources. Reprinted material is quoted with permission, and sources are indicated. A wide variety of references are listed. Reasonable efforts have been made to publish reliable data and information, but the author and the publisher cannot assume responsibility for the validity of all materials or for the consequences of their use.

No part of this book may be reprinted, reproduced, transmitted, or utilized in any form by any electronic, mechanical, or other means, now known or hereafter invented, including photocopying, microfilming, and recording, or in any information storage or retrieval system, without written permission from the publishers.

For permission to photocopy or use material electronically from this work, please access www.copyright.com (http://www.copyright.com/) or contact the Copyright Clearance Center, Inc. (CCC) 222 Rosewood Drive, Danvers, MA 01923, 978-750-8400. CCC is a not-for-profit organization that provides licenses and registration for a variety of users. For organizations that have been granted a photocopy license by the CCC, a separate system of payment has been arranged.

Trademark Notice: Product or corporate names may be trademarks or registered trademarks, and are used only for identification and explanation without intent to infringe.

Library of Congress Cataloging-in-Publication Data

Engineering electromagnetics : applications / edited by Rajeev Bansal.
p. cm.
Includes bibliographical references and index.
ISBN 0-8493-7363-8 (alk. paper)
1. Electromagnetism. I. Bansal, Rajeev.

QC760.E54 2006
621.3--dc22 2006040452

informa
Taylor & Francis Group
is the Academic Division of Informa plc.

Visit the Taylor & Francis Web site at
http://www.taylorandfrancis.com

and the CRC Press Web site at
http://www.crcpress.com

To the memory of my parents

Preface

AIM

This volume, derived from the *Handbook of Engineering Electromagnetics* (2004), is intended as a desk reference for the broad area of engineering electromagnetics. Since electromagnetics provides the underpinnings for many technological fields such as wireless communications, fiber optics, microwave engineering, radar, electromagnetic compatibility, material science, and biomedicine, there is a great deal of interest and need for training in the engineering applications of electromagnetics. Practicing engineers in these diverse fields need to understand how electromagnetic principles can be applied to the formulation and solution of actual engineering problems. As technologies wax and wane and engineers move around, they find themselves learning new applications on the run.

Engineering Electromagnetics: Applications should serve as a bridge between standard textbooks in electromagnetic theory and specialized references such as a handbook on wireless antenna design. While textbooks are comprehensive in terms of the theoretical development of the subject matter, they are usually deficient in the application of that theory to practical applications. Specialized handbooks on the other hand often provide detailed lists of formulas, tables, and graphs, but do not provide the insight needed to appreciate the underlying physical concepts. This application-oriented volume will permit a practicing engineer/scientist to:

- **Review** the necessary electromagnetic **theory** in the context of the application he/she is interested in.
- Gain an appreciation for the key electromagnetic **terms and parameters**.
- Learn how to apply the theory to **formulate engineering problems**.
- Obtain **guidance to the specialized literature** for additional details.

SCOPE

Since *Engineering Electromagnetics: Applications* is intended to be useful to engineers engaged in electromagnetic applications in a variety of professional settings, the coverage of topics is correspondingly broad in scope (as can be inferred from the Table of Contents). In terms of *electromagnetic technologies*, radar, wireless communication, satellite communication, and optical communication are covered (Chapters 1–4). Chapter 5 provides an introduction to various numerical techniques being used for computer-aided solutions to complex electromagnetic problems. Given the ubiquitous nature of electromagnetic fields, it is important to consider their biological effects and safety standards (Chapter 6). Chapter 7 presents a concise survey of current and evolving biomedical applications while Chapter 8 is a review of the techniques used for measuring the electromagnetic properties of biological materials. Pertinent *data* in the form of tables and graphs has been provided within the context of the subject matter. In addition, Appendices A and B are brief compilations of important electromagnetic constants and units, respectively. Finally, Appendix C is intended as a convenient tutorial on vector analysis and coordinate systems.

Rajeev Bansal

Contents

Acknowledgments

First and foremost, I would like to thank all the contributors, whose hard work is reflected in the pages of this volume. My editors at Taylor & Francis, specially Mr. Taisuke Soda, have provided valuable help and advice throughout the project. I would like to thank Mr. Anthony Palladino for his help in preparing the manuscript of Appendix C. Finally, I would like to express my gratitude to my family for its unfailing support and encouragement.

Editor

Rajeev Bansal received his Ph.D. in Applied Physics from Harvard University in 1981. Since then he has taught and conducted research in the area of applied electromagnetics at the University of Connecticut where he is currently a professor of electrical engineering. His technical contributions include the edited volume *Handbook of Engineering Electromagnetics* (2004), two coauthored book chapters on submarine antennas (2005) and semiconductor dipole antennas (1986), two patents (1989 and 1993), and over 75 journal/conference papers. Dr. Bansal has served on the editorial boards of *International Journal of RF and Microwave Computer-Aided Engineering, Journal of Electromagnetic Waves and Applications, Radio Science, IEEE Antennas and Propagation Magazine*, and *IEEE Microwave Magazine*. He is a member of the Electromagnetics Academy and the Technical Coordinating Committee of the IEEE Microwave Theory & Techniques Society. He has served as a consultant to the Naval Undersea Warfare Center, Newport, RI.

Contributors

Nathan Blaunstein *Ben-Gurion University of the Negev, Beer Sheva, Israel*

Afshin Daryoush *Drexel University, Philadelphia, Pennsylvania*

Riadh Habash *University of Ottawa, Ottawa, Ontario, Canada*

Randy L. Haupt *Utah State University, Logan, Utah*

James C. Lin *University of Illinois at Chicago, Chicago, Illinois*

Joseph C. Palais *Arizona State University, Tempe, Arizona*

Matthew N. O. Sadiku *Prairie View A&M University, Prairie View, Texas*

Levent Sevgi *Dogus University, Istanbul, Turkey*

Mohammad-Reza Tofighi *Drexel University, Philadelphia, Pennsylvania*

1

Radar

Levent Sevgi
Dogus University
Istanbul, Turkey

Electronic sensors are being used in a variety of applications in our modern life, from security and defense to public health, education to transportation, science to sports. The sensors may be electromagnetic, acoustic, thermal, chemical, biological, etc. A radar (an acronym for *radio detection and ranging*) is commonly used for an electromagnetic sensor. In this chapter, fundamentals of radar are presented. Starting from the historical background, the theory, the signal environment, the radar equation, and applications are outlined.

1.1. INTRODUCTION AND HISTORICAL BACKGROUND

Radar is about using electromagnetic waves to detect the presence of objects and to extract as much information as possible from the interaction of electromagnetic waves with objects. The concept can be traced back to the pioneering studies on radio transmission and reception; to the works of Hertz in 1886, Hulsmeyer in 1903, and Marconi in 1922. Radar development studies accelerated in the United Kingdom, France, Germany, and the United States during 1935–1940 and particularly in the United States during 1940–1945. The period 1950–1960 corresponds to the introduction of new techniques in radar applications, especially coherent techniques, such as Doppler processing and pulse compression. The principles, technology, and applications of radar were publicized by fundamental books, such as those written by Skolnik and Barton (see, for example, Refs. 1–3, their latter editions) during 1960–1970. The solid-state technology, integrated circuits, microprocessors, etc., accelerated its development during 1970–1980, and finally, the period 1980–1990 corresponds to the mature age of the radar theory and technology. A brief historical overview is given in Ref. 4.

As given in the applicable IEEE standard [5], a radar is

> A device for transmitting electromagnetic signals and receiving echoes from objects of interests (targets) within its volume of coverage. Presence of a target is revealed by detection of its echo or its transponder reply. Additional information about a target provided by a radar includes one or more of the following: distance (range), by the elapsed time between transmission of the signal and reception of the return signal; direction, by use of directive antenna patterns; rate of change of range, by measurement of Doppler shift; description or classification of target, by analysis of echoes and their variation with time.

This simple and clear definition of radar shows that

Radar is a device that transmits and receives electromagnetic signals.
There are objects of interest (targets) and noninterest (clutter, interference, jammer).

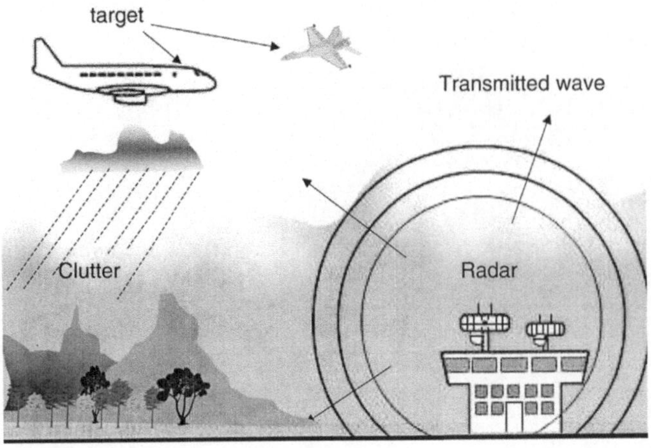

Figure 1.1 A typical radar scenario.

Information is extracted from the echo signal initially by detection.

Target information includes, but not limited to range, range rate (velocity), direction, description, etc.

A typical radar scenario is pictured in Fig. 1.1, where an airport surveillance radar is in operation. The scenario includes two air targets (a fighter and a commercial airplane), mountains and trees on the ground, clouds in the sky, rain, etc. The radar transmitted signal interacts with the environment, and its receiver receives echoes from possible targets, unwanted echoes from mountains, trees, clouds in the sky, rain, etc. The total received echo is a signal, which contains signatures of different components, generally categorized as target, noise, clutter, and interference.

1.2. TERMS AND CONCEPTS

A radar *target* is the object of interest that is embedded in noise and clutter together with interfering signals. *Noise* is a floor signal which limits the smallest signal that can be measured in the receiver. Noise is present in all electronic circuits, although it is often quite small compared with useful signals. *Clutter* is a radar (background) echo or group of echoes from ground, sea, rain, birds, chaff, etc., that is operationally unwanted in the situation being considered. There is no single definition for clutter, and clutter or target may interchange depending on the duty of the radar. For example, an echo from rain is clutter for an airport surveillance radar but is the target for a weather radar. Similarly, ground echo is clutter for a ground surveillance radar but is itself the target (useful signal) for a ground imaging radar.

Radars can be classified according to purpose, application, type, installation, operating frequency, transmit waveform, receiver processing techniques, etc., as illustrated in Table 1.1 Major purposes of a radar may be detection, tracking, classification, identification, surveillance, imaging, or guidance, as listed.

Target detection is the ability to distinguish target at the receiver. The total radar echo signal at the receiver consists of the target (wanted signal), the noise (unwanted, uncorrelated signal), the clutter (unwanted but correlated echoes from unwanted targets), and the interference (unintentional such as radio and TV broadcast signals and/or intentional jamming signals).

Target tracking is the process of following the moving target continuously, i.e., to monitor its range, direction, velocity, etc. Tracking may be done mechanically (i.e., by steering the receiver

Table 1.1 Types of Radars According to Different Parameters

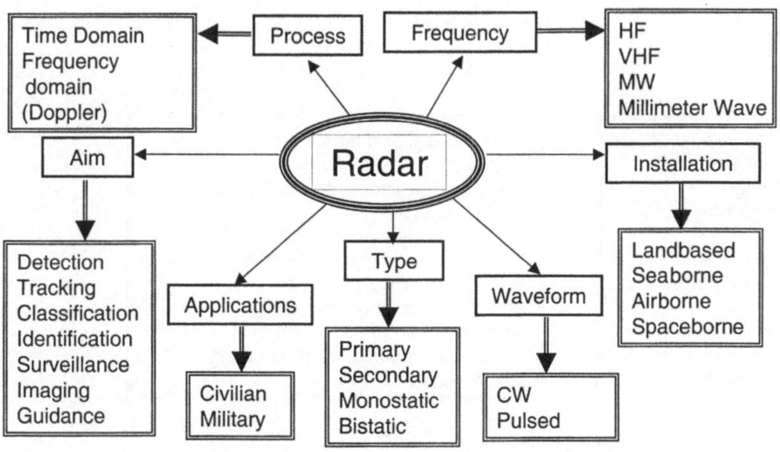

antenna in a way to hold the target inside the receiver beam) or electronically (i.e., by digital beam steering techniques at the receiver processor). It may be single-target tracking or multitarget tracking, where the latter requires target discrimination.

Target classification is to distinguish certain types of targets and group them according to certain characteristics called *features*. For example, to group the detected sea surface targets into frigates, boats, tankers, or air targets into fighters, cargo planes, etc., is the process of classification. Grouping them as military and civilian targets is also a form of classification. Possible distinguishing features may be their size, speed, onboard electronic devices, electromagnetic reflectivity, maneuvers, etc.

Target identification is the process of finding out "who" the target is. This knowledge of a particular radar return signal that is from a specific target may be obtained by determining size, shape, timing, position, maneuvers, or rate of change of any of these parameters by means of coded responses through secondary radar or by electronic counter measures (i.e., by listening to and recording active communication and radar systems onboard of the target).

Imaging radars are microwave (MW) radars, which can provide high resolution in range and cross-range to obtain "radar picture" of surface and air targets and earth's surface. They are range profiling (RP) radars, synthetic aperture radars (SAR), and inverse synthetic aperture radars (ISAR).

An RP radar is a high-resolution active instrument, which has range resolution cell sizes much smaller than typical dimensions of the observed targets so that multireturns from different range cells along the target can be used to have a longitudinal reflectivity profile.

SAR is an airborne or spaceborne active instrument that produces high-resolution imagery of surface targets and earth's surface (ocean and terrain), which achieves its mission by tracing the target via the motion of the platform. The high cross-range resolution is obtained via a synthetic aperture or, equivalently, via Doppler processing. The term *synthetic aperture* refers to the distance the radar travels during data collection for Doppler processing.

ISAR is a land-based and/or airborne active instrument that produces high-resolution imagery of surface and air targets, and it uses the motion of the target as information.

Surveillance is systematic observation of a region (aerospace, surface, or subsurface areas) by a different number of different sensors, primarily for the purpose of detecting, tracking, classifying, and identifying activities of interest. Surveillance may be air to air (A/A), air to ground (A/G), air to surface (A/S), surface to air (S/A), surface to surface (S/S), etc.

Basic radar applications are listed in Table 1.2. They may be grouped in two ways: monostatic/bistatic and primary/secondary. A monostatic radar is a radar where its transmitter and

Table 1.2 Some Applications of Current Radars

Civil Appl.	Military Appl.
Air traffic control and flight management	Land, ocean, and air surveillance
Intelligent traffic management systems	Detection and tracking
Precision approach and landing	Classification and identification ballistic
Vessel traffic management (harbors, waterways, straits)	Missile defense
Navigation and collision avoidance	Airborne early warning
Weather radar and ocean monitoring	Fire control and missile guidance
Search and rescue	Mortar and artillary location
Ground surveillance and intruder alarms	Search and rescue operations
Ground probing and subsurface imaging	Ground probing and subsurface
Vehicle speed sensors and altimeters	Detection simulation and modeling
Wide-area surveillance	Multifunction
Multifunction	

receiver are colocated. When the transmitter and the receiver sites are separated, the radar is said to be bistatic. In a primary radar system, subsystem, or mode of operation, the return signals are the echoes of its own emitted signals obtained by reflection from the target. On the other hand, a secondary radar extracts target information from a target transmit signal sent by any IFF (identify friend or foe) transponder. It is a radar technique or mode of operation in which the return signals are obtained from a transponder or a repeater carried by the target.

Radars may be installed at fixed locations (land based) or mobile on a truck (land based), aircraft (airborne), ship (seaborne), or satellite or space shuttle (spaceborne). They may use signals with different frequencies (such as HF, 3–30 MHz; VHF, 30–300 MHz; or microwave (MW), 300 MHz up to tens of GHz) with different waveforms [continuous wave (CW), frequency modulated CW (FMCW), FM interrupted CW (FMICW), pulsed, etc.]. Their receiver processor may perform detection either in the time domain (TD) or in the frequency domain (FD).

CW radars are simple and occupy minimal spread in the frequency spectrum. Its transmitted power level is much less than the peak power level of a pulsed radar. It is used to measure the speed of a target (i.e., a traffic radar) by using the Doppler effect (i.e., the shift in the frequency of CW signal caused by the radial speed of a target moving toward or away from the radar). It is also used to detect moving targets in a region (e.g., an intruder alarm). On the other hand, it is not capable of measuring the range of a target, unless the CW signal is frequency modulated. In the frequency modulated CW (FMCW) radars, the frequency of the CW wave is periodically modulated by applying a frequency shift that varies linearly with time in the range of $f_0 \pm f_m$, where f_0 is the frequency of the carrier wave and f_m is the frequency deviation. CW and FMCW radars are also attractive because of their low-level transmitter power requirements, which are within the capability of current solid-state power amplifiers (low cost). However, CW and FMCW radars are bistatic since their transmitter (with a relatively high-level transmitted power) and receiver (with a relatively low-level echo signal) are on at the same time. This results in a direct arrival of the transmitted signal with noise to the receiver, where it competes with the target echo.

Frequency is a basic radar parameter that determines not only the design and construction of a radar but also the application and performance. Table 1.3 lists military and commercial radar bands. Although there are many in the table, we group radar frequencies mainly into three; HF radars, VHF radars, and MW radars, according to EM propagation characteristics and target interaction properties in these regions. The MW radars find wide areas of applications, some of which are listed in Table 1.4 together with the assigned frequency ranges.

Table 1.3 Military and Commercial Radar Bands

VHF	30–300 MHz	Very high frequency	138–144 MHz 890–942 MHz
S	2–4 GHz		2.3–2.5 GHz
C	4–8 GHz		5.25–5.925 GHz
Ku	12–18 GHz		13.4–14.0 GHz
K	18–27 GHz		24.05–24.25 GHz
mm	40–300 GHz	Millimeter waves	33.4–36.0 GHz

Table 1.4 Frequency Ranges Assigned to Certain Radar Applications

Frequency range	Spectrum allowance
1.35–1.4 GHz	Military comms/radar
1.435–1.535 GHz	L-band telemetry
2.45–2.69 GHz	Commercial comms/radar
2.9–3.7 GHz	Miscellaneous radar
4.2–4.4 GHz	Radar altimeter
5.25–5.925 GHz	Miscellaneous radars
8.5–10.55 GHz	Miscellaneous radars
9.3–9.5 GHz	Weather radar and maritime navigation radar
13.25–14 GHz	Miscellaneous radars and satellite comms
15.7–17.7 GHz	Miscellaneous radars
24.25–25.25 GHz	Navigation radar
33.4–36 GHz	Miscellaneous radar

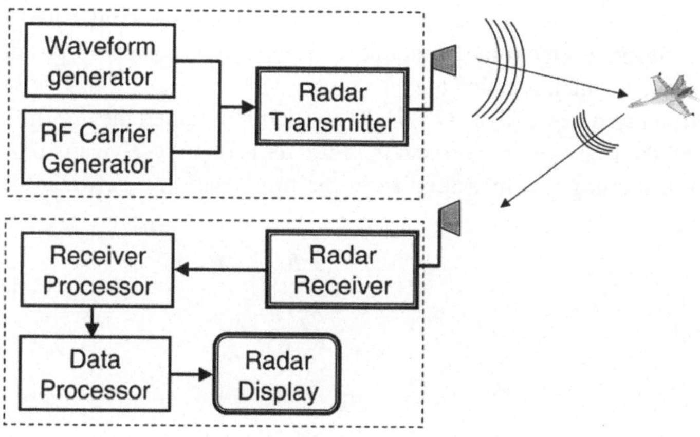

Figure 1.2 A simple block diagram of a radar.

Pulsed radars find more applications than CW radars; therefore, most of the chapter is devoted to pulsed radar systems. Unless otherwise stated, the information included here refers to pulsed radars.

A typical block diagram of a radar is shown in Fig. 1.2, which consists of a transmitter block and a receiver block. The carrier signal is generated from a local oscillator and modulated by a suitable radar waveform (best suited for the operational purposes) in the transmitter, and this signal is transmitted via the transmit antenna system. All the echoes are received by the receive antenna system and processed first in the receiver processor unit [which includes all electronic processing stages, such as filtering, amplifying, mixing, and analog-to-digital converting (ADC)] and then in the data processor unit (the hardware units where digital data is processed by computer algorithms). Finally, the output is displayed as graphics in the video display unit.

1.2.1. Resolution and Accuracy

Radar is a measuring device that measures target's range (distance between the radar and the target), range rate (velocity of the target), direction (angular position of the target), and reflected power [radar cross section (RCS) of the target]. Because of imperfections in any measuring instrument, some amount of error will always be introduced. The errors of a radar are characterized by two performance parameters: resolution and accuracy. *Resolution* is the radar's ability to distinguish two targets in close proximity of each other, mostly in a three-dimensional space: (1) range, (2) bearing, and (3) velocity (Doppler). *Accuracy* is the ability of the radar to measure the true value (i.e., the true range, velocity, direction, etc.) to within some stated error specification, and intuition tells us that it must be related to the received power level of the target (i.e., sharpness of the target signal above the noise level). Any measurement made in a gaussian type noise environment and with a signal-to-noise ratio (SNR), using a system with a basic resolution Δ, will have an rms error, δ, which can generally be expressed as $\delta = \Delta/(2\mathrm{SNR})^{1/2}$ [1]. It has to be noted that this definition of accuracy corresponds to measurement errors which are bounded by one standard deviation. Hence, assuming a gaussian-like noise distribution, one can conclude that the probabilities for the actual result to be within δ and 2δ vicinity of its measured value is approximately 70% and 95%, respectively.

1.2.2. PRF and Maximum Range

Ideal transmitted and received time series of a pulsed radar (without carrier signal) are pictured together in Fig. 1.3. Here, pulses marked as 1, 2, 3, ..., are the transmitted pulses and the rest are the received pulses. When the transmitter is on during the transmission of a pulse having a pulse width τ, the receiver is off (to secure its sensitive electronic components from high transmitted power effects). This is repeated every T seconds, which is called the *pulse repetition interval* (PRI), inverse of which is equal to the pulse repetition frequency ($\mathrm{PRF} = 1/T$). The ratio τ/T is called the *duty factor*. During the time interval ($T - \tau$), the transmitter is off and the receiver is on to receive any possible target echoes. The range (R) of the target is measured from the time delay (t_d) between the transmit and received signals as

$$R = \frac{ct_d}{2}\ \mathrm{m} \tag{1.1}$$

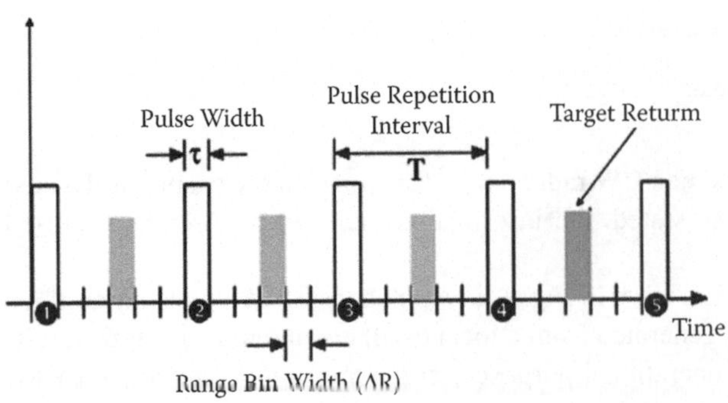

Figure 1.3 Radar transmit and receive pulse definitions.

where c is the speed of light ($c = 3 \times 10^8$ m/s). The factor 2 in Eq. (1.1) arises because the distance traveled by the signal is $2R$, i.e., to the target and back. The maximum useful range is determined by the PRI (or PRF) as

$$R_{max} = \frac{cT}{2} = \frac{c}{2\text{PRF}} \text{ m} \tag{1.2}$$

Usually, the radar receiver samples the received time echo signal every τ seconds and each sample represents a distance ΔR called a *range bin* or *range gate*:

$$\Delta R = \frac{c\tau}{2} \text{ m} \tag{1.3}$$

which is also called a *range resolution*. The number of range bins N is then equal to the maximum range divided by the range resolution ($N = R_{max}/\Delta R$). The narrower the pulse width τ, the better the range resolution, and in turn, the higher the number of range bins. On the other hand, the narrower the pulse width, the wider the bandwidth B [Hz] of the signal (i.e., $\tau \sim 1/B$). It should be noted that there are different definitions for both pulse width and bandwidth of a signal when it is not an ideal rectangular pulse. Here, they are both defined as the distance between half-power points of the pulse in TD and of the spectrum in FD. For a pulse width τ, the range accuracy can be given as [1]

$$\delta R \cong \frac{c\tau}{2\sqrt{2 \times \text{SNR}}} = \frac{c}{2B\sqrt{2 \times \text{SNR}}} \text{ m} \tag{1.4}$$

1.2.3. Pulse Integration and Doppler Frequency

A radially approaching (receding) target causes a slight increase (decrease) in the carrier frequency of the radar that is proportional to its radial speed. This is called *Doppler effect* and is used in radar systems to detect and/or discriminate targets in FD. Targets that cannot be discriminated in TD, because of strong unwanted echoes, may easily be discriminated in FD by using radial velocity differences. Radars that use the Doppler effect are called *pulse Doppler* or *moving-target-indicator* (MTI) radars.

Measurements of the position of a target can be made quickly (with a unique received pulse), but it takes some time to estimate velocities and to distinguish differences in velocities of targets. The smaller the velocity differences, the longer the time needed to estimate. This is clearly understood from the Fourier theory. It is known that analytical Fourier transform is defined for infinite time ($-\infty < t < \infty$) and infinite frequency ($-\infty < f < \infty$) range. Observing a signal for an infinite duration in time yields a zero frequency resolution, that is, one can get frequency information from an infinite time series at any particular frequency. Similarly, one needs to know infinite frequency range behavior to rebuild the signal *exactly* in TD. As for all other discrete real signal processing cases, radar signals have a finite duration in time. From a finite duration time series (let's call it observation time T_{obs}) with Δt sampling interval, one can obtain FD response via discrete Fourier transform (DFT), or mostly fast fourier transform (FFT), with a maximum frequency f_{max} and a minimum frequency resolution Δf as

$$f_{max} = \frac{1}{2\Delta t} \text{ Hz} \quad \text{and} \quad \Delta f = \frac{1}{T_{obs}} \text{ Hz} \tag{1.5}$$

It is obvious from Eq. (1.5) that one needs to obverse the target (illuminate the target, collect consecutive pulses at the receiver and maintain a time series) longer if a better frequency resolution is required. This process is called *integration* and is done to enhance detectability, to reduce measurement errors, to improve resolution or some other performances. The length of time taken to make an observation with a radar is called the *integration time*. Coherent (incoherent) integration is the process of collecting consecutive pulses in TD, where the receiver is tuned to the same carrier frequency with the transmitter with (without) phase locking to it. In coherent radars, a complex time series is formed (with target echo amplitude and phase) and Fourier transformed (via FFT), and detection is done in FD, where a moving target with a radial speed component appears as an impulse like signal along the frequency axis, far from the zero frequency. This Doppler shift of a target depends on the radial velocity v_r and the carrier signal wavelength λ_0 of the radar as

$$f_d = \frac{2v_r}{\lambda_0}\ \text{Hz} \tag{1.6}$$

Finally, the velocity accuracy of a target can be calculated as [1]

$$\delta v_r \cong \frac{\Delta f_d}{\sqrt{2 \times \text{SNR}}} \approx \frac{\lambda}{2T_{\text{int}}\sqrt{2 \times \text{SNR}}}\ \text{m/s} \tag{1.7}$$

1.2.4. Angular and Elevation Scan

Location of a target with a radar system requires determining its range (radial distance), azimuth (angular position), and height. In a pulsed radar, range information is extracted by directly time gating in the receiver. Azimuth and elevation information is obtained by the scanning characteristics of the receive antenna system, as either a mechanical scanning or an electronic scanning system. In mechanical scanning (see Fig. 1.4), the directive antenna with a narrow beam characteristics is rotated mechanically with a constant speed, which is adequate to illuminate each angular sector for a while and receive the number of required echoes. If the antenna has beam widths of $\Delta\varphi$ (rad) and $\Delta\theta$ (rad) in azimuth and elevation, respectively, the radar azimuth and elevation resolutions will be $\Delta\varphi$ and $\Delta\theta$, respectively. Its azimuth and elevation accuracies are

$$\delta\varphi \cong \frac{\Delta\varphi}{\sqrt{2 \times \text{SNR}}}\ \text{rad} \qquad \delta\theta \cong \frac{\Delta\theta}{\sqrt{2 \times \text{SNR}}}\ \text{rad} \tag{1.8}$$

respectively.

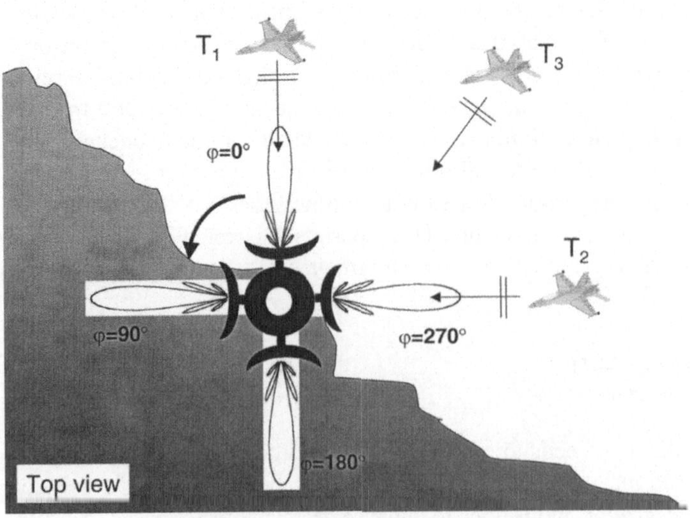

Figure 1.4 Mechanical scanning.

MW radars with mechanical scanning use dish type (parabolic reflector) antennas and their directivity (i.e., lossless) gain is given as [2]

$$G = \frac{4\pi}{\Delta\varphi\Delta\theta} \tag{1.9}$$

For a solid beam width $\Omega = \varepsilon\Delta\varphi\Delta\theta$, the number of beams required to scan the hemisphere (4π steradians being the entire sphere) is

$$\text{Number of beams} = \frac{2\pi}{\Delta\varphi\Delta\theta} = \frac{G}{2} \tag{1.10}$$

Mechanical scanning is used mostly in MW radars, especially in tracking radars, where the target's path is determined and its route is predicted. Although tracking can be carried out using range, angle, or Doppler information, angular tracking is the characteristic feature in tracking radars. It should be noted that all surveillance radars may also be considered as tracking systems (to some extent) since they keep track of many targets simultaneously. The process of mechanical tracking is called *track-while-scan.*

In HF and most of VHF radars, mechanical scanning is not possible because of the large apertures of the receiving antenna systems; therefore, electronic scanning is the only way to obtain angular information of a target. A typical transmit and receive antenna site of a HF surface wave (HFSW) radar (that operates in the frequency range of 3–6 MHz) is pictured in Fig. 1.5a, where a 24 × 4 vertical monopole array over the earth's surface is used as the receive antenna system [6,7]. Here, the aim is to cover a coastal region, so a quadlet (four element) end-fire array is used as the receive array element. Its mission is to direct radar energy toward the ocean with a high front-to-back ratio. As narrow as 4°–5° azimuth beam widths are obtained by using 16 to 24 quadlets and as much as 100°–120° azimuthal coverage can be obtained. The width and length of this receive antenna array is 300 m × 600 m, which may extend to more than a kilometer. In a HFSW radar, angular locations of the targets are obtained solely by digital beam forming (i.e., electronic scanning). Typical beams formed digitally (with computer simulations [6]) are given in Fig. 1.5b. As easily seen from the figure, the shapes of the antenna beams are quite different and become distorted as the beams leave the boresight azimuthally. For example, it is easy to locate the angular position of target T_1 in beam A in the figure, since the main beam is far stronger than the other lobes (side and back lobes), where echo signals from T_2 and T_3 are easily suppressed. The angular discrimination is not that good in beam B (for 30° angular scan) and gets worse in beam C (for 55° angular scan), since strong unwanted side and back lobes appear, where targets T_2 and T_3 may appear to be in the same angular direction.

Both mechanical and electronic scanning have advantages and limitations, which forces the radar engineer to do optimization depending on the data flow and processing speed in the radar computer. Good coverage and good resolution are payoffs. Users usually desire to have a radar with good coverage (long in range and wide in azimuth/elevation) and good resolution (narrow range and angular resolutions), which means a high number of beams to scan or form. The higher the number of beams, the higher the scan rate and the lower the dwell time (pulse integration period).

1.2.5. Analog-to-Digital Conversion (ADC) Process

The power of today's radar systems comes from both its electronic subsystems and intelligent software. The return echo signal is processed by high-speed computers via powerful algorithms, and this is accomplished first by digitizing (sampling) the time signal via ADC. Sampling a signal in TD makes its spectrum periodic in FD. Mathematically, it corresponds to multiplying the time signal by an infinite extend impulse train. A Fourier transform of an infinite-extent impulse train with impulse

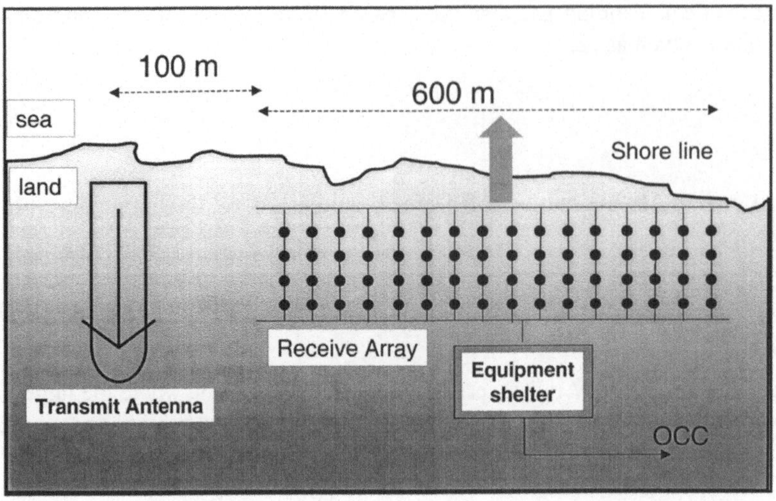

(a)

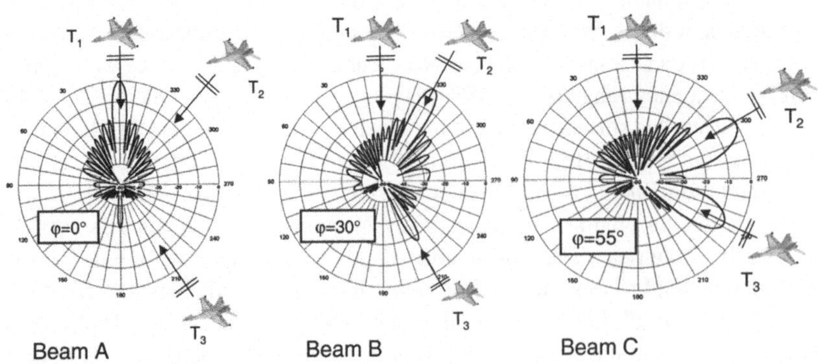

(b)

Figure 1.5 (a) A transmit and receive antenna site of an HFSW radar. (b) Electronic scanning.

separation τ is another infinite-extent impulse train with a separation in the FD of $1/\tau$. Also, multiplication in TD corresponds to convolution in FD, which makes the spectrum periodic. Finally, if a signal reconstruction in TD is required a low-pass filtration in FD (that is equivalent to multiplication of the discrete time signal with a Zinc function in TD) is applied, which are illustrated in Fig. 1.6. This is a well-established theory (sampling theory) in digital signal processing.

ADC translates the input voltage of the receiver to binary numbers that computer hardware can process. The radar receiver may have very large echoes from nearby huge targets (and often from clutter) and at the same time very weak echoes from distant small targets. Therefore, a radar receiver must have a high *dynamic range* (the dynamic range of a receiver is the range between maximum and minimum detectable signals), which is an obvious consequence of the two-way path loss. The higher the dynamic range the greater the number of bits in the ADC. In most radar receiver systems, the echo signal undergoes some electronic and digital processes, such as RF filtering, IF converting and amplifying, and finally video (baseband) filtering and information extracting. At which stage ADC will be used depends on the speeds of the ADC. With today's ADC speeds (which reach up to Gbit/s) ADC may be used almost anywhere in the radar receiver.

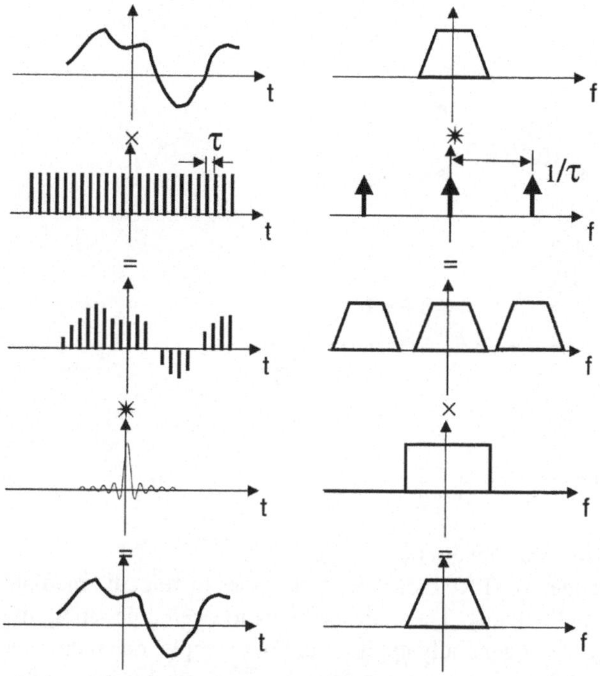

Figure 1.6 Sampling and DFT effects (×: multiplication, *: convolution).

1.2.6. Range-Doppler Ambiguities

Equation (1.2) relates the maximum range of a radar to its PRF. The lower the PRF (the higher the PRI), the longer the maximum range. If longer ranges (longer T) with good resolution (shorter τ) are to be obtained, the radar needs to operate at very high peak powers, since the transmitted average and peak powers, P_t and P_p, are related via the duty factor as

$$P_t = P_p \frac{\tau}{T} \text{ W} \tag{1.11}$$

It is not easy to obtain high peak powers in solid-state radar technology; therefore, low PRF radars with high resolution are not easily realizable.

What happens if longer ranges (longer than R_{max} determined by T) are to be covered? As illustrated in Fig. 1.7, the receiver may not distinguish whether the second received echo (Rx 2) belongs to the first transmitted pulse (Tx 1) or the second one (Tx 2). In this case the received echo amplitude may give a clue (as the range of a target increases its echo amplitude decreases because of propagation losses) but there may still be an ambiguity. For example, a huge target (with strong RCS) at range 3 may have the same order echo amplitude with a small target (with weak RCS) at range 2.

PRF (i.e., $1/T$) also determines maximum target speed in Doppler FD, since the spectrum become periodic with PRF. The lower the PRF, the lower the maximum target speed in a Doppler radar. Therefore, designing low PRF radar to avoid range ambiguity problems causes ambiguities in the Doppler domain for today's high speed targets. Also, targets with radial velocities of integer multiple of radar PRF cannot be detected. These are called *blind velocities* (v_b) and are calculated as [2]

$$v_b = \text{PRF} \frac{n\lambda}{2} \text{ m/s} \tag{1.12}$$

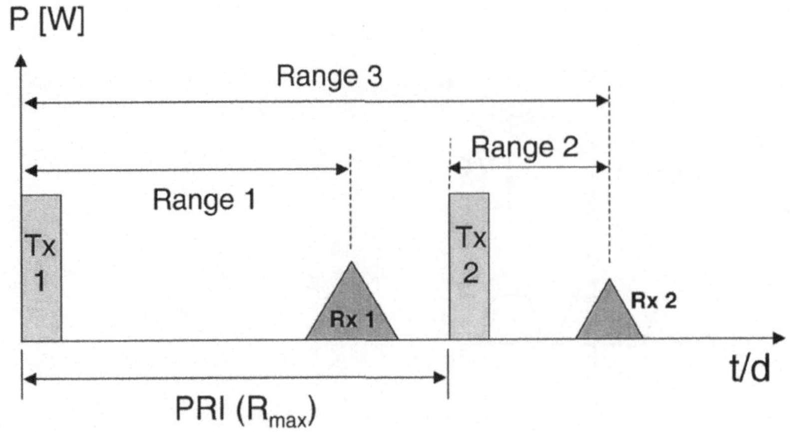

Figure 1.7 PRF, maximum range, and ambiguities.

where n is an integer and λ is the radar carrier wavelength ($\lambda = c/f_0$).

High PRF radars are ambiguous in range; low PRF radars are ambiguous in velocity; medium PRF radars are ambiguous in both range and velocity. There are methods to overcome ambiguity problems without changing PRF, but they are not generally applicable. For example, one method is to label each pulse (transmitting them with different frequencies, phases, polarizations, etc.) so that echoes can easily be assigned to their transmitted pulses. But, this does not work for targets when their RCS fluctuates from pulse to pulse. Labeling the transmitted pulse also causes difficulties in Doppler processing. Another way of overcoming ambiguities is to use a burst of pulses with variable PRFs, which in turn may result in the occurrence of blind ranges and blind speeds at different places.

1.2.7. Pulse Compression and Matched Filter

The detectability of an echo is dependent on the total energy of the radar signal. The higher the transmitted energy, the longer the range and the higher the detectability. The aim of a radar is (1) to detect the target, (2) to measure its range and velocity as accurately as possible, and (3) to discriminate targets in range, angle, and velocity, which are not compatible with each other. For example, transmitted energy must be increased to increase detectability, which may be achieved by using a longer pulse duration τ. On the contrary, to increase range resolution, a short pulse duration τ (i.e., a large bandwidth) must be used, which also results in poor Doppler resolution. To achieve both, short pulses with high energy must be used, which requires very large peak powers. Very large peak powers are not desirable for long-term operation with minimal cost or may not be available in the solid-state design. The solution is pulse compression, that is, to use long pulses during transmission for long range coverage and better detection performance (also for good Doppler resolution) requiring a reasonable peak power. During reception, short pulses are needed to achieve a high-range resolution. This can be accomplished by designing a suitable waveform, normally by using frequency modulation and by using a correlation receiver, the matched filter (Fig. 1.8).

The radar echo at the receiver consists of a possible target, noise, clutter, and interference signals. The higher the target signal among the others, the better the detection process. If one assumes the noise floor as the threshold signal, the fundamental criterion that determines the ability to detect a target is the signal-to-noise ratio (SNR). As will be discussed later, noise power depends on the receiver bandwidth and is independent of the radar signal. Therefore, for a fixed noise power SNR may be maximized by using a correlation receiver (whose frequency domain implementation is the matched filter). As shown in Fig. 1.8, the matched filter is a filter, whose transfer bandwidth is equal to the radar pulse

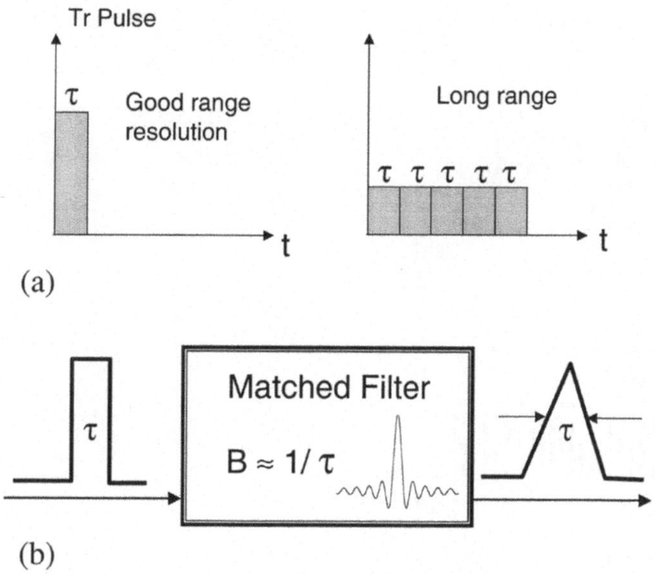

Figure 1.8 (a) Pulse compression and (b) matched filter.

bandwidth. Physically, this corresponds to minimizing the noise power, which results in maximization of SNR. Geometrically, it corresponds to sharpening the peak of the received pulse (as shown in the figure) which in turn maximizes the SNR.

1.2.8. CFAR Detection and Decision Making

At the radar receiver, target echoes are contaminated with other unwanted echoes (such as noise and clutter); therefore, a threshold decision is required for target detection. Constant-false-alarm-rate (CFAR) detection in a radar is a technique to extract target signals from the noise and clutter, which, in general, are all random processes [8]. As a random signal, the total radar echo (target, noise, and clutter) fluctuates with time. A random variable can be represented by a probability density function (PDF) of its amplitude fluctuation, plus its frequency content, the power spectrum function. The important parameters of a PDF are mean and median values, standard deviation, and cumulative probability function.

A typical radar echo is pictured in Fig. 1.9. Here, analog and digitized (sampled) radar echoes are given on top and bottom, respectively. The vertical axis is the power or voltage at the receiver, the horizontal axis may be time, range, or frequency (velocity). The horizontal dotted line represents the mean value of the noise floor. The horizontal dashed line is the CFAR detection threshold, which is well above the noise floor determined by a given SNR value. The main problem in CFAR detection is to determine the most suitable SNR value, which allows best decision making among the following four cases:

1. Target is present, and its echo is above the threshold (*true detection*).
2. Target is present, but its echo is below the threshold (*missed detection*).
3. Target is not present, but an echo is above the threshold (*false alarm*).
4. Target is not present, and no echo is above the threshold (*no action*).

For example, peak A, B, C, and D in Fig. 1.9 correspond to a true detection, a missed detection, a false alarm, and no action cases, respectively. The reliability of the CFAR detection unit

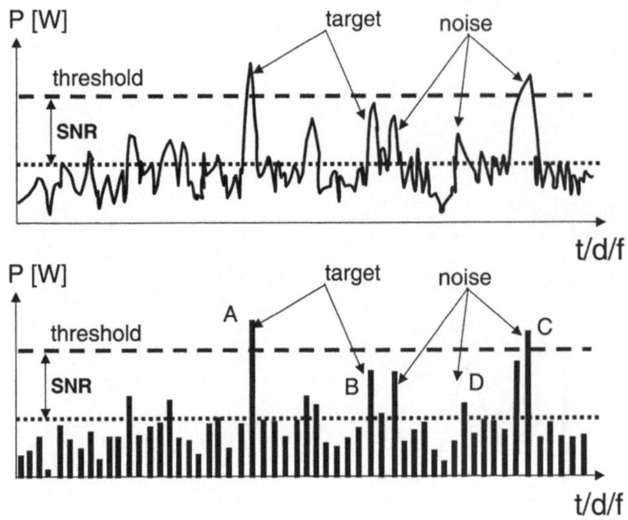

Figure 1.9 CFAR detection and decision making.

depends on understanding the stochastic characteristics of random processes target, noise, and clutter, which are discussed in the following sections.

1.2.9. False Alarm Probability and ROC

For any threshold setting, there will be a corresponding probability that noise alone (and/or clutter) may exceed this threshold. When this happens, the radar will erroneously report a detected target at the corresponding position. The associated probability is therefore called *the false alarm probability*, P_{fa}. In a similar way, the probability of making a correct detection P_d can be associated for each chosen threshold. Since they both depend on the same threshold setting, it is clear that they are somehow related to each other.

If one assumes that at decision making instants the probability density functions of noise and noise plus target associated with the amplitude (y) distributions are $p_n(y)$ and $p_{s+n}(y)$, respectively, these probabilities are determined as

$$P_{fa} = \int_Y^\infty p_n(y)\,dy \qquad P_d = \int_Y^\infty p_{s+n}(y)\,dy \tag{1.13}$$

Above relations indicate that

Increasing the detection threshold Y will decrease both P_{fa} and P_d.
Decreasing the detection threshold Y will increase both P_{fa} and P_d.

which show that there is always a trade-off between improving detections and reducing false alarms. Since both P_{fa} and P_d change with the detection threshold level, it is possible to calculate a value of SNR which is required to achieve a certain detection probability P_d, for a given value of P_{fa}. In practice, P_{fa} is chosen first and then P_d is calculated or determined by using receiver operating characteristics (ROC) of the radar receiver. A typical ROC is illustrated in Fig. 1.10, where detection probability vs. SNR are plotted for different false alarm rates. Very low false alarm probabilities are used in radar systems as shown in the figure: as low as $P_{fa} = 10^{-6}$ to 10^{-8}. If the noise amplitude

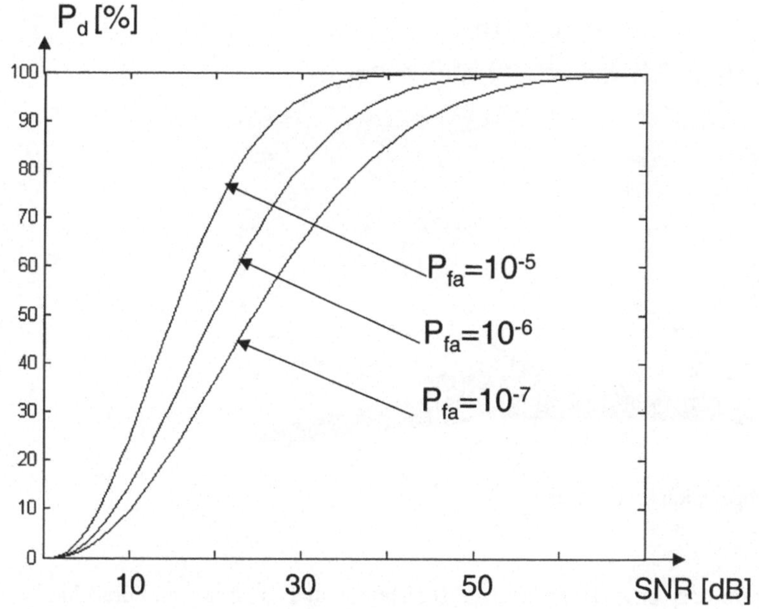

Figure 1.10 A typical radar ROC graph.

distribution is assumed to be gaussian, the ROC of a radar can be obtained analytically. If not, or if clutter determines CFAR detection threshold, then these graphs must be prepared either experimentally or numerically.

The determination of the ROC of a radar depends on a number of factors, including the statistical behavior of the target fluctuations and of the noise plus clutter, the number of pulse integrated coherently and incoherently. All these are extensively studied and well documented in the literature [9].

1.3. PROPAGATION ENVIRONMENT AND PATH LOSS PREDICTION

Radars operate above the earth's surface through the atmosphere and understanding EM scattering properties and propagation effects are critical [10–14]. A typical propagation environment is pictured in Fig. 1.11. Here, surface (on land and ocean) and air targets, land-based and airborne radars, which may operate at different frequencies from HF, VHF to MWs are illustrated. The propagation scenario may include the spherical earth's surface with nonflat terrain (mountains, valleys, etc.), rough surfaces (e.g., ocean surface or land irregularities and vegetation) and atmospheric variations (e.g., clouds, rain), all of which behave quite differently in different frequency regimes.

In general, propagation occurs via ground waves and sky waves. Ground waves have three components: direct waves, ground-reflected waves, and surface waves. Sky waves use high altitude atmospheric layers (i.e., D, E, and F layers of the atmosphere, called the *ionosphere*). The model environment is a spherical earth with various ground characteristics, above which exists a radially inhomogeneous atmosphere. Since the radar and the target may be anywhere on or above the ground, the propagation model may have different canonical features. The physical characteristics of propagation depend on many parameters, such as the operating frequency, medium parameters, transmitter and receiver locations, and the geometry (boundary conditions, BC) between them.

Depending on the mission and allowed frequency bands, propagation characteristics may be totally different. In Table 1.4, some of the frequency ranges and assigned missions are listed. The

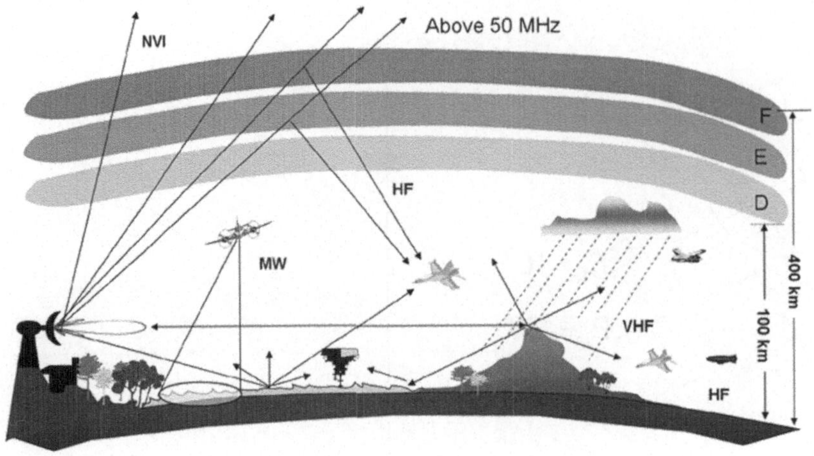

Figure 1.11 Radio wave propagation environment.

mission may be to defend a military base or airport against airborne attacks, to assist aircraft landing in a critical region surrounded by high hills, to assist in weather or oceanographic surveillance, a maritime navigation along a narrow waterway, or to measure the speed of a vehicle (police Doppler radar), height of a vehicle (altimeter), etc. The mission may also be to prepare agricultural maps (SAR), to identify ships (ISAR) or subsurface imaging (GPR). Propagation requirements for these missions are quite different from each other. Some applications need a range of hundreds of kilometers, others work over less than a meter. Generally speaking, radars may be classified into three groups in terms of their propagation characteristics: HF radars, VHF radars, and MW radars (Infrared and optical sensors are not mentioned here). Some of the propagation characteristics in terms of this classification can be listed as follows [10,11]:

For A/A and/or S/A microwave radars, it is essential to understand free-space path loss, first. This is not actual loss in fact and is nothing but a decrease in power density because of the spherical wave spread. *Free-space path loss* is defined as the power density of a unit power isotropic radar transmitter captured by an isotropic (unit gain) receive antenna with an effective aperture of $A_e = \lambda^2/4\pi$ and given by

$$Lfree = \left(\frac{4\pi d}{\lambda}\right)^2 = \left(\frac{4\pi df}{c}\right)^2 \tag{1.14}$$

where d, f, and λ are the distance between the transmitter and receiver, the frequency, and the wavelength of the radar carrier, respectively. By using d in kilometers and f in MHz, Eq. (1.14) can be rearranged in logarithmic form as

$$Lfree = 32.4 + 20\log_{10} d_{\text{km}} + 20\log_{10} f_{\text{MHz}} \text{ dB} \tag{1.15}$$

For A/G, A/S, and S/S microwave radars the total propagation loss is more than the free-space path loss and may include

Absorption loss (due to atmospheric, ionospheric gases and precipitation)

Earth's curvature loss (due to the spherical nature of the ground, which diverges the waves)

Reflection, multipath, and scattering loss (due to ground irregularities)
Diffraction loss (due to the nonflat terrain along the propagation path)
Refractivity loss (due to ducting, guiding, and antiguiding effects of atmospheric layers, especially in the first few kilometers above the ground)
Depolarization loss (due to tropo-scattering effects), etc.

These are defined as *additional losses*, and the *total path loss* is usually defined as the free-space path loss plus additional losses.

For MW radars only line-of-sight (LOS) propagation is possible. *Line of sight* is defined as the distance when there is no obstacle between the transmitter and the target. Within LOS is the interference region where ground waves have all three components. Beyond the LOS is the diffraction region, and propagation is possible only by means of surface waves and/or sky waves (via ionospheric reflections).

Since S/S MW radars are limited by the LOS, the radar needs elevated platforms to overcome the earth's curvature effects. The LOS (in kilometers) of a MW radar with platform height h (m) can be calculated via [2]

$$\text{LOS} = 4.12\sqrt{h}\ \text{km} \qquad (1.16)$$

For example, a S/S MW radar on a 40-m tower may cover ranges up to only 25–26 km.

For S/S, HF radars the frequency range of interest is between 3–30 MHz depending on the area of surveillance and the mode of propagation. Frequencies between 3–6 MHz are used for wide-area ocean surveillance (up to 400–500 km in range), while 5–15 MHz are good for 10–30 km (may reach up to 25 MHz for ranges of a few kilometers) when the surface-wave mode of operation is used. On the other hand, sky-wave mode of propagation may also be used for S/S wide-area surveillance up to a few thousands of kilometers.

For S/S, SWHF radars the transmitter and the receiver are close to surface; so direct and ground-reflected waves cancel each other, and only surface wave remains. The earth's electrical parameters are important in reaching longer ranges. Sea surface is a good conductor, but ground is a poor conductor at these frequencies. For example, with the same transmitter and receiver characteristics, a 5-MHz signal, which reaches the 400-km range over the sea can only reach up to 40–50 km in range over poor ground. Typical electrical parameters (σ, conductivity, ε_r, relative permittivity) of ground are listed in Table 1.5 [3].

In the lower HF band (3–10 MHz), propagation well beyond the LOS is possible via surface waves. Surface waves are hardly excited and coupled to the surface, when the transmitter is many wavelengths (e.g., 3λ–4λ) above the ground. When excited and coupled, they exponentially decay with height. Surface waves are rarely used above 10–15 MHz.

At higher HF frequencies (15–30 MHz), beyond the LOS coverage is only possible by means of the sky waves. The lower layer of the ionosphere (D layer) absorbs EM waves and causes extra loss. The higher layers (E and F) bend EM waves toward the earth's surface, act as a reflecting upper boundary and form a kind of earth–ionosphere waveguide. This

Table 1.5 Typical Ground Characteristics at MF and HF

Ground	σ (S/m)	ε_r
Sea	5.0	80.0
Medium ground	0.01	15.0
Poor ground	0.001	7.0

waveguide can be used at most up to the frequencies of 45–50 MHz. Beyond 50-MHz EM waves are not bent and escape into outer space.

At lower VHF frequencies (50–150 MHz), propagation beyond the LOS is still possible by means of diffracted EM wave components (typically, 5–20 km beyond obstacles).

At upper VHF frequencies and above, (i.e., for frequencies 200 MHz and above) propagation is limited by the LOS, because surface waves are negligible at these frequencies.

Ground-wave propagation through atmosphere (up to nearly 100 GHz) is affected by oxygen and water-vapor molecules. The air can be considered as a nondispersive medium and can be represented by its refractive index ($n = \sqrt{\varepsilon_r}$). The refractivity of the propagation medium should be well understood, since nonflat terrain and/or earth's curvature may also be implemented via refractivity in most of the analytical as well as numerical approaches. Refractive index of the air is very close to unity (e.g., 1.000320); therefore, it is customary to use the refractivity N, defined as

$$N = (n-1) \times 10^6 \tag{1.17}$$

N is dimensionless, but is measured in "N units" for convenience. N depends on the pressure P (mbar), the absolute temperature T (K) and the partial pressure of water vapor e (mbar) as [12]

$$N = 77.6\frac{P}{T} + 3.73 \times 10^5 \frac{e}{T^2} \tag{1.18}$$

which is valid in earth–troposphere waveguides and can be used in ground-wave propagation modeling. If the refractive index were constant, radio waves would propagate in straight lines. Since n decreases with height, radio waves are bent downward toward the earth, so that the radio horizon lies further away than the optical horizon (i.e., LOS). It should be noted that the radio horizon effect is taken into account either by using N with the effective earth radius a_e or by introducing a fictitious medium where N is replaced by the modified refractivity M [12],

$$M = N + \frac{x}{a} \times 10^6 = N + 157x \tag{1.19}$$

with the height x given in kilometers. In Eq. (1.19)

$$a = 6378\text{ km} \qquad \frac{10^6}{a} = 157 \quad \text{and} \quad \frac{\partial M}{\partial x} = \frac{\partial N}{\partial x} + 157 \tag{1.20}$$

For the standard atmosphere (i.e., for a vertical linearly decreasing refractive index), N decreases by about 40 Nunit/km, while M increases by about 117 Nunit/km. Subrefraction (superrefraction) occurs when the rate of change in N with respect to height (i.e., $\partial N/\partial x$) is less (more) than 40 Nunit/km.

A linearly decreasing (increasing) vertical refractive index variation forces the waves to be trapped near the earth's surface while propagating. Similar effects are also caused by concave and convex surfaces. Therefore, there is an analogy between refractive index and surface geometry in terms of propagation effects. By using this analogy, earth's curvature effect is included into the refractive index of the air. Earth's curvature effect is equivalent to a vertical refractivity gradient of 157 Nunit/km (i.e., linear vertical increasing refractivity profile).

Characteristic features of the EM wave propagation are graphically illustrated in Figs. 1.12–1.14. In Fig. 1.12, both refractivity and nonflat terrain effects are shown at 30 MHz as a range–height field strength color map, when the transmitter is on the ground [11]. The ducting effect is clearly observed in the second graph under a trilinear refractivity variation. In Fig. 1.13, surface-wave

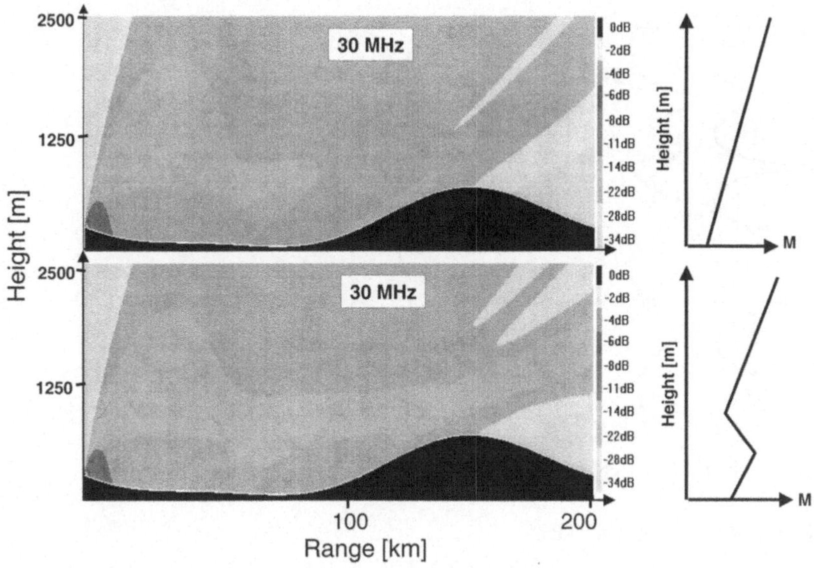

Figure 1.12 Propagation over earth's surface.

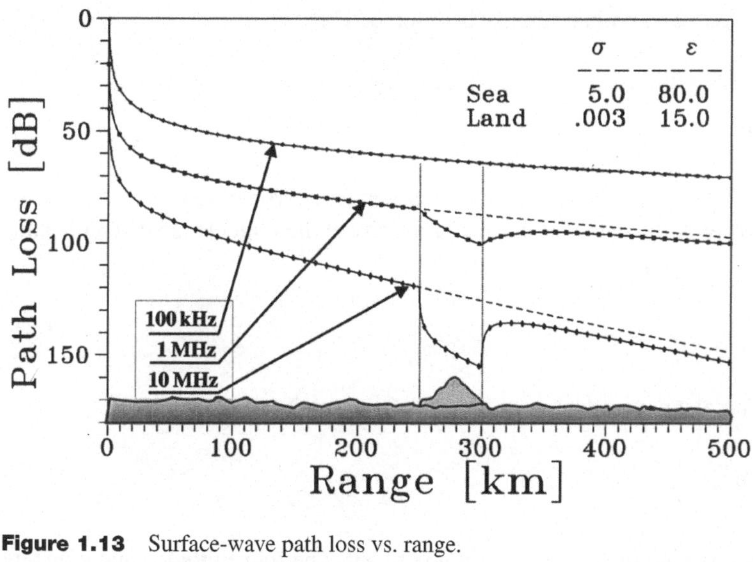

Figure 1.13 Surface-wave path loss vs. range.

path loss vs. range is plotted at three different frequencies along the ocean surface with a 50-km length island at a distance of 250 km from the radar. The radar transmitter and the receiver heights are zero. The dashed curves are for a homogeneous path. At 100 kHz, the island hardly affects the smooth path loss variation. But in the HF band, the path loss increases substantially over land, and signal recovery is observed behind the island. Finally, normalized vertical fields (propagation factor) vs. height are plotted in Fig. 1.14. This corresponds to propagation over a spherical earth with standard atmosphere with a 3-GHz transmitter that is located 31 m above the surface. The five curves correspond to ranges of 20, 30, 40, 50, and 60 km, respectively. The ground is assumed to be a perfect electric conductor (PEC). A 10-dB scale is also given in the figure. The 31-m transmitter height places the first four range curves into the region of interference between the direct and ground

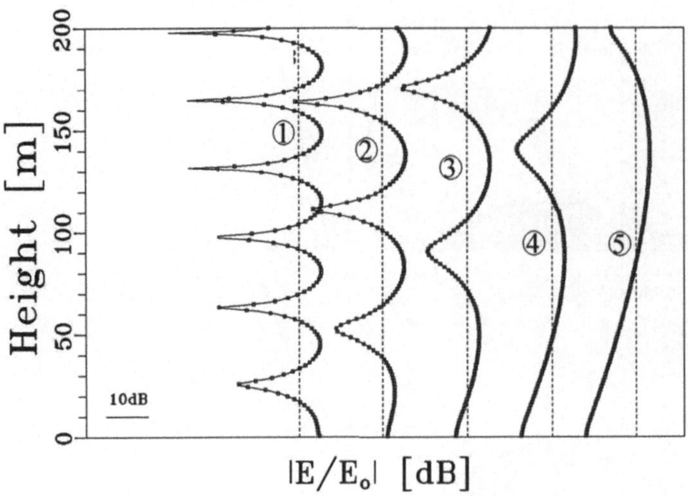

Figure 1.14 Propagation factor vs. height at 3 GHz.

reflected waves (within LOS) as evident in the plots. At the 60-km range, heights up to 200 m are in the shadow region below the LOS and no interference is observed [11].

1.4. RADAR EQUATION

A derivation of radar equation is best understood with the case of an isotropic transmitting antenna. If the average transmitted power is P_t (W), then the power density at the target located at a distance R (m) is

$$\text{PD} = \frac{P_t}{4\pi R^2}\ \text{W/m}^2 \tag{1.21}$$

If one uses a directive antenna, the power density at the target is given by

$$\text{PD} = \frac{P_t}{4\pi R^2} \times G_t \text{W/m}^2 \tag{1.22}$$

where G_t is the transmit antenna gain in the direction of the target. The transmitted signal interacts with the target and is reflected back to the radar. The power density of this echo signal back at the radar is

$$\text{PD} = \frac{P_t}{4\pi R^2} \times G_t \times \sigma \times \frac{1}{4\pi R^2}\ \text{W/m}^2 \tag{1.23}$$

where σ (m^2) is the radar cross section of the target. The echo is received by the receive antenna of the radar and the power received is then given by

$$P_r = \frac{P_t}{4\pi R^2} \times G_t \times \sigma \times \frac{1}{4\pi R^2} \times A_e \text{W} \tag{1.24}$$

where A_e (m^2) is the effective receiving antenna aperture. The effective antenna aperture depends on the operating frequency and the antenna gain, mostly as

$$A_e = \frac{G_r \lambda^2}{4\pi} \text{ m}^2 \tag{1.25}$$

where G_r is the receive antenna gain in the target direction. As given in Eq. (1.11), range is inversely proportional to power. If the minimum detectable (received) power $P_{\min}$ is defined as

$$P_{\min} = P_t \times G_t \times G_r \times \sigma \times \frac{\lambda^2}{(4\pi)^3 R^4} \text{ W} \tag{1.26}$$

then maximum range can be obtained as

$$R_{\max} = \left[\frac{P_t G_t G_r \sigma \lambda^2}{(4\pi)^3 P_{\min}} \right]^{1/4} \text{ m} \tag{1.27}$$

which is called the simplest form of *radar equation*. If Eq. (1.26) is rewritten in terms of free-space path loss L_{free}

$$P_{\min} = \left[\frac{P_t \times G_t \times G_r \times \sigma \times 4\pi}{\lambda^2 L_{\text{free}}^2} \right] \text{W} \tag{1.28}$$

is obtained. It should be remembered that the basic radar equation is derived for S/A and/or A/A surveillance (i.e., in free space). When A/S or S/S is of interest surface wave attenuation factor A is introduced and may be combined with the free-space path loss, yielding a total one way propagation loss, L_p as [3]

$$L_p = \left(\frac{4\pi R}{\lambda A} \right)^2 \tag{1.29}$$

The signal at the radar receiver fluctuates for a variety of reasons, which makes the total radar echo a random process. The total radar echo contains targets, noise, clutter, and other interfering signals (such as jamming signals, intentional radio and communication broadcast signals). The minimum required target signal is defined via signal-to-threshold ratio, where the threshold is usually determined either by noise or by clutter. Therefore, the radar equation is generally given as signal-to-noise ratio (SNR) or signal-to-clutter ratio (SCR).

1.4.1. Noise-Limited Detection

Electronic circuits and receivers are affected by a variety of noise sources. Typical noise sources and their frequency variations are illustrated in Fig. 1.15. Here, the horizontal axis is the frequency in log scale and the vertical axis is the electric field in dBμV/m for a 1-kHz bandwidth. In terms of radar engineering, noise can be classified into two groups: thermal noise (also called *internal noise*, caused by electronic devices themselves) and the environmental noise (atmospheric, cosmic, man-made, etc.). As shown in the figure, thermal noise dominates the others for frequencies above a few hundred MHz. Therefore, noise limited radar detection can be grouped into thermal noise limited detection (e.g., MW radars) and environmental noise limited detection (e.g., HF radars).

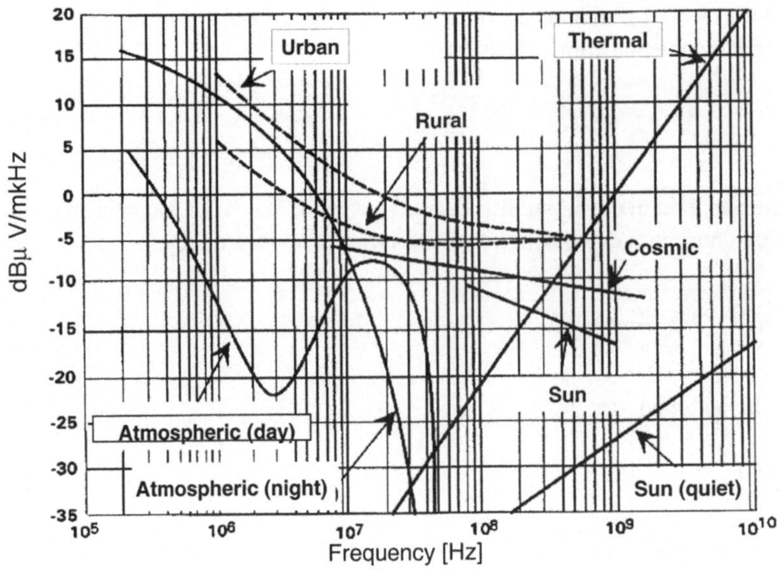

Figure 1.15 Characteristic noise types and frequency regions.

Thermal noise (which is also called *white noise*) is directly proportional to the receiver bandwidth and can be calculated as

$$N_t = kTBW \tag{1.30}$$

where B (Hz) is the noise bandwidth, k is Boltzmann's constant ($k = 1.38 \times 10^{-23}$ J/K), and T is the temperature in Kelvin. Thermal noise can be considered to be white noise (i.e., gaussian amplitude distribution and flat power spectral density, $S(f) = kT$). Since $kT = -204$ dBW/Hz at 300 K, thermal noise can also be calculated as

$$N_t = -204 + B_{dB_{Hz}}\,\text{dB} \tag{1.31}$$

On the other hand, the environmental noise (daytime and nighttime atmospheric noise and man-made noise in rural and urban areas) determines the noise floor at HF frequencies. As shown in the figure, nighttime atmospheric noise level is much higher than the daytime level, especially at lower HF frequencies, which degrades the performances of HF radars at night. Environmental noise can be expressed by the empirical formula [6]

$$N_e = \left(N_0 - 12.6 \ln \frac{f_{MHz}}{3} \right) B\text{dBW} \tag{1.32}$$

where, N_0 is equal to man-made noise level of the selected radar site (typically, -136, -148, -164 dBW/Hz in residental, rural, or remote sites, respectively) and the term inside the parentheses is the noise density given as dB/Hz.

As a result, radar equation for a noise limited detection case can be written as

$$\text{SNR} = \frac{P_t \times G_t \times G_r \times \sigma \times 4\pi}{\lambda^2 L_p^2 N} \tag{1.33}$$

where N is N_t or N_e, depending on whether the radar is a MW (thermal noise limited) or an HF (environmental noise limited) radar, respectively. It is important to note that noise is present at the input of the radar receiver, is range and operating frequency independent, and is proportional to the receiver bandwidth. The narrower the receiver bandwidth, the lower the noise threshold and higher the SNR. It should be noted that Eq. (1.33) is a basic radar equation and may change due to the usage of peak power, coherent and/or incoherent integration, etc.

In order to find out the maximum range from Eq. (1.33), the first step is to determine the detection probability and the false alarm rate. These two determine the minimum SNR, which can be extracted form the ROC of the radar receiver prepared for different target types, number of coherent and incoherent integration, etc. Once SNR is determined, $R_{\max}$ can be calculated from Eq. (1.33) together with Eq. (1.29).

1.4.2. Clutter-Limited Detection

Unlike noise, which is present inherently in the receiver, clutter is a target like echo signal that comes from many small scatterers, such as rain droplets, birds, ocean waves, terrain irregularities, vegetation, aurora, meteors, etc.; therefore, it is a radar parameter and range dependent. When clutter power dominates over noise (i.e., when SNR $\gg$ SCR) the radar is said to operate in a clutter limited condition. In this case SCR is calculated as

$$\text{SCR} = \frac{\sigma_t}{\sigma_c} \tag{1.34}$$

where σ_t and σ_c are the RCS of a target and clutter, respectively. Clutter may occur as distributed clutter, which increases with radar resolution, and as point clutter, which does not. Clutter characteristics and surface and volume clutter calculations are discussed in the next section.

1.5. RADAR SIGNAL ENVIRONMENT

A total radar echo usually consists of (1) target, (2) noise, (3) clutter, and (4) interference signals, all of which randomly fluctuate with time. This means, a radar signal environment is a stochastic environment. Usually, the target signal is embedded within a background (noise + clutter + interference), its power level is much less than the others and it is extremely difficult to extract it. The process of extracting useful information (generally the target) from the total echo is called (stochastic) *signal processing* and performed via powerful, intelligent algorithms. The power of these algorithms arise from the physical understanding of the target, noise, clutter, and interference signals.

1.5.1. Target

A radar target is characterized by its EM reflectivity. This reflectivity is called the *radar cross section* (RCS) σ and is defined as

$$\sigma = \text{RCS} = \lim_{R\to\infty} 4\pi R^2 \frac{|E_s|^2}{|E_i|^2}\ \text{m}^2 \tag{1.35}$$

where R, E_i, and E_s are the distance, electric fields of the illuminating and target-scattered waves, respectively. Spherical coordinates are of interest in RCS calculations, so both incident and scattered fields may be either E_θ or E_φ. Also, RCS is a far field concept and has to be measured and/or simulated at ranges sufficiently far from the radar transmitter. In other words, the illuminating wave has to be a plane wave. σ has the dimension of area in m^2, or in dB m^2 (referred to 1 m^2).

RCS of a target is classified according to the types of radars (monostatic and bistatic radars), and the polarization of the transmitter and the receiver. In most of practical cases, RCS of a target is

Table 1.6 Co- and Cross-Polarized RCS Cases

Linear[a]	VV	VH
	HH	HV
Circular[b]	RL	RR
	LR	LL

[a]V: vertical, H: horizontal.
[b]R: right, L: left.

mentioned for the four cases listed in Table 1.6, when RCS of a target is of interest these parameters should be given:

Angle of incidence (θ_i,φ_i) and angle of scatter (θ_s,φ_s)
Incident and scattered field polarizations
Frequency and target geometry (size)
RCS value (in m^2 or dB)

Depending on the radar operating frequency (i.e., wavelength) and size of a target, RCS of a radar target falls into one of three characteristic regimes (where qualitative as well as quantitative differences occur) [15];

Low frequencies (*Rayleigh region*) where target dimensions (l) are much less than the radar wavelength ($1 \ll \lambda$). In this region a radar target acts as a point reflector, and its RCS is proportional to the fourth power of frequency ($\sigma \approx f^4$).

Medium frequencies (*resonance region*) where target dimensions and the radar wavelength are of the same order ($l \approx \lambda$). In this region, the target contributes to its RCS as a whole, (which is called *bulk RCS*), therefore mathematical RCS calculations are almost impossible for targets with complex geometries. Fortunately, there are powerful time and frequency domain RCS tools in this regime [16]. Aircraft and ships have dimensions (typically tens of meters) that put them in the resonant scattering regime for HF radars, for example (where radar wavelengths are also of the order of tens of meters).

High frequencies (*optical region*) where target dimensions are very large compared to the radar wavelength ($l \gg \lambda$). For example, aircraft and ships mentioned above fall in this region for the microwave radars (where radar wavelengths are of the order of centimeters). In this region, RCS is roughly the same size as the real area of target. Local characteristics within the area of illumination usually dominate the target RCS and high-frequency asymptotic techniques, such as geometric optics (GO), geometric theory of diffraction (GTD), physical optics (PO), physical theory of diffraction (PTD), and uniform theory of diffraction (UTD), can be applied [17] in this regime.

Spatial and temporal RCS fluctuations differ from target to target. The fluctuations may be slow (correlated within seconds, e.g., a tanker ship navigating on a calm sea) or fast (correlated only in miliseconds, e.g., a maneuvering high speed fighter aircraft); the fluctuating RCS values may be distributed or may accumulate around a few dominant scatterers. These are categorized in terms of probability distribution functions and are grouped into three as

Steady targets like mountains, buildings, etc. (TYPE 0)
Group of small scatterers (without dominant contribution) (slow, TYPE 1 and fast, TYPE 2)
Group of scatterers with a few dominant ones (slow, TYPE 3 and fast, TYPE 4)

These types are also called *Swerling types* (SW) and determine radar waveform as well as echo integration process in the radar receiver. For example, since SW 1 and SW 3 targets (which

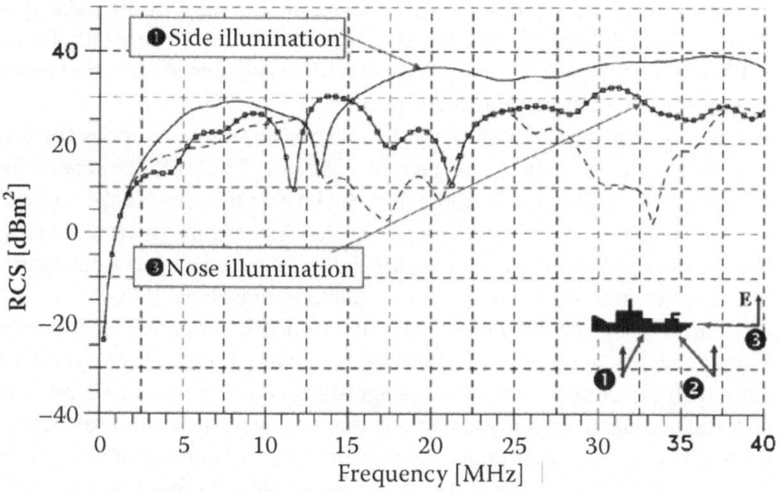

Figure 1.16 RCS vs. frequency of a 45 m-long, 20 m-wide navy frigate model.

have a decorrelation time of seconds) are stationary from pulse-to-pulse (with PRI of miliseconds) pulse-to-pulse coherent integration can be applied. Target echo fluctuations act upon detectability factor and, in general, are considered a loss factor. They are included as a fluctuation loss in the radar equation. Fluctuation loss is small (large) for low (high) P_d and heavy for SW 1 and SW 2 type targets.

In general, RCS of a target cannot be given in terms of simple mathematical functions, and either powerful numerical simulation methods or real measurements are used to obtain RCS values of different targets. On the other hand, simple relations can be used for simple geometries in optical RCS regime. For example, the RCS of a perfectly electrical conductor square plate is given as

$$\text{RCS} = \frac{4\pi a^2}{\lambda^2}\ \text{m}^2 \qquad \text{(for vertical illumination)} \tag{1.36}$$

where a is the edge dimension and λ is the radar wavelength. At 3 GHz (with S-band radar), a 1-m^2 PEC square plate yields nearly 1200 m^2 (~31 dB) RCS and increases to more than 10,000 m^2 (~40 dB) at 10 GHz (X-band radar).

RCS of a target may be given as a figure (or table), where, for example, RCS vs. frequency is given, or, as radial plots, where mono- and/or bistatic RCS variations at different frequencies are plotted. A typical example is given in Fig. 1.16, for a navy frigate model calculated via the FDTD technique [15,16].

1.5.2. Noise

Sensitivity of a radar receiver shows the lowest signal level that is measurable with the device (which is called floor signal). Usually, the receiver sensitivity, i.e., the floor signal, is determined by either internal or external electromagnetic disturbances. Electromagnetic disturbances caused by internal and/or external, man-made or natural are generally called noise. For example, man-made sources, such as power lines, electric razors, hair-dry machines, etc., and, natural sources, such as, lightning, electrical storms, galactic effects, etc. are typical external noise sources. On the other hand, power amplifiers, mixers, diodes, and transistors are some of the internal noise sources, because of random collisions of the electrons. Various noise sources and their effective frequency ranges are shown in Fig. 1.15.

Roughly speaking, noise can be classified into internal and external in most of the radar systems. Internal noise, which is also called *thermal noise* (as explained in Sec. 1.4.1), generally limits the detection threshold in MW radars (typically at frequencies 100 MHz and above). External noise determines the detection threshold at HF and partially VHF radars.

The amount of noise that is present at the input of a radar receiver depends on various factors, such as receiver bandwidth, antenna radiation pattern, antenna side lobes, or objects of illumination. In practical radar receivers, the noise level is found to be more than the level at the input of the receiver by a factor known as *noise figure*. The ratio of the noise present at the output of the radar receiver to the noise due to thermal effects alone is called the *receiver's noise figure*. Noise figure is also an important parameter that determines the maximum range in the radar equation given above.

Noise is a random signal and must be handled in a stochastic manner. Its amplitude distribution characteristics, average value, deviation, frequency characteristics (power spectrum), etc. must be well understood when dealing with detection theory. The distinguishing characteristic of the noise is that it is a pulse-to-pulse uncorrelated signal (usually called as white noise or gaussian noise). Signal correlation in time domain and power spectrum in frequency domain forms a Fourier pair. An uncorrelated (a delta type) function in time domain corresponds to a constant value in frequency domain. Therefore, the larger the receiver bandwidth, the higher the noise level in radar receivers.

The sensitivity of a radar is determined by its ability to maximize the SNR of the received echo. In other words, noise must be minimized while amplifying the target signal. The probability that a noise spike will reach a certain level at the output terminals of the receiver is given by a Rayleigh distribution function. The probability of a target signal imbedded in a noise is given by a gaussian distribution. Using these two distributions (under noise limited detection conditions), detection probabilities and false alarm rates can be calculated mathematically.

Noise elimination in a radar receiver can be achieved by integration, either in predetection or postdetection stages [2].

1.5.3. Clutter

Clutter is a word used to describe all unwanted echoes in a radar receiver. Clutter can be characterized as a distributed nondirectional source. Depending on the "mission" of a radar, what is clutter in one application may not be so in another. For example, a radar designed to detect aircraft includes the echoes from land, sea, clouds, rain, birds, insects, etc., which are all called *clutter.* On the other hand, aircraft is one of the clutters for an HF radar designed for oceanographic surveillance. Similarly, backscatter echoes from land can degrade the performance of many radars (as land clutter) but represent the target of interest for a ground-mapping radar.

Unwanted echoes usually occur as distributed clutter, as surface clutter (such as land and sea echoes) or as volume clutter (such as rain, chaff). Because of its distributed nature, clutter is characterized in terms of RCS density, rather than the RCS as described for conventional radar targets (ships, aircraft, etc.). For surface clutter, the average RCS density σ_0, the RCS per unit area, is given by the ratio

$$\sigma_0 = \frac{\sigma_c}{A_c}\,\mathrm{m}^2/\mathrm{m}^2 \tag{1.37}$$

where σ_c is the RCS of the area A_c. Similarly, volume clutter RCS density is given as the average RCS of a unit volume V_c as

$$\eta_0 = \frac{\sigma_c}{V_c}\,\mathrm{m}^{-1} \tag{1.38}$$

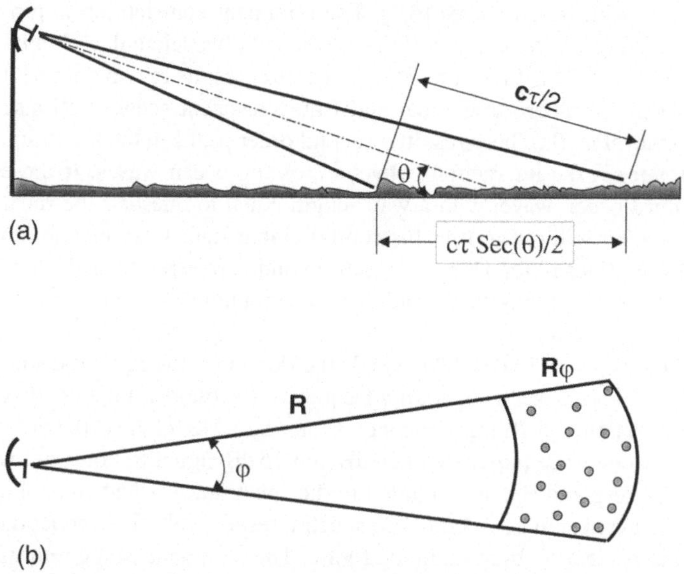

Figure 1.17 Geometry of a radar clutter: (a) elevation view and (b) plan view.

where σ_c is the RCS of a unit volume V_c. The values of both surface and volume RCS densities depend on many factors, such as, the type of terrain observed, the direction of illumination and observation, radar wavelength, polarization. For example, the ocean RCS density depends on ocean wave height, wind characteristics (direction and speed), and wave direction. Ground RCS density depends on soil type, surface roughness, foliage cover, etc. Typical RCS densities, for example, for ocean are -80 dB at S band for $0.1°$ grazing angle for vertical polarization, but increases to -45 dB at X band for $3°$ grazing angle for the same polarization. On the other hand, values up to -30 dB and -35 dB are used for HF frequencies.

The geometry of surface clutter is pictured in Fig. 1.17. With the parameters mentioned in the figure (i.e., pulse length τ, range R, azimuth beam width φ, elevation angle θ) the clutter RCS can be given as

$$\sigma_c = \sigma_0 R\varphi \frac{c\tau}{2} \sec\theta \ \mathrm{m}^2 \tag{1.39}$$

and the signal-to-clutter ratio in Eq. (1.34) reduces to

$$\mathrm{SCR} = \frac{\sigma_t}{\sigma_0 R\varphi (c\tau/2) \sec\theta} \tag{1.40}$$

Ocean and land clutters are basic surface clutters that determine the performances of surveillance radars. They are also stochastic processes and are characterized by temporal as well as spatial distribution characteristics. Both are pulse-to-pulse coherent signals, so elimination by just averaging (postdetection integration as done for noise) is not possible. Usually, clutter elimination is performed in FD. Land clutter occupies a very low frequency range around zero Doppler frequency. On the other hand, ocean clutter has different Doppler characteristics at different radar frequencies.

A typical example is given in Fig. 1.18a, where (synthetically produced) Doppler characteristics of the ocean clutter are pictured for a HFSW radar. Ocean clutter is the result of the interaction of the radiated electromagnetic wave with ocean waves [6,7]. The dominant contribution is produced by scatter from ocean waves having a wavelength half that of the radar wavelength and moving radially to and away from the radar site. This first-order resonant scatter results in two dominant peaks called *Bragg lines* [1,2]. Ocean waves are trochoidal and Bragg resonant scatter will also occur at harmonics of the principal wavelength. These result in second order peaks in the spectrum. Another source of second-order scatter is the interaction between crossing ocean waves. If these crossing ocean waves generate a third ocean wave, with a wavelength equal to one-half the radar wavelength, then Bragg resonance scatter will occur. It is this condition that leads to an increase in the continuum level between the Bragg lines in the Doppler spectrum and is referred to as the second-order continuum. The energy contained within the second-order continuum is related to the sea state and hence surface wind speed and duration.

In Fig. 1.18a, the operating frequency is 5 MHz. This yields the dominant Bragg frequencies of ± 0.228 Hz that are normalized to ± 1 Hz. For this operating frequency, the blind velocities (they correspond to ocean waves speed resonating at Bragg frequencies) are ± 13.7 kn (1 kn $\approx$ 0.5 m/s). The clutter to noise ratio is 30 dB. Two surface targets with 20 dB and 15 dB signal to noise ratios and 14 kn and 20 kn radial velocities, respectively, are included in the spectrum. Dashed and solid lines correspond to one spectrum and average of consecutive 20 spectra, respectively. The reduction in noise floor by spectrum averaging is clearly observed in the figure. The dominant Bragg returns at ± 1 Hz, second-order continuum and the target with 20 kn velocity are clearly seen in the figure. The other target with 14 kn radial velocity is obscured by the dominant Bragg return at 1 Hz.

Usually, the real Doppler spectra are not as pure as the simulated ones, which means detection decisions are really hard to give. Typical real Doppler spectra recorded between 1998 and 1999 is given in Fig. 1.18b (recorded at Cape Bonavista with a HFSW radar operated at 3.6 MHz). On top, Doppler spectra at two different radial ranges along a chosen beam is shown. At the bottom, range profile at fixed radar beam, at two different times are plotted. A threshold level of 20 dB below the thermal noise floor is chosen for the vertical axis for the peak signal power level. The complexity of ocean clutter spectrum is clearly observed in these plots.

There are many differences between ocean and land clutter. When compared to ocean clutter, land clutter is less time dependent, but backscatter from land is significantly greater than ocean clutter in most cases.

1.5.4. Interference

Interfering signals received within the total radar echo may be classified into different groups:

Intentional broadcast and communication systems and their harmonics
Local and remote cochannel signals
Intentional jamming signals (ECM, electronic counter measures)
Unintentional multipath arrivals (e.g., ground reflected target echo or signals from mutual interactions between different nearby targets)
Scattered signals from atmospheric discontinuities, etc.

These interfering signals have different characteristics, which determine the techniques that may be applied in interference cancellation. For example, radio broadcast signals may be narrow band and direction dependent, therefore, the interfering signal may appear in one azimuth beam along all the range gates and may not in nearby beams. Applying a correlation process between the cells of different beams and/or along range cells may be an effective interference cancellation technique. On the other hand, jamming signals are high power, broad band signals; therefore, angular and/or range correlation may not be effective. In this case, correlation in the frequency domain may be a solution.

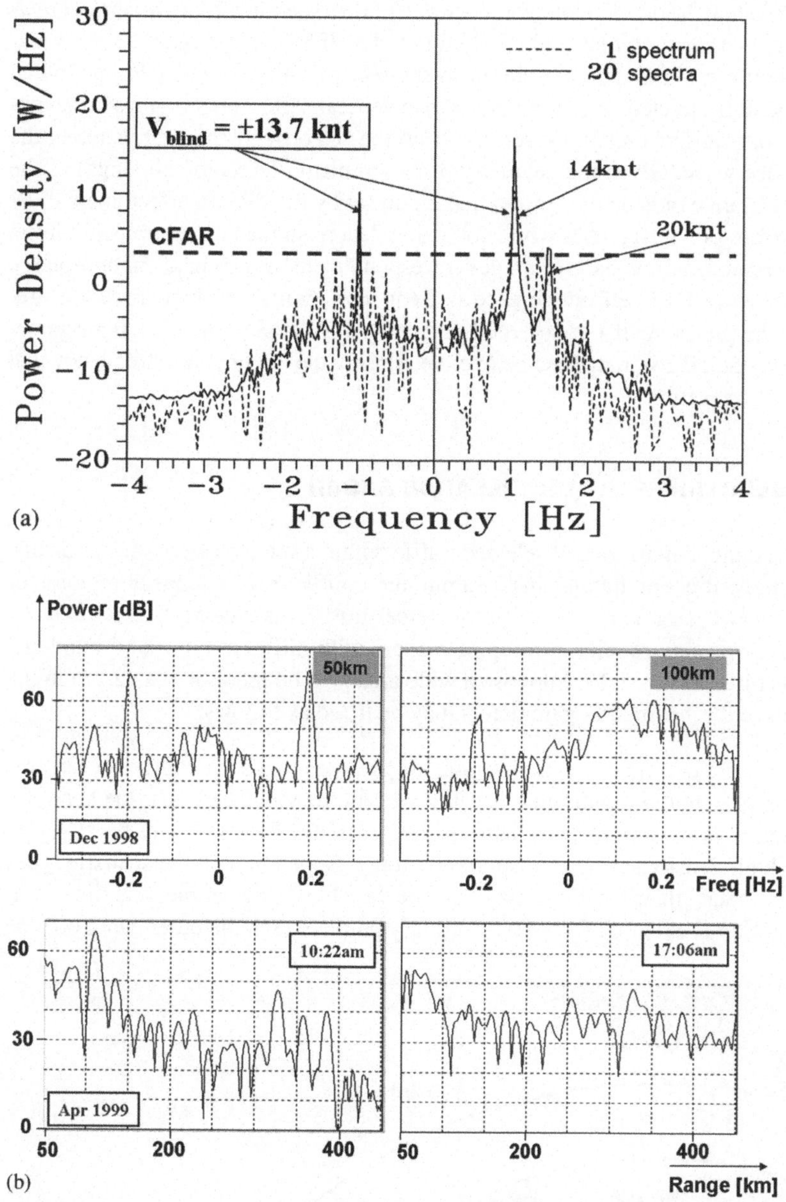

Figure 1.18 (a) Typical (synthetic) HF ocean spectra at 5 MHz. (b) Typical (real) HF ocean spectra at 3.6 MHz.

Cochannel interference, coming from local or remote sources, may also be a problem in the detection process. Local interfering signals are generally from known sources and interference can be avoided by choosing alternate frequencies. Interference from distance sources poses a more serious problem in that it is more random in time and frequency.

Interfering signals depend also on the type of the radar in operation. For example, ionosphere is a severe interference path for HFSW radars. Not all the energy emitted by the HFSW radar propagates along the surface. Some energy is directed upward and, as with short-wave radio, may, under certain conditions, reflect from the ionosphere. In some cases, the energy reflected from the ionosphere returns to the radar. This signal may be viewed as multipath clutter or self-interference.

Ionospheric self-interference may be divided into two main categories, specifically, *near vertical incidence (NVI) clutter* and *range folded clutter.* With NVI clutter, the HFSW radar signal travels vertically from the radar and is reflected from an ionospheric layer directly back to the radar. Range folded clutter occurs when the signal is directed at an angle other than vertical. After reflecting from the ionosphere the signal travels outward whereupon it reflects from the sea or land and returns along the same path, or via the surface wave. Given the geometry of the problem, the total path length of the returned signal places it at a range outside the system maximum set by the PRI. In effect, the HFSW radar will receive returns from previous pulses while collecting data from the current transmit pulse. NVI self-interference appears at a narrow band of ranges corresponding to the height of an ionospheric layer. One option for combating NVI self-interference is frequency agility. By increasing the frequency, the layer-critical frequency will be exceeded and the HFSW radar signal will penetrate through the layer. Similarly, the HFSW radar can be operated during the daytime at a frequency that does not support skywave propagation.

1.6. PARAMETER SELECTION FOR SURVEILLANCE RADAR

Radar surveillance is to maintain cognizance of selected traffic within a selected area, such as an airport terminal area, air route, critical mountains inside a military conflict region, coastal regions for offshore security, and waterways or narrow straits for vessel traffic management. This may be achieved with a single radar or may require a group of sensors with different types and numbers. One typical scenario is pictured in Fig. 1.19. Suppose a wide area is to be monitored as given in the figure. Basic requirements and fundamental parameters may be listed as follows:

The requirement may be a tactical coverage for military purposes in a high-conflict ocean area or cruise and tanker traffic monitoring in a heavy traffic region. There may be beautiful islands, fishing regions, and/or petroleum drilling regions. The area may be along an international high-density surface transportation route. Depending on the scenario types and number of radars, their locations may be quite variable. For example, real time, continuous monitoring is essential for most of military purposes. On the other hand, off-line

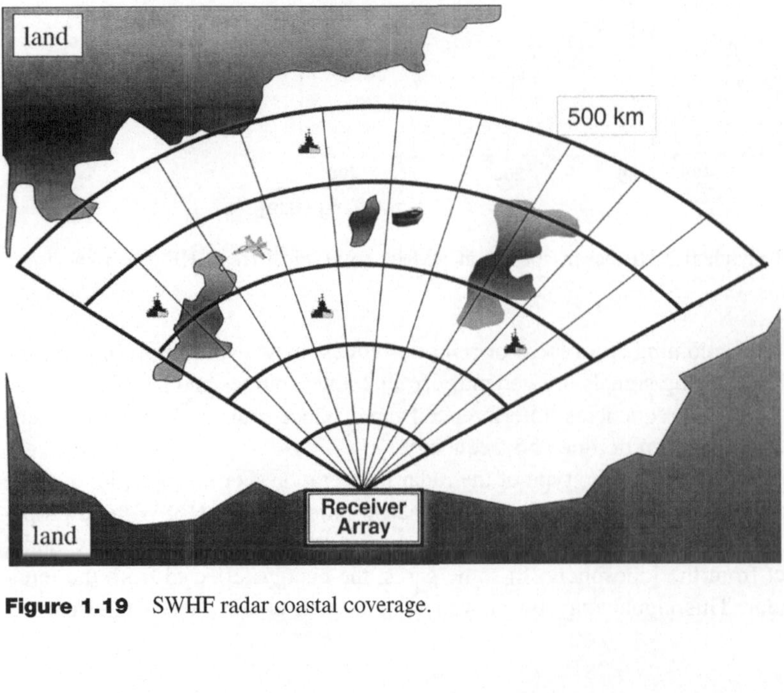

Figure 1.19 SWHF radar coastal coverage.

monitoring with a few hours update may be adequate for monitoring illegal fishing or cruise transportation, etc.

In the scenario in Fig. 1.19, suppose the region, up to 500 km in range and 120 degrees in azimuth is to be covered and both surface and air targets are to be monitored continuously. This is a typical scenario for countries with long coastal regions. The United Nations Convention on the Law of the Sea (UNCLOS) gives coastal nations sovereign rights over 200 nautical miles (nm) of sea known as the Exclusive Economic Zone (EEZ). In return countries are required to establish and maintain Administration, Law Enforcement, and Environmental Protection over this new frontier that is many times larger than their previous 12 nm territorial limits.

What sensors are available to monitor these typical regions? Traditional land-based MW radars are limited to operate within LOS. Even by elevating the radar platform the maximum range is limited to 50–60 km. The EEZ can be covered by a number of airborne radars, but these provide only a snap shot in time of activity within the EEZ. Sky-wave high-frequency (HF) radars can be used for this purpose, but they need large installations, are expensive and detection of surface targets is still limited. Satellites have neither the spatial nor the temporal resolution to provide the necessary level of real-time surveillance.

An optimal solution may be an integrated maritime surveillance (IMS) system that uses multitype multisensors. Effective surveillance also requires the integration of data from a number of complementary sensors. The primary sensor for this scenario may be one or two HFSW radars that are capable of tracking both surface and airborne targets at ranges in excess of 200 nm. The radar data may be enhanced with target identification obtained from automatic identification systems, such as Automatic Dependent Surveillance (ADS) systems [6,7], IFF, as well as information obtained from patrol vessels, communications, mandatory reporting procedures, etc. This information is associated with the radar data to provide a complete, real time, picture of activities within the EEZ.

Once types and number of radars are determined, power requirements, receive and transmit antenna systems, optimal radar waveform, signal and data processing hardware, and software, display utilities, etc., are to be taken into account for the types of targets to be monitored.

1.7. HFSW RADAR-BASED WIDE-AREA SURVEILLANCE

The region in Fig. 1.19 may be covered with one SWHF, located along a shore as pictured. It should be long range (up to 500 km) radar with wide azimuth coverage capability. HFSW radars use the lower end of HF frequency band (3–6 MHz) to provide the required coverage. A typical HFSW radar site is pictured in Fig. 1.5a. A broad band transmit antenna is located 100–200 m away from the receiver array that consist of 16 subarrays [7]. The arrays are located parallel to the shoreline with a clear field of view of the desired coverage area and may be located on a beach or cliff. The receive array yields a nominal beam width, at boresight, of approximately 5–10 degrees. For optimum performance the radar system should be located at an electrically quite area (where environmental noise level is as low as required) as defined by International Radio Consultative Committee (CCIR) [6,7].

The antenna system for an HFSW radar is designed to satisfy a number of criteria. The transmit antenna must provide a high gain over the specified band. Energy must be distributed equally and only over the desired surveillance area. The receive array must be parallel to the shore line, have high and equal array gain over the entire surveillance area with minimum sensitivity to signals arriving from other directions. Both transmit and receive arrays must also provide a deep, broad, null at NVI. When predicting the performance of both the transmit and receive arrays the effect of local site topography and ground conditions must be considered. Operating at the low end of the HF band requires that the receive array occupies a significant shoreline area, with the aperture of the array inversely proportional to frequency. For example, at 3 MHz, a 5° azimuthal beam width requires an array aperture of approximately 1 km.

The HFSW detects targets in three-dimensional space: range, bearing, and velocity (Doppler):

1. The *range resolution* is directly proportional to the bandwidth of the transmitted waveform. HFSW radar typically operates with a maximum bandwidth of 10–20 kHz, for which the resolution is about 8–15 km (note that several MHz bandwidths are used in MW radars where a few centimeters of range resolutions are of interest).
2. The *azimuthal resolution* is directly proportional to the aperture size of the antenna. The aperture is measured in terms of the radar wavelength. For an antenna array with element separation of half a wavelength, 16-element to 24-element antenna arrays produce beam widths at boresight of approximately 5–10 degrees. This translates to a cross range of 50–60-km at a distance of 400–500 km. As the beam is steered away from boresight, the beam width increases (azimuth beam width is usually less than a degree in MW radars).
3. The *velocity resolution* is directly proportional to the coherent integration time. HFSW radar employs three simultaneous integration intervals corresponding to 20 s for air targets, 164 s for ship targets and 1200 s for near stationary targets. These correspond to velocity resolutions of 4, 0.5, and 0.06 kn, respectively (CIT of miliseconds are used in MW radars).

Even though the resolution capabilities of HFSW radars seem at first glance to be rather moderate, in practice it is not a serious issue since two targets can be resolved provided that they are separable in one of the three dimensions. The probability of having two targets in close proximity in all three dimensions is not high. In the event that two targets are not resolvable, the radar will track either the larger or the composite return of the two until such time that they can be resolved.

Another performance parameter is the accuracy of the estimate of target position. Although the resolution capabilities of HFSW radars are moderate, an accuracy of better than one-tenth of the basic resolution can be achieved with even moderate signal-to-noise ratios.

A typical HFSW radar receiver is pictured in Fig. 1.20. The receiver obtains echoes from the operational area and two-step gating is applied: range gating followed by digital beam forming. As shown in the figure time histories of data of N × M resolution cells are accumulated (coherently integrated) and Fourier transformed (via FFT), and detection is achieved by Doppler processing. Target detection using a CFAR algorithm, follows beamforming. Different CFAR variants are used for surface and air targets as well as constant-velocity and manoeuvring targets. Because of the

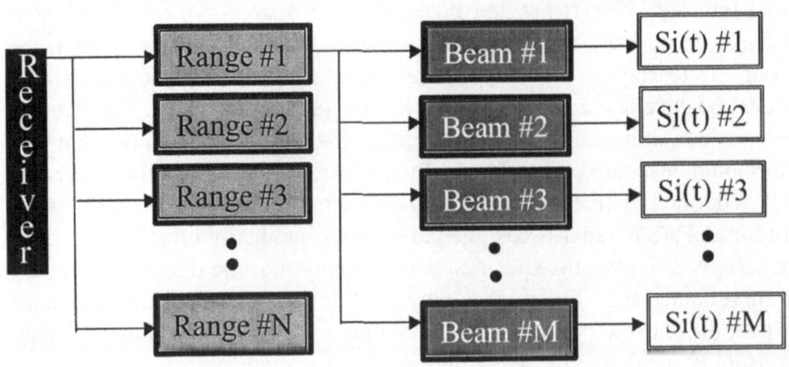

Figure 1.20 SWHF Receiver stages.

complexity of the real signal environment, it is desirable to employ adaptive parameters in the detection process to accommodate clutter, noise, and interference levels that vary from a CIT to CIT as well as from cell to cell.

HFSW radars are coherent radars and detection is based on the SNR at a given Doppler where the noise bandwidth is determined by the inverse of the coherent integration time. A consequence of the much reduced noise bandwidth is that the PD, even at a relatively low SNR, is extremely high and for the noise-limited case, the PFA at a given Doppler bin is very low. The number of false alarms for a given CIT is equal to the probability of false alarm multiplied by the number of independent range/azimuth Doppler cells (approximately a few million). Consequently even a moderate probability of false alarm translates into a high number of false detections. These false detections are dominated by ionospheric clutter and are characterized as rings of detections corresponding to the vertical range of the various ionospheric layers. The high number of high false alarm associated with HFSW radars is typical as illustrated in Fig. 1.21. Here, detections associated with 30 consecutive CITs (approximately 75 min) have been plotted on a range-azimuth scale. The figure is derived from real data collected in February 1999 [7]. The 10 to 20 surface target tracks among the large number of false detections (as many as thousands) may be observed in the figure. The spoking of the data can be attributed to detections that occurred in a single beam. This high false alarm rate is almost unavoidable if the number of missed detections is to be minimized. Therefore, the Tracker must be designed to accommodate these false alarms but ensure that they do not propagate through the system to generate false tracks.

For a target that travels at a constant velocity within a CIT interval, the echo is characterized by an impulse in the Doppler spectrum. Surface targets appear in the vicinity of the Bragg peaks, while air targets are generally far removed from this sea clutter region and are typically detected against a noise background. Separate, optimized, detection processes are used to accommodate both the clutter and noise limited detection scenarios.

A typical HFSW radar picture is given in Fig. 1.22, which was obtained in Cape Race on June 9, 1999 (around noon). The data set are for days when a ground truthing aircraft was used to verify both the targets under track and to search for any targets not seen by the radar. In the figures a "?" corresponds to an independently observed target. The suffix V indicates a visual observation and the

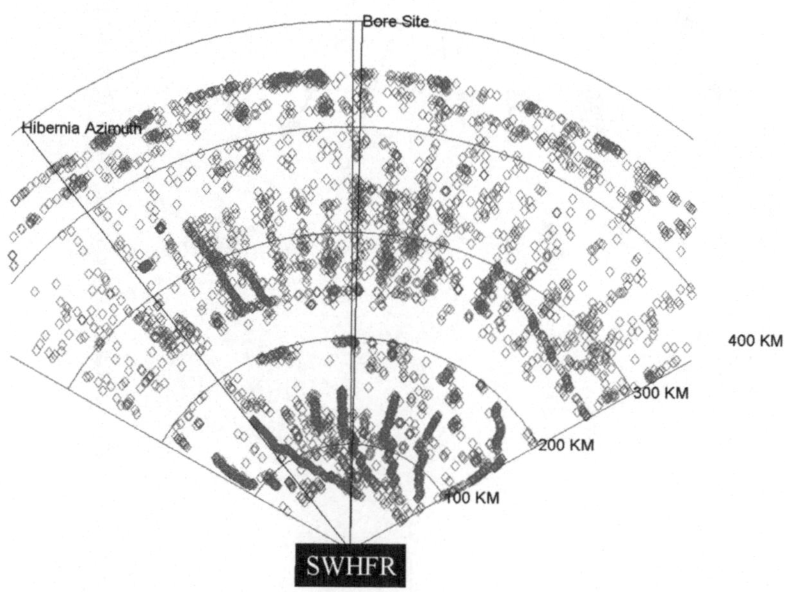

Figure 1.21 Detection of a SWHFR in 30 CIT.

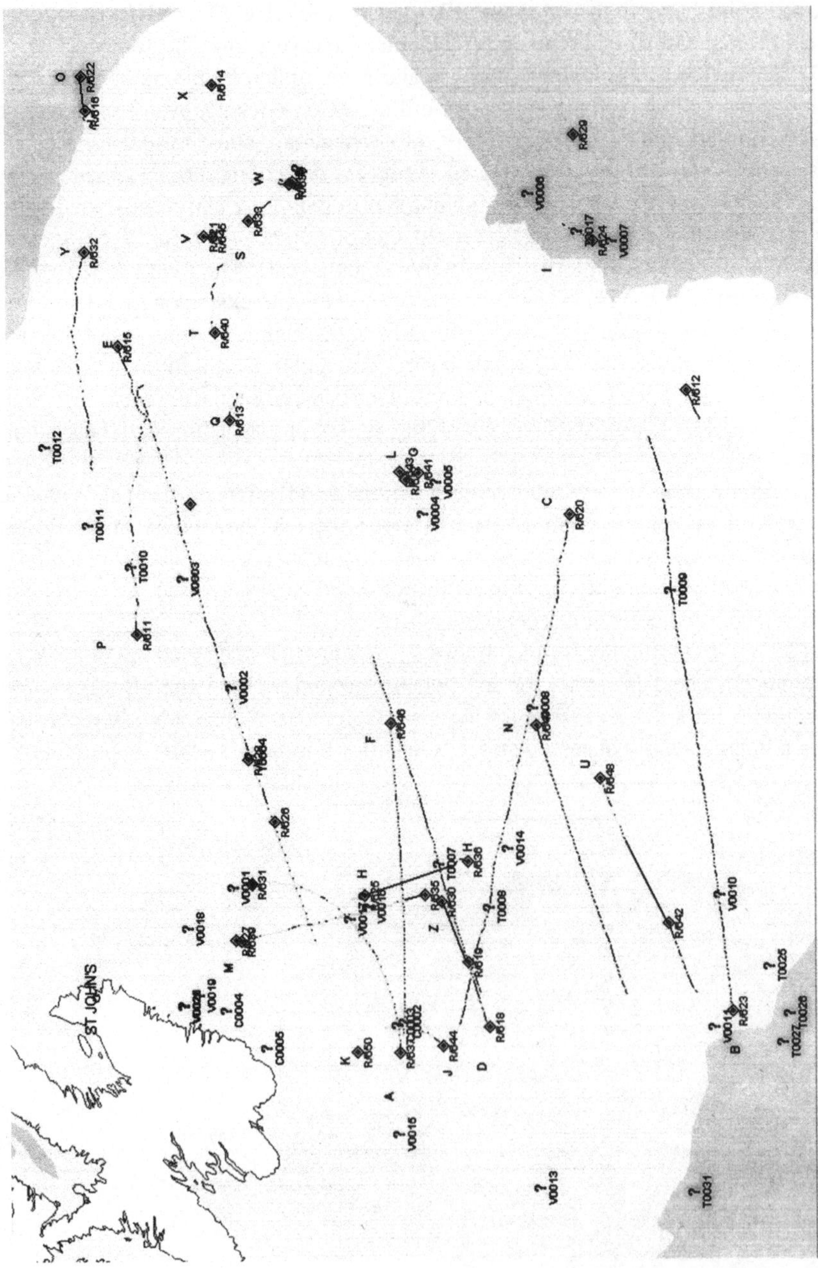

Figure 1.22 Typical (real) HFSWR system picture (Cape Race, June 2000).

suffix T represents an airborne radar observation. It can be observed that the HFSWR successfully tracked all targets observed by the aircraft. Those targets that were tracked by the radar but were not observed by the aircraft entered the coverage area after the aircraft had finished searching that zone.

1.8. CONCLUSION

Radars are electronic sensors that extract information from EM wave–object interaction. From simple detections back in 1940s to today's complex multisensor integrated systems, radars have had a great impact on modern life. They are used as monitoring, guiding, controlling, etc., instruments in a variety of applications. Since electronic sensing information will always be required, radar will continue to play essential roles in modern societies in the future.

REFERENCES

1. Skolnik, M.I. *Introduction to Radar Systems;* McGraw Hill: New York, 1985.
2. Skolnik, M.I. *Radar Handbook;* McGraw Hill: New York, 1990.
3. Barton, D.K. *Modern Radar System Analysis*; Artech House: Norwood, MA, 1988.
4. Skolnik, M.I. Fifty years of radar. Proc. IEEE **1985**, *73*.
5. IEEE Standard Radar Definitions, std-686-1990.
6. Sevgi, L.; Ponsford, A.M.; Chan, H.C. An integrated maritime surveillance system based on surface wave HF radars, Part I—Theoretical background and numerical simulations. IEEE Antennas Propagation Mag. **2001**, *43*(4), 28–43.
7. Ponsford, A.M.; Sevgi, L.; Chan H.C. An integrated maritime surveillance system based on surface wave HF radars, Part II—Operational Status and System Performance. IEEE Antennas Propagation Mag. **2001**, *43*(5), 52–63.
8. Popoulis, A. *Probability, Random Variables and Stochastic Processes*; McGraw Hill: New York, 1985.
9. Farina, A. (Ed.). *Optimized Radar Processors*; IEE Publication: London, U.K., 1987.
10. Sevgi, L.; Akleman, F.; Felsen, L.B. Ground wave propagation modeling: problem-matched analytical formulations and direct numerical techniques. IEEE Antennas Propagation Mag. **Feb. 2002**, *44*(1), 55–75.
11. Sevgi, L.; Felsen, L.B. A new algorithm for ground wave propagation problems based on a hybrid ray-mode approach. Int. J. Numerical Modeling **1998**, *11*(2), 8–103.
12. Hall, M.P.M.; Barclay, L.W.; Hewitt, M.T. *Propagation of Radiowaves*; IEEE Publication: London, U.K., 1996.
13. Fock, V.A. *Electromagnetic Diffraction and Propagation Problems*; Pergamon: Oxford, 1965.
14. Kerr, D.E. (Ed.). *The Propagation of Short Radio Waves, Radiation Lab. Series*; McGraw-Hill: New York, 1951.
15. Shaeffer, J.F.; Tuley, M.T.; Knot E.F. *Radar Cross Section*; Artech House: Norwood, MA, 1985.
16. Sevgi, L. Target reflectivity and RCS interaction in integrated maritime surveillance systems based on surface wave HF *radar radars*. IEEE Antennas Propagation Mag. **2001**, *43*(1), 36–51.
17. Balanis, C.A. *Advanced Engineering Electromagnetics*; Wiley: New York, 1989.

Other Fundamental Radar Books

18. Meikle, H.D. *Modern Radar Systems*; Artech House: Norwood, MA, 2001.
19. Sullivan, R.J. *Microwave Radar: Imaging and Advanced Concepts*; Artech House: Norwood, MA, 2001.
20. Ince, N.; Topuz, E.; Panayirci, E.; Isik C. *Principles of Integrated Maritime Surveillance Systems*; Kluwer Academic: Boston, 2000.
21. Kingsley, S.; Quegan, S. *Understanding Radar Systems*; McGraw Hill Co.: London, U.K., 1992.
22. Nathanson, F.E. *Radar Design Principles*; McGraw Hill: New York, 1991.
23. Eaves, J.L.; Reedy, E.K. (Eds.). *Principles of Modern Radar*; Van Nostrand Reinhold: New York, 1987.
24. Eaves, L.; Reedy, E.K. *Principles of Modern Radars*, Van Nostrand Reinhold: New York, 1987.
25. Meeks, M.L. *Radar Propagation at Low Altitudes*; Artech House: Norwood, MA, 1982.

2
Wireless Communication Systems

Nathan Blaunstein
Ben-Gurion University of the Negev
Beer Sheva, Israel

2.1. INTRODUCTION

This chapter provides a basic tutorial on key aspects of wireless communication systems. Because the optical communication systems are also "wireless," we must declare at the outset that such systems are not a subject of this chapter despite the fact that radio waves and optical waves are both independent parts of the electromagnetic spectrum. Therefore, this chapter will focus on antennas as transducers of radio waves, radio propagation in the wireless communication channels with emphasis on land communication channels. Specific propagation models for various land environments (rural, forested, hilly, built-up) will be presented with a view to understanding the main propagation characteristics in such environments, such as path loss, and slow and fast fading effects. The cellular concept for wireless systems and a strategy for cell design will also be discussed briefly.

2.1.1. Definition of the Wireless Communication System

Although wire-based communication systems, such as telephony which connects each telephone with a central operator station through a pair (or more) of copper wires or cables (usually called the *local* or *subscriber loops*), have been successfully employed for more than a century, during recent decades wireless communication systems have been developed to satisfy continually increasing demands for personal, local, mobile and satellite communications, by enhancing and replacing wire-media loops with wireless communication media.

Depending on the specific application of wireless communications, these media include the subsoil layers, water and ground surface, atmosphere, ionosphere, and cosmic space. We will treat each medium as a radio propagation channel across which radio signals are sent. These signals are created by transmitting antennas whose dimensions, as we will show later, are approximately the same as the wavelength of the radio signal generated. According to the Huygens principle, these waves are similar to the light rays in optics, which can be generated by a light bulb or a spot light. These radio waves are captured by the receiving antenna connected to a receiver.

Thus, a basic wireless communication system consists of a transmitter (T), a receiver (R), and the radio propagation channel (Fig. 2.1 according to Ref. 1). As follows from the simple scheme depicted in Fig. 2.1, there are *three* main independent electronic and electromagnetic design tasks related to this communication system. The *first* task is the specification of the electronic equipment that controls all operations within the transmitter, including the transmitting antenna operation. The radio propagation channel, denoted as a *second* element in the scheme presented in Fig. 2.1, plays a separate independent role. Its main output characteristics depend on the conditions of radio wave propagation in the various operational environments. The *third* task concerns the same operations

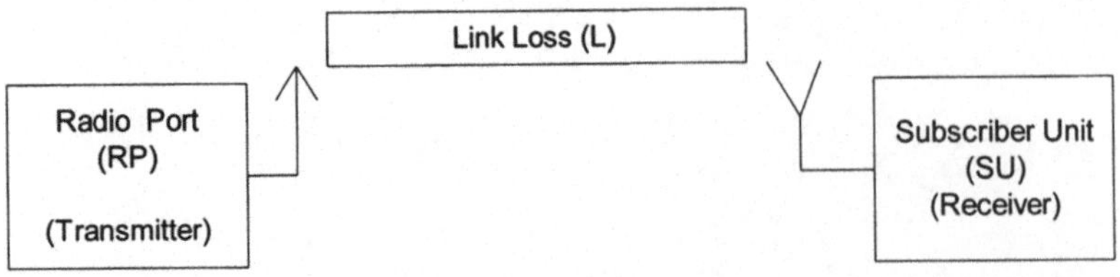

Figure 2.1 The simple scheme of three main independent electronic and electromagnetic design tasks related to the wireless communication channels.

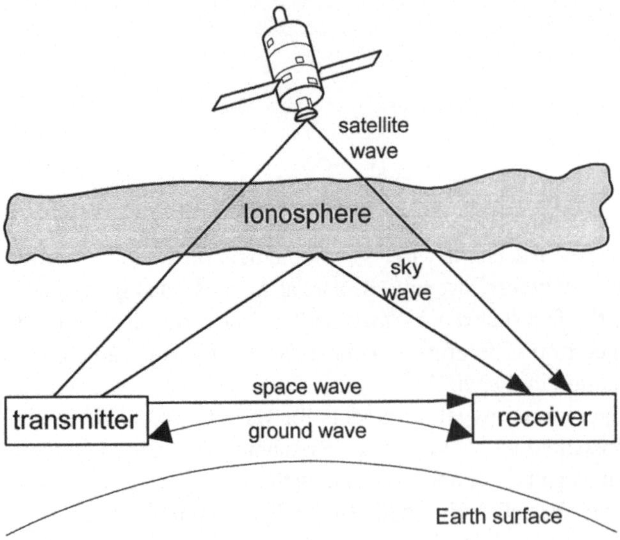

Figure 2.2 Different kinds of radio waves.

and signals, but for the receiver, with its own peculiarities. For both kinds of antennas, the transmitting and the receiving, an important issue is the influence of different kinds of obstacles located around the antennas and the environmental conditions.

2.1.2. Frequency Spectrum for Wireless Communications

The optimal *frequency band* for each propagation channel is determined and limited by the technical requirements of each communication system and by the conditions of radio propagation through each channel.

Extremely low and *very low frequencies* (ELF and VLF) are frequencies below 3 kHz and from 3 kHz up to 30 kHz, respectively. The VLF band corresponds to waves, which propagate through the waveguide formed by the earth's surface and the ionosphere at long distances with low attenuation [0.1–0.5 decibel (dB) per 1000 km]. Frequencies lower than 3 kHz (ELF band) are effective for underwater communication channels and for mines and subterranean communication.

Low frequencies (LF) and *medium frequencies* (MF) are frequencies from 30 kHz up to 300 kHz and 300 kHz to 3 MHz, respectively. They are useful for radio navigation of ships and aircrafts, and for broadcasting. Such radio waves propagate along the earth's surface by following the curvature of the earth, as shown in Fig. 2.2, and in the literature are called *surface* waves. Because of the long wavelengths of surface waves (for example, a 100-kHz signal has a wavelength of 3000 m), ground features, such as buildings, hills, trees, and built-up topography, do not affect the radiosignal propagation significantly.

High frequencies (HF) are those which are located in the band from 3 MHz up to 30 MHz. Signals in this spectrum propagate by means of reflections caused by the ionosphere and, therefore, are called the *sky* waves. This type of radio signals is used for long-distance land communications by use of broadcasting stations ("short-wave radio").

Very high frequencies (VHF) are located in the band from 30 MHz up to 300 MHz. They are used in a line-of sight (LOS) mode for TV communications, in long-range radar systems and in radio-navigation systems.

Ultra-high frequencies (UHF) are those that are located in the band from 300 MHz up to 3 GHz (in some literature its upper part from 0.5 GHz up to 3 GHz is also divided into P, L, S bands). This frequency band is very effective for wireless microwave links for cellular systems (fixed and mobile) and for satellite communication channels (since these frequencies penetrate the ionosphere). In the literature, these waves are sometimes called the *satellite* waves.

In recent decades radio waves with frequencies higher than 3 GHz (C, X, K bands, up to several hundred GHz, which are also loosely described as *microwaves*) have begun to be used for constructing new kinds of wireless communication channels.

2.1.3. Noise

The effectiveness of each wireless communication system depends on noise inside it, which in the literature is separated into the *additive* (or *white*) and the *multiplicative* noise [2–9]. Let us consider briefly the sources of such kinds of noise.

The *additive noise* arises from [1]

Noise in the receiver antenna
Noise within the electronic equipment that communicates with antenna
Background and ambient noise (galactic, atmospheric, man-made, etc.)

Now let us consider each type of noise, which exists in a complete communication system. Noise is generated within each element of electronic communication channel because of the random motion of electrons within the various components of the equipment. The noise power inside the transmitter–receiver electronic channel at a given system bandwidth B_w is given by [10,11]

$$N_F = k_B T_0 B_w \tag{2.1}$$

where $k_B = 1.38 \times 10^{-23}\,\mathrm{Ws\,K^{-1}}$ is Boltzmann's constant, $T_0 = 290\,\mathrm{K}$ (17°C). Taking also into account the *noise figure* F of the receiver [2,4]

$$F = 1 + \frac{T_e}{T_0} \tag{2.2}$$

where T_e is the effective noise temperature at the receiver, we can express the total effective noise power at the receiver input Eq. (2.1) as

$$N_F = k_B T_0 B_w F \tag{2.3}$$

The *multiplicative noise* arises from the processes encountered by transmitted radio waves during their travel from the transmitter to the receiver, such as (Fig. 2.3):

Multireflections from ground surface, walls, and hills
Multiscattering from rough surfaces such as the sea, rough terrain, buildings, and trees
Multidiffraction from the edges of walls, building rooftops, and hilltops

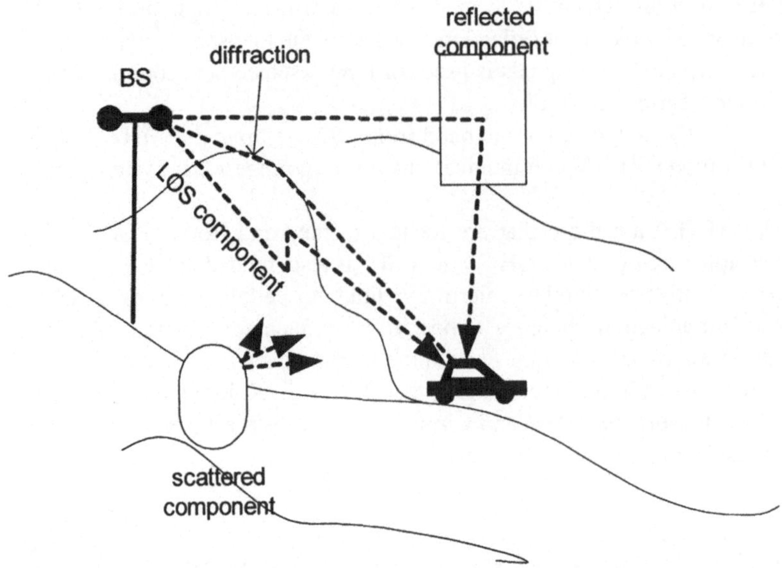

Figure 2.3 Presentation of the triple nature of fading phenomena.

To gain a better understanding of the multiplicative noise, it is very important to define the *propagation characteristics* of the radio communication channel.

2.1.4. Main Propagation Characteristics

In real communication channels, the radio waves reach the receiver in a multipath situation in which the various waves arrive with different radiopaths and time delays. At the receiver, such waves are combined to give an oscillating resultant signal, the variations of which depend on the distribution of phases amongst the incoming component waves. The signal amplitude variations are known as *fading* [1–9]. Fading is basically a spatial phenomenon, but spatial signal variations are experienced as temporal variations by a receiver and/or transmitter moving through the multipath field or due to moving scatters, such as a truck passing the area between two terminal antennas. Thus we can talk here about space-domain and time-domain variations of EM field in land environments. Moreover, if one deals with mobile communication systems, one observes the effects of random fading in the frequency domain, i.e., the complicated interference picture of the received signal caused by receiver/transmitter movements, which is known in literature as the Doppler effect [2–9].

Numerous theoretical and experimental investigations of spatial and temporal variations of radio waves in conditions of built-up areas have shown that the urban propagation channel is approximately stationary in time, but the spatial variations of signal level have a *triple nature* (see Fig. 2.4 according to Ref. 5).

The *first* one is the *path loss*, which can be defined as an overall decrease in the signal strength with distance between two terminals, the transmitter and the receiver, when the signal is expressed in decibels. The physical processes, which cause this phenomenon are the spreading of electromagnetic wave radiated outward in space by the transmitter antenna and the obstructing effects of any natural and man-made object surrounding this antenna. The spatial and temporal variations of the signal path loss are large and slow.

Large-scale (in the space domain) and *long-term* (in the time domain) *fading* is the *second* one, which is usually called in the literature a *shadow* or *slow fading* [5,8], because it is caused by diffraction from the buildings' corners and their rooftops, or from the hills' tops located along the

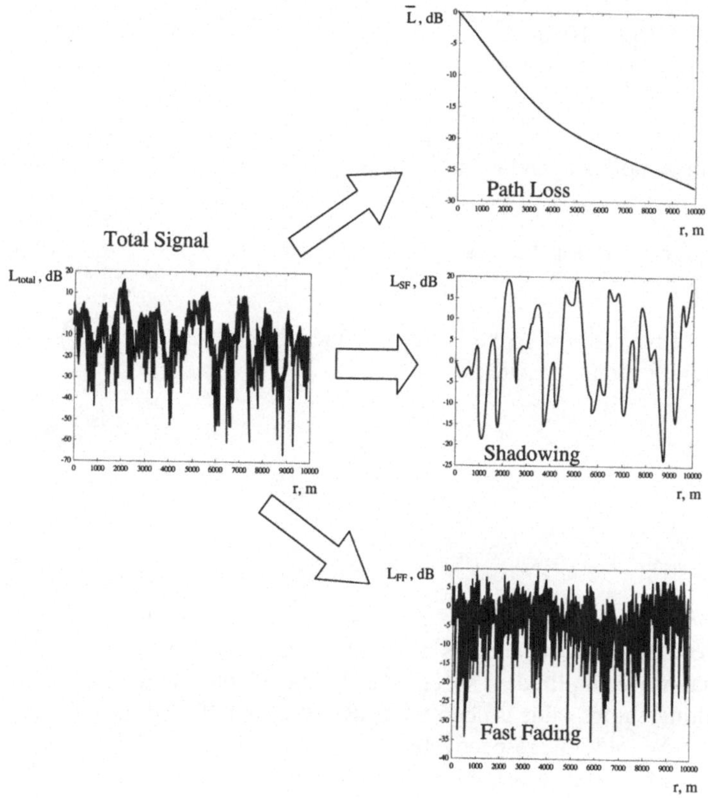

Figure 2.4 Illustration of Doppler effect.

radio link surrounding the terminal antennas. The spatial scale of large-scale variations is of the order of the obstructions' dimensions, that is, from several to several tens of meters.

The *third* one is the *small-scale fading* in the space domain and *short-term* or *fast* signal variations in the time domain, which are caused by the mutual interference of the wave components of the multiray field. The characteristic scale of such waves in the space domain is changed from half wavelength to three wavelength [3,8–12]. Therefore they are usually called *fast-fading* signals in the literature (see also bibliography in Refs. 5–9).

Path Loss

This is a principal characteristic that determines the effectiveness of the propagation channel in various kinds of environment. It defines variations of the signal amplitude or field intensity along the propagation trajectory (*path*) from point to point within the communication channel. For its quantitative evaluation we will assume that the signal–wave amplitude at the point $\mathbf{r}_1$ along the propagation path is $A_1(\mathbf{r}_1)$ or the signal–wave intensity is $J(\mathbf{r}_1) = A_1^2(\mathbf{r}_1)$. In the process of propagation along the path, at any next point $\mathbf{r}_2$ the signal–wave amplitude is $A_2(\mathbf{r}_2)$ or intensity $J(\mathbf{r}_2) = A_2^2(\mathbf{r}_2)$. In the literature the *path loss* is defined as a logarithmic difference between the amplitude or the intensity (sometime it is called *power*) at the points $\mathbf{r}_1$ and $\mathbf{r}_2$ along the propagation path in the medium.

In other words, *path loss*, which is denoted by L and measured in decibels (dB), can be evaluated as

For signal amplitude $A(\mathbf{r}_j)$ at two points $\mathbf{r}_1$ and $\mathbf{r}_2$ along the propagation path [1]

$$L = 10\log\frac{A^2(r_2)}{A^2(r_1)} = 10\log A^2(r_2) - 10\log A^2(r_2)$$
$$= 20\log A(\mathbf{r}_2) - 20\log A(\mathbf{r}_1)\,\mathrm{dB} \tag{2.4}$$

For signal intensity $J(\mathbf{r}_j)$ at two points $\mathbf{r}_1$ and $\mathbf{r}_2$ along the propagation path

$$L = 10\log\frac{J(\mathbf{r}_2)}{J(\mathbf{r}_1)} = 10\log J(\mathbf{r}_2) - 10\log J(\mathbf{r}_1)\,\mathrm{dB} \tag{2.5}$$

If we take point $\mathbf{r}_1$ as the origin of the radiopath (the transmitter location) and assume $A(\mathbf{r}_1) = 1$, then the loss L at any arbitrary point $\mathbf{r}$ along the path is

$$L = 20\log A(\mathbf{r})\,\mathrm{dB} \tag{2.6a}$$

and

$$L = 10\log J(\mathbf{r})\,\mathrm{dB} \tag{2.6b}$$

The next question is: What are the units in which the received power is measured at the receiver? According to Refs. 1–8, the resulting output value is denoted in dB/(V/m), dB/(mV/m), and dB/(μV/m), if the reference signal–wave amplitude is specified as 1 V/m, mV/m, and μV/m, respectively. In the same way, the resulting output value is denoted in dB, dBm, and dBμ, if the reference signal/wave power is 1 W (watt), mW, and μW, respectively.

Taking into account relations between measured power units, that is, $1\,\mu\mathrm{W} = 10^{-3}\mathrm{mW} = 10^{-6}\mathrm{W}$, one can easily obtain

$$0\ \mathrm{dB}\mu = -30\,\mathrm{dBm} = -60\,\mathrm{dBW} \tag{2.7}$$

For example, if the received power level is $-15\,\mathrm{dBm}$, we have $10\log P_{\mathrm{mW}} = -15\,\mathrm{dBm}$, from which it immediately follows that $P_{\mathrm{mW}} = 10^{-1.5} = 0.0316\,\mathrm{mW}$.

The second main characteristic of communication channels is the *signal-to-noise ratio* (SNR or S/N). In decibels this characteristic can be presented as follows: for the receiver (output) channel where noise (artificial and natural) is significant

$$\mathrm{SNR} = P_R - N_F\ \mathrm{dB} \tag{2.8}$$

where P_R is the power at the receiver and N_F is described by Eq. (12.3). Both P_R and N_F are assumed to be in the same dB units, e.g., dBW.

Example. A receiver of the wireless communication system has a bandwidth of 250 kHz and requires that its input SNR should be not more than 10 dB when the input signal is $-105\,\mathrm{dBm}$. The background temperature at the input of the system is $T_0 = 300\,\mathrm{K}$. Find the maximum value of noise figure and the corresponding effective noise temperature at the input of such a receiver.

Solution. Using expression (2.3) in dB, that is,

$$N_F = 10\log(k_B T_0 B_w F)$$

we finally get

$$\mathrm{SNR} = P_R - N_F = P_R - F_{\mathrm{dB}} - 10\log(k_B T_0 B_w)$$

So, rewriting P_R in dBW according to Eq. (2.7), the unknown noise figure can be obtained as follows

$$F_{\text{dB}} = P_R - 10\log(k_B T_0 B_w) = (-105-30)\,\text{dBW} - 10\,\text{dB}$$
$$-10\log(1.38\times10^{-23}\,\text{W Hz}^{-1}\text{K}^{-1}\times300\,\text{K}\times250\times10^{3}\,\text{Hz})$$
$$= -135-10+170-10\log103.5 = 4.85\,\text{dB}$$

From Eq. (2.2) we finally get

$$T_e = T_0(F-1) = 300(10^{4.85/10}-1) = 616\,\text{K}$$

2.1.5. Multipath Characteristics of the Multiplicative Noise

We will start, first of all, with a qualitative description of *slow* and *fast* fading following Refs. 2–9.

Long-Term or Slow Fading

As was shown [2–9], because the *slow* spatial signal variations (expressed in decibels, dB) tend to normal or gaussian distributions, the average signal power variations, as a result of their averaging within some individual small area, tend to the log-normal distribution (expressed in dB) with the standard deviation that depends on the relief of the terrain and on the type of built-up area [3–8,12].

Short-Term or Fast Fading

As follows from Fig. 2.4, the *fast* fading (expressed in dB) is observed over distances of about half or one wavelength. When talking about this phenomenon, we must contrast two main situations in the cellular propagation channel: the first one is when the subscribers' antennas are stationary with respect to the base station, which can be formally termed a *static multipath* situation [2–6]; the second one is when subscribers' antennas are in motion relative to the base station, which can be formally termed a *dynamic multipath* situation [2–6]. For the case of stationary receiver and transmitter (*static multipath channel*), due to multiple reflections and scattering from various obstructions around the arbitrary transmitter and receiver, the narrowband radio signals travel along different paths of varying lengths. In the case of a *dynamic multipath* situation, either the subscribers' antennas are in movement or the objects surrounding the stationary antennas move, the spatial variations of resultant signal at the receiver can be seen as temporal variations at the receiver as it moves through the multipath field. Moreover, in such a dynamic multipath situation a signal fading at the mobile receiver occurs in the time domain. This temporal fading relates to a shift of frequency radiated by the stationary transmitter. In fact, the time variations, or dynamic changes of the propagation path lengths are related to the Doppler effect, which is due to relative movements between a stationary transmitter and a moving receiver.

To illustrate the effects of phase change in the time domain due to Doppler frequency shift (called the *Doppler effect*), let us consider a mobile receiver moving at a constant velocity v, along the path X_1X_2, as it is shown in Fig. 2.5. As follows from the geometry presented in Fig. 2.5, the difference in path lengths traveled by a signal from source S to the mobile at points X_1 and X_2 is $\Delta\ell = \ell\cos\theta = v\Delta t\cos\theta$, where Δt is the time required for the moving receiver to travel from point X_1 to X_2 along the path, and θ is the angle between the mobile direction along X_1X_2 and direction to the source at the current point X_i, that is, X_iS, $i = 1, 2$. The phase change in the resultant received signal due to the difference in path lengths is therefore [8]

$$\Delta\Phi = k\Delta\ell = \frac{2\pi}{\lambda}\ell\cos\theta = \frac{2\pi v\Delta t}{\lambda}\cos\theta \tag{2.9}$$

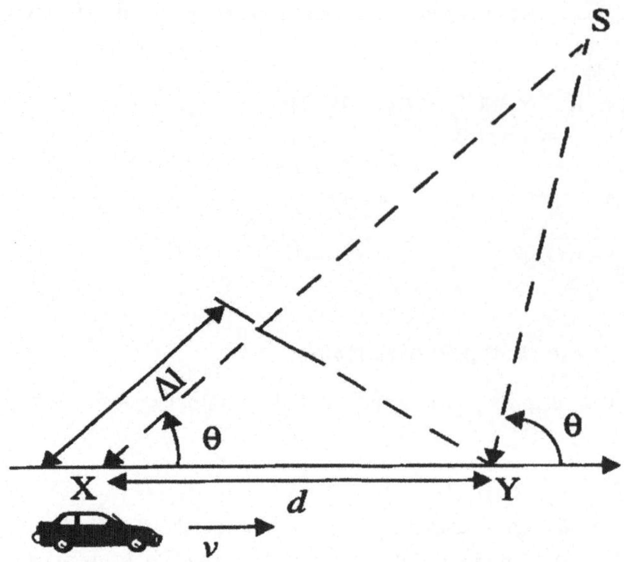

Figure 2.5 Relationship between representations of both signals' spectra according to [8].

Hence the apparent change in frequency radiated, or Doppler shift, is given by f_D, where

$$f_D = \frac{1}{2\pi}\frac{\Delta\Phi}{\Delta t} = \frac{v}{\lambda}\cos\theta \tag{2.10}$$

Because the Doppler shift relates, according to Eq. (2.10), to the mobile velocity and the spatial angle between the direction of mobile motion and the direction of arrival of the signal, it can be positive or negative depending on whether the mobile receiver is moving toward or away from the transmitter. In fact, as follows from Eq. (2.10), if the mobile moves *toward* the direction of arrival of the signal then $f_D > 0$, i.e., the apparent received frequency is increased, while if it moves away from the direction of arrival of the signal then $f_D < 0$, i.e., the apparent received frequency is decreased. Signals arriving from directly ahead of or directly behind the mobile correspond to the maximum rate of phase changes, giving $f_{D\,\max} = v/\lambda$.

2.1.6. Narrowband and Wideband Signal Representations

Now we will consider a question: what kinds of radio signals propagate in wireless communication channels. First of all we will consider a CW or *narrowband* signal representation. A voice modulated CW signal transmitted at the carrier frequency, f_c, which in the literature is called the transmitted *band-pass* or *RF* signal [2–8], can be expressed in the following form:

$$s(t) = A(t)\cos[2\pi f_c t + \varphi(t)] \tag{2.11}$$

where $A(t)$ is the signal envelope and $\varphi(t)$ is its phase.

For example, if a 3-kHz voice signal amplitude modulates a carrier at $f_c = 900\,\text{MHz}$, its fractional bandwidth is very narrow, that is, $6 \times 10^3\,\text{Hz}/9 \times 10^8\,\text{Hz} \approx 7 \times 10^{-6}$ or $7 \times 10^{-4}\%$.

Since all information in the signal is contained within the phase and envelope time variations, usually in the literature one encounters an alternative form of band-pass signal $s(t)$

$$u(t) = A(t)\exp(j\varphi(t)) \tag{2.12}$$

which is called a *complex base-band* representation of $s(t)$. It is clear from Eqs. (2.11) and (2.12) that the relation between the *band-pass* (*RF*) and the *complex base-band* signal representations is

$$s(t) = \mathrm{Re}[u(t)\exp(j2\pi f_c t)] \tag{2.13}$$

The relationship between the representations of both signals, Eqs. (2.11) and (2.12), in the frequency domain is shown schematically in Fig. 2.6, from which follows that the complex base-band signal is a frequency-shifted version of the band-pass signal with the same spectral shape but centered in the close proximity of zero frequency despite the carrier f_c. Moreover, the mean power of the base-band signal is

$$\langle P_s(t)\rangle = \frac{\langle |u(t)|^2\rangle}{2} = \frac{\langle u(t)u^*(t)\rangle}{2} \tag{2.14}$$

which is the same result as the mean-square value of the real, band-pass signal $s(t)$.

The complex envelope of the received *CW* (*narrowband*) signal can be presented according to Eq. (2.12) within the multipath channel as the phasor sum of N baseband individual multiray components arriving at the receiver with the corresponding time delay, τ_i, $i = 0, 1, 2, \ldots$ [8],

$$r(t) = \sum_{i=0}^{N-1} u_i(t) = \sum_{i=0}^{N-1} A_i \exp(j\varphi_i(t, \tau_i)) \tag{2.15}$$

If we assume that during the vehicle movements over a local area the amplitude A_i variations are small enough, whereas phases φ_i vary greatly due to changes in propagation distance over the space,

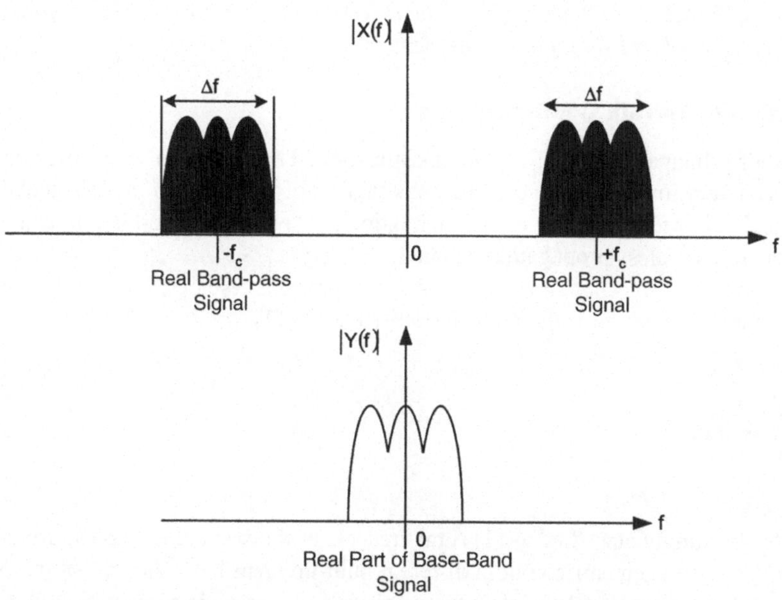

Figure 2.6 Schematic illustration of Doppler effect [according to Ref. 5].

then as a result we obtain great random oscillations of the total signal $r(t)$ during the receiver movement over small distances. Finally, since $r(t)$ is the phasor sum Eq. (2.15) of the individual multipath components, the instantaneous phases of the multipath components cause the large fluctuations which typifies small-scale fast fading for CW signal $s(t)$. The average received power over a local area is then given by [8]

$$\langle Pcw \rangle \approx \sum_{i=0}^{N-1} \langle A_i^2 \rangle + 2\sum_{i=0}^{N-1} \sum_{i,j \neq i} \langle A_i A_j \rangle \langle \cos(\varphi_i - \varphi_j) \rangle \tag{2.16}$$

Here the time averaging was done for *CW* measurements made by a mobile vehicle, as the receiver, over the local measured area [8].

For a *wideband* (or *pulse*) probing signal the total received power is simply related to a sum of the powers of the individual multipath components Eq. (2.15), where each component has a random amplitude and phase at any time t, and, then, the average small-scale received power can be presented as [8]

$$\langle P_{\text{pulse}} \rangle \approx \sum_{i=0}^{N-1} \langle A_i^2 \rangle \tag{2.17}$$

Hence, in the multipath wideband propagation channel the small-scale received power is simply the sum of the powers received in each multipath component. Because in practice, the amplitudes of individual multipath components do not fluctuate widely in a local area, then the received power of the wideband (pulse) signal does not fluctuate significantly when a vehicle moves over a local area. Comparison between the CW and pulse small-scale power presentation, Eqs. (2.16) and (2.17), shows that when $\langle A_i A_j \rangle = 0$ and/or $\langle \cos(\varphi_i - \varphi_j) \rangle = 0$, the average power for a CW signal is equivalent to the average received power for a pulse signal in a small-scale region. This can occur either when the path amplitudes are uncorrelated, that is, each multipath component is independent after reflection or scattering, or when multipath phases are independently and uniformly distributed over $[0, 2\pi]$. Thus we can conclude that in UHF–microwave bands, when the multipath components traverse differential path lengths, having hundreds of wavelengths: *The received local ensemble average powers of wideband signal and narrowband signal are equivalent.*

2.1.7. Characterization of Terrain Configurations

Now let us consider another principal question: terrain classification. The process of classification of terrain configurations is a very important stage in the construction of propagation models above the ground surface and finally, in predicting the signal attenuation (or "path loss") and fading characteristics within each concrete wireless propagation channel.

The simple classification of *terrain configuration* follows from practical research and experience of designers of such communication systems. It can be presented as [1–11]

Open area
Flat ground surface
Curved, but smooth terrain
Hilly terrain
Mountains

The *built-up areas* can also be simply classified as (1) rural areas, (2) suburban areas, and (3) urban areas. Many experiments that have been carried out in different built-up areas have shown that there are many specific factors, which must be taken into account to describe specific propagation phenomena in built-up areas, such as [1–11]

Buildings' density or terrain coverage by buildings (in percents)
Buildings' contours or their individual dimensions
Buildings' average height
Positions of buildings with respect to the base station and fixed or mobile receivers
Positions of both antennas, receiver and transmitter, with respect to the rooftops' level
Density of vegetation; presence of gardens, parks, lakes etc.
Degree of "roughness" or "hilliness" of a terrain surface

Using these specific characteristics and parameters, we can easily classify various kinds of terrain by examining topographic maps for each deployment of a wireless communication system.

2.1.8. Various Propagation Situations in Built-up Areas

As remarked earlier, a very important characteristic of the propagation channel is the location and position of both antennas with respect to the obstacles placed around them. Usually there are three possible situations shown in Fig. 2.7a–c, respectively [1]:

1. Both antennas, receiver and transmitter, are placed above the tops of obstacles (in a built-up area this means that they are above the rooftops' level).

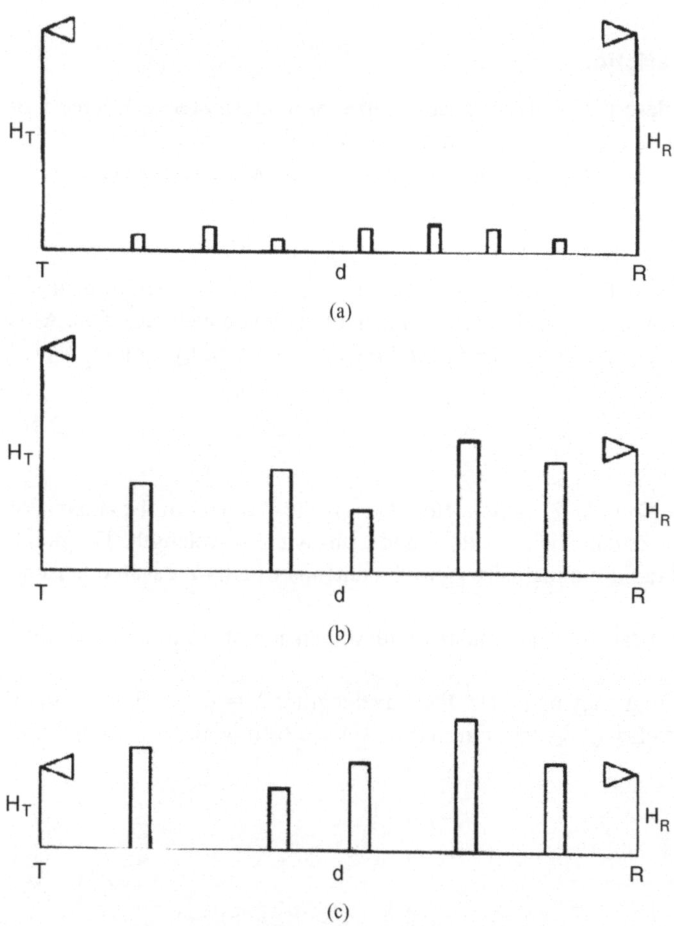

Figure 2.7 Three possible situations with receiving and transmitting antennas.

2. One of the antennas is higher than the tops of the obstacles (namely, the roofs), but the second one is lower.
3. Both antennas are below the tops of the obstacles.

In the first situation they are in *direct visibility* or LOS conditions. In the last two situations, one or both antennas are in *clutter* or obstructive conditions, which call non-line-of-sight (NLOS) conditions. In all these cases the profile of terrain surface is also very important and may vary from flat and smooth, with curvature, up to rough and hilly terrain.

2.2. ANTENNA BASICS

As was mentioned above and shown in Fig. 2.1 according to Ref. 13, a radio antenna, transmitting or receiving, is an independent element of the wireless communication system, which converts the current and/or voltage generated by the wire-based circuit, such as a transmission line, a waveguide or coaxial cable, into electromagnetic field energy propagating through space. In unbounded free space, the fields propagate in the form of spherical waves, whose amplitude, as will be shown below, is inversely proportional to the distance from the antenna. Each radio signal can be represented as a progressive electromagnetic wave [1], which propagates along a given direction.

2.2.1. Main Antenna Characteristics

The principal characteristics used to describe an antenna acting either as a transmitter or as a receiver are *radiation pattern, polarization, directivity, gain, efficiency*, and antenna *impedance*. We will define them briefly. More information about antennas and their characteristics can be found in Refs. 13–16.

Radiation Pattern

The radiation pattern of any antenna is defined usually as the relative distribution of electromagnetic power in space. Here we must differentiate a near-field and a far-field region of such radiation. As is shown in Fig. 2.8, the near-field region, called also the *Fresnel region*, is defined by a radius R [13]

$$R = \frac{2l^2}{\lambda} \tag{2.18}$$

beyond which lies the far field or the *Fraunhofer region*. Here l (m) is the diameter of the antenna or area of the smallest sphere where the antenna is embedded and λ(m) is the wavelength. The radiation pattern is a plot of the far-field radiation intensity from the antenna usually measured per unit solid angle (Fig. 2.8).

Example. Find the far-field distance for an antenna with dimension of 0.9 m and operating frequency of 1 GHz.

Solution. For the operating frequency of 1 GHz, the wavelength is $\lambda = c/f = (3 \times 10^8\,\text{m/s})/(10^9\,\text{Hz}) = 0.3\,\text{m}$. Then, according to Eq. (2.8), the minimum distance, from which the Fraunhofer zone has already begun, is

$$r_F \geq R = \frac{2l^2}{\lambda} = \frac{2 \times 0.81}{0.3} = 5.4\,\text{m}$$

Mathematically the radiation intensity can be presented as the product of the power density (or the time-averaged Poynting vector S in watts per square meter [1,13–16]) and the square of the distance

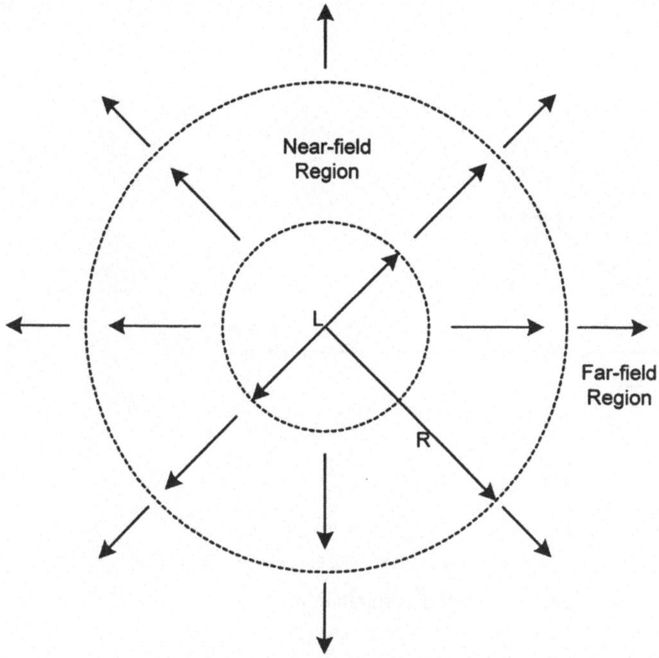

Figure 2.8 Definition of field regions surrounding the antenna [13].

from the antenna r, that is, [13–16]

$$I = r^2 S = r^2 E_\theta H_\phi \tag{2.19}$$

where E_θ and H_ϕ are the components of the electrical and magnetic fields of antenna radiation in the spherical coordinate system shown in Fig. 2.9. The shape of the radiation pattern defines a type of the antenna: *isotropic*, *directional*, and *omnidirectional.*

An *isotropic antenna* refers to an antenna radiating equally in all directions, that is, its radiation power uniformly distributed in all directions. For such an antenna with a total power P, which spreads uniformly over a sphere of radius r, the power density at this distance in any direction equals [13–16]

$$S = \frac{P}{\text{area of a sphere}} = \frac{P}{4\pi r^2} \tag{2.20}$$

Then, according to Eq. (2.19), the radiation intensity of an isotropic antenna equals

$$I = \frac{P}{4\pi} \tag{2.21}$$

A *directional antenna* transmits (or receives) waves more efficiently in certain directions than in others. A radiation pattern plot for directional antenna is shown in Fig. 2.10 according to Ref. 6, illustrating the *main lobe*, which includes the direction of maximum radiation intensity, a *back lobe* with radiation in the opposite direction of the main lobe, and several *side lobes* separated by nulls where no radiation occurs. For such kind of antennas, some unified parameters are usually used

The *half-power beam width* (see Fig. 2.10), or simply the beam width, is the solid angle that bounds the area of the main lobe where the half-power points are located.

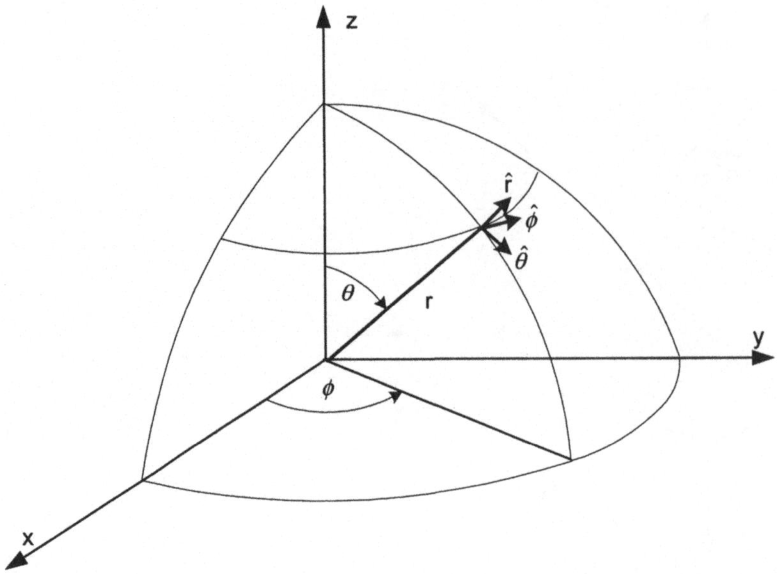

Figure 2.9 Spherical coordinate system for antenna parameters computation.

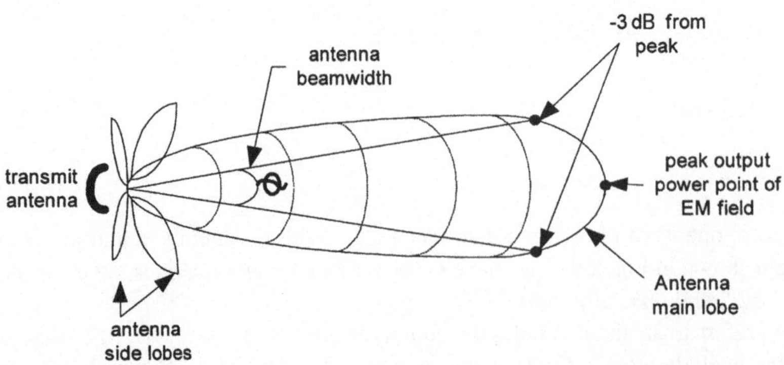

Figure 2.10 Radiation pattern of the directive antenna.

The *front–back ratio* is the ratio between the peak amplitudes of the main and back lobes, usually expressed in decibels.

The *side-lobe level* is the amplitude of the biggest side lobe, usually expressed in decibels relative to the peak of the main lobe.

A more practical type of the antenna, as compared with the idealized isotropic one, is an *omnidirectional antenna*, whose radiation is constant in the plane of azimuth but may vary in the vertical plane.

The parameter "directivity" is used to describe non-isotropic antennas and the variation of its signal intensity in all directions. The concept of directivity indicates that an antenna concentrates the field energy in specific directions. If so, we can define the directivity D as the ratio between the power of the transmitting radiation in the specific direction determined by spherical coordinates (ϕ, θ), according to Fig. 2.9, to that of the equivalent isotropic antenna [13–16]:

$$D(\phi, \theta) = \frac{P(\phi, \theta)}{P_{\text{isotropy}}} \tag{2.22}$$

The use of an isotropic antenna as a reference in Eq. (2.22) allows one to measure the directivity in units dBi, i.e., relative to an isotropic antenna, D dBi $= 10 \log D_{\text{isotropy}}$.

Because of power dissipation within the antenna circuit itself, a new parameter is introduced to describe the *radiation efficiency*, denoted by η and defined as the ratio between the power actually radiated by the antenna to the power accepted by the antenna.

The *power gain, G,* or *antenna gain,* expressed in dB, is defined as 4π times the ratio of the radiation intensity in a given direction to the total power accepted by the antenna. The gain is related to the radiation efficiency and the directivity by [13–16]

$$G(\phi, \theta) = \eta D(\phi, \theta) \tag{2.23}$$

Usually in practice the terms "directivity" and "gain" refer to the maximum value of $D(\phi, \theta) = D_{\max}$, denoted simply by D, and $G(\phi,\theta) = G_{\max}$, denoted simply by G. If so, we can rewrite Eq. (2.23) as a simple relation $G = \eta D$.

Using these notations we can rewrite now the formula Eq. (2.20) for radiated power density of the transmitting directive antenna as

$$S = \frac{P}{4\pi r^2} G = \frac{P}{4\pi r^2} \eta D \tag{2.24}$$

For a *receiving* antenna, a parameter called the *effective area* or *aperture* (also called *antenna cross-section*) is usually used, and denoted by A_e. The *effective aperture* of the receiving antenna is defined as the ratio of the power P_R, which is delivered to a matched receiver, to the power density S of electromagnetic radiation of the transmitter antenna, according to Eq. (2.20), which arrives at the receiver antenna, i.e.,

$$A_e = \frac{P_R}{S} \tag{2.25}$$

It should be noted that this value differs from the real geometrical area of the receiving antenna, which collects the arriving wave energy. The maximum antenna gain G is also related to the effective antenna aperture as follows [13–16]:

$$G = \frac{4\pi}{\lambda^2} A_e \tag{2.26}$$

As follows from Eq. (2.8), the power collected at the receiver antenna must be greater than the power of noise there. An improvement in the SNR can be obtained through use of antennas with high directivity or gain. In fact, for an isotropic antenna with $G = 4\pi (D = 1)$, we have $A_e = \lambda^2/4$ according to Eq. (2.26). By comparison, a short dipole antenna (about which we will talk later) has directivity $D = 1.5$, i.e., $A_e = 1.5(\lambda^2/4) = 3\lambda^2/8$. In this case the power collected at the receiving dipole antenna is 1.5 times greater than that for the isotropic antenna.

Polarization is also one of the important parameters of the antenna. It is defined by the orientation of the electric field component E of the radiated electromagnetic wave.

2.2.2. Antennas in Free Space

Formulas obtained above allow us to obtain the relation between power at the transmitter and the receiver antennas located in free space. This relation in called the *Friis transmission formula.* Let us present it below. According to formulas (2.24) and (2.25), for two antennas separated by a distance r, great enough to take into account only a far-field regions of both antennas, we get [13–16]

$$P_R = SA_{eR} = \frac{P_T}{4\pi r^2} G_T A_{eR} \tag{2.27}$$

Since the effective aperture of the receiver antenna equals [see Eq. (2.26)] $A_{eR} = \lambda^2 G_R/4\pi$, we finally get

$$P_R = SA_{eR} = P_T \left(\frac{\lambda}{4\pi r}\right)^2 G_T G_R \tag{2.28}$$

or introducing a new parameter, the path gain PG, for antennas in free space we get

$$PG = \frac{P_R}{P_T} = \left(\frac{\lambda}{4\pi r}\right)^2 G_T G_R \tag{2.29}$$

Here G_T and G_R are the maximum gains of the transmitting and the receiving antennas, respectively.

2.2.3. Types of Antennas

There is a wide variety of available antenna systems used in different areas of wireless communications. We describe briefly the dipole antenna, which is widely used in practical applications and will refer the reader to the literature [13–16] where all types of antennas are fully described.

Dipole Antennas

The basic structures and the current distributions of the hertzian and $\lambda/2$-dipole antennas are shown in Figs. 2.11a,b and 2.12a,b, respectively [6]. It is seen that for a hertzian dipole $l/\lambda \ll 1$, i.e., it can be considered as a "short" antenna with a uniform current distribution along its length and with maximum gain $G = 1.5$, which corresponds to angle $\theta = 90°$ (normal to the dipole axis). As for the $\lambda/2$-dipole antenna, presented in Fig. 2.12, the maximum of field radiation also occurs at $\theta = 90°$; the maximum gain is now $G = 1.64$ [5,6]. So, the gains of the two kinds of dipole antennas are not very different. But, as was mentioned in [5,6,13–16], the radiation resistance of the $\lambda/2$-dipole antenna is much higher with respect to that of a Hertzian one (making the longer dipole easier to match to the feedline), and the loss resistance is small (higher efficiency).

Using now the Friis formula Eq. (2.29), we can compare the path gain for the isotropic and the $\lambda/2$-dipole antennas. In fact, according to Eq. (2.29), if two terminals, the transmitter and the receiver, are separated by a distance of 1 km, the ratio of their path gain in these two cases is [5,6]

$$\frac{(PG)_{\text{isotropic}}}{(PG)_{\lambda/2\text{-dipole}}} = \frac{0.57\times10^{-9}}{1.53\times10^{-9}} = 0.37 \tag{2.30}$$

that is, the path gain of the $\lambda/2$-dipole antenna approximately in 3 times greater than that for the isotropic antenna.

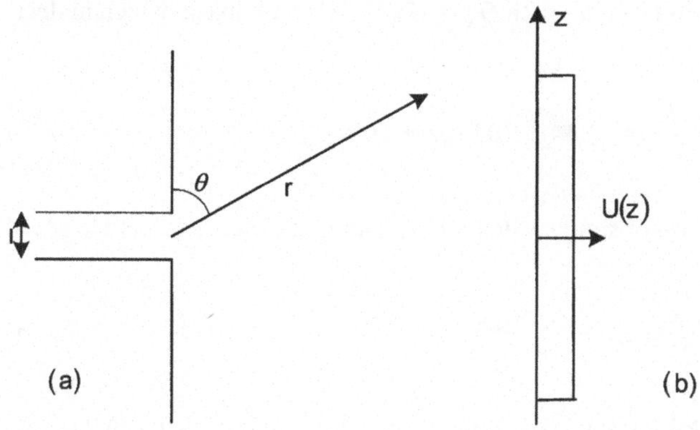

Figure 2.11 Hertzian dipole: (a) antenna and (b) current distribution according to [6].

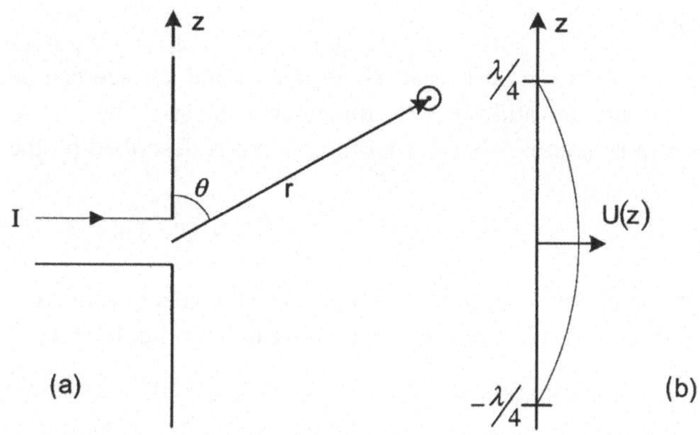

Figure 2.12 $\lambda/2$ dipole: (a) antenna and (b) current distribution according to [6].

2.3. PATH LOSS PREDICTION MODELS IN VARIOUS OUTDOOR COMMUNICATION LINKS

As was mentioned in Sec. 2.1, the path loss is a main characteristic of the radio propagation channel, which is investigated in more detail in the literature [1–12]. Below, we will present briefly the propagation models suitable for practical applications in wireless communications.

2.3.1. Free-Space Path Loss

Let us consider a nonisotropic antenna placed in free space as a transmitter of P_T watts and with a directivity gain G_T. At an arbitrary large distance r ($r > r_F$, where r_F is the Fraunhofer far-field range) from the source, the radiated power is uniformly distributed over the surface area of a sphere of radius r. If P_R is the power at the receiver antenna, which is located at distance r from the transmitter antenna and has a directivity gain G_R, then the *path loss* in decibels according to definition Eq. (2.5) and Friis formula Eq. (2.29) can be determined as [1–11]

$$L = 10\log\frac{P_T}{P_R} = 10\log\frac{(4\pi r/\lambda)^2}{G_T G_R} = L_0 + 10\log\frac{1}{G_T G_R} \tag{2.31}$$

Here L_0 is the path loss for an isotropic point source (with $G_R = G_T = 1$) in free space, which in decibels can be presented as

$$L_0 = 10\log\left(\frac{4\pi fr}{c}\right)^2 = 20\log\frac{4\pi fr}{c} = 32.44 + 20\log r + 20\log f \tag{2.32}$$

where the value 34.44 is obtained by the use of simple calculations, taking into account that the speed of light $c = 3 \times 10^8$ m/s:

$$32.44 = 20\log\frac{4\pi \times 10^3\text{m} \times 10^6(1/\text{s})}{3 \times 10^8\,\text{m/s}} = 20\log\frac{40\pi}{3} \tag{2.33}$$

In expression (2.32) the distance r is in kilometers (km), and frequency f is in megahertz (MHz). As the result, the path loss for both directive antennas, in free space, is

$$L_F = 32.44 + 20\log d_{\text{km}} + 20\log f_{\text{MHz}} - 10\log G_T - 10\log G_R \tag{2.34}$$

2.3.2. Path Loss over a Flat Terrain

The simplest case of radio wave propagation over terrain is that where the ground surface can be assumed as flat. The assumption of "flat terrain" is valid for radio links between subscribers up to 10–15 km apart [1–7]. The main process is a reflection from flat terrain, which is described by the reflection coefficients.

Reflection Coefficients

Following Refs. 1–6, we will present now the expressions for the complex coefficients of reflection (Γ) for waves with vertical (denoted by index V) and horizontal (denoted by index H) polarization, respectively.

For *horizontal* polarization,

$$\Gamma_H = |\Gamma_H| e^{-j\varphi_H} = \frac{\sin\psi - (\varepsilon_r - \cos^2\psi)^{1/2}}{\sin\psi + (\varepsilon_r - \cos^2\psi)^{1/2}} \tag{2.35a}$$

For *vertical* polarization,

$$\Gamma_V = |\Gamma_V| e^{-j\varphi_v} = \frac{\varepsilon_r \sin\psi - (\varepsilon_r - \cos^2\psi)^{1/2}}{\varepsilon_r \sin\psi + (\varepsilon_r - \cos^2\psi)^{1/2}} \tag{2.35b}$$

Here $|\Gamma_V|$, $|\Gamma_H|$ and φ_V, φ_H are the magnitude and the phase of the coefficients of reflection for vertical and horizontal polarization, respectively; $\psi = (\pi/2) - \theta_0$ is the grazing angle; θ_0 is the angle of wave incidence. The knowledge of reflection coefficient amplitude and phase variations is a very important factor in the prediction of path loss for different situations in the land propagation channels. In practice, for wave propagation over terrain, the ground properties are determined by the conductivity and the absolute dielectric permittivity (dielectric constant) of the subsoil medium, $\varepsilon = \varepsilon_0\varepsilon_r$, where ε_0 is the permittivity of vacuum and ε_r is the complex relative permittivity of the ground surface.

Line-of-Sight (LOS) Two-Ray Model

The two-ray model was first proposed for describing the process of radio wave propagation over flat terrain [1,5–7], which is based on the superposition of a direct ray from the source and a ray reflected from the flat ground surface, as shown in Fig. 2.13. Following [1,5–7], we can present relation

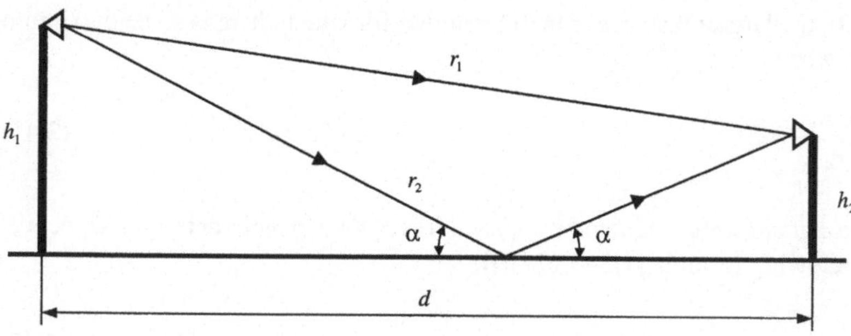

Figure 2.13 Geometry of two-ray model.

between the filed strength and power at the transmitter as

$$E = \sqrt{\frac{30 G_T G_R P_T}{r_1}} \tag{2.36}$$

where r_1 is the trajectory of the direct wave as presented in Fig. 2.13. Then the total field at the receiver is the sum of direct and received waves, that is,

$$E_R = E_T\left(1 + \frac{r}{r_1}|\Gamma|e^{-jk\Delta r}\right) \tag{2.37}$$

Here $\Gamma(\psi)$ is the reflection coefficient described by formulas (2.35a) and (2.35b) for horizontal and vertical polarization, respectively, $\Delta r = r_2 - r_1$ (see Fig. 2.13) is the difference in the radio paths of the two waves, $\Delta\varphi = k\Delta r$ is the phase difference between the reflected and direct waves which can be presented as [1–7]

$$\Delta\varphi = k\,\Delta d = \frac{2\pi}{\lambda} r\left\{\left[\left[1 + \left(\frac{H_R + H_T}{r}\right)^2\right] - \left[1 + \left(\frac{H_R - H_T}{r}\right)^2\right]\right]\right\} \tag{2.38}$$

where h_R and h_T are the receiver and transmitter antenna heights, respectively, and r is the distance between them. For $r_1 \gg (h_T \pm h_R)$ and $r \gg (h_T \pm h_R)$, with the assumption that $r_1 \approx r_2 \approx r$, Eq. (2.38) can be rewritten as

$$\Delta\varphi = \frac{4\pi h_R h_T}{\lambda r} \tag{2.39}$$

Furthermore, if we now assume that $G_R \approx G_T = 1$ (omnidirectional antennas) and that $\Gamma(\psi) \approx -1$ for the farthest ranges from transmitter (when the grazing angle is small), we will finally obtain the absolute value of the power at the receiver

$$\begin{aligned} |P_R| &= |P_T|\left(\frac{\lambda}{4\pi r}\right)^2 \left|1 + \cos^2 k\,\Delta r - 2\cos k\,\Delta r + \sin^2 k\,\Delta r\right| \\ &= |P_T|\left(\frac{\lambda}{4\pi r}\right)^2 \sin^2\frac{k\,\Delta r}{2} \end{aligned} \tag{2.40}$$

As follows from Eq. (2.40), the largest distance from transmitter, for which there is some maximum of received power, occurs when

$$\frac{k\,\Delta r}{2} \approx \frac{\pi}{2} \qquad \sin\frac{k\,\Delta r}{2} \approx 1 \tag{2.41}$$

This distance is called the *critical range*, denoted by r_b, and it is approximately determined according to Eq. (2.41) by the following formula [1,5–7,17–19]:

$$r_b \approx \frac{4h_R h_T}{\lambda} \tag{2.42}$$

In other critical case of small incident angles, that is, when $\sin^2(k\,\Delta r/2) \approx (k\,\Delta r/2)^2$, $\Delta r = 2h_T h_R/r$, which is valid for large distances between antennas relative to antenna heights ($r \gg h_T, h_R$), from Eq. (2.40) an approximate formula of the path loss, called *path loss in the model of flat terrain*, can be obtained

$$L_{FT} = 10\log\frac{|P_T|}{|P_R|} = 10\log\frac{r^4}{h_T^2 h_R^2} = 40\log r_{\mathrm{m}} - 20\log(h_{T\mathrm{m}} h_{R\mathrm{m}}) \tag{2.43}$$

Using now the *critical range* definition (2.42), we can present the path loss over flat terrain in the following form [17]

$$L = \begin{cases} L_B + 20\,\log\dfrac{r}{r_B} & r \le r_B \\ L_B + 40\,\log\dfrac{r}{r_B} & r > r_B \end{cases} \tag{2.44}$$

where L_B is the path loss in free space at the distance that equals the critical range, i.e., $r = r_B$, which can be calculated from the following expression [17]:

$$L_B = 32.44 + 20\log r_{B\mathrm{km}} + 20\log f_{\mathrm{MHz}}$$

As follows from Eq. (2.44), there are two modes of field intensity decay at distances r less than the *break point* $r = r_B$, and beyond this point, that is, $\sim r^{-q}$, $q = 2$ for $r \le r_B$, and $\sim r^{-q}$, $q = 4$ for $r > r_B$.

2.3.3. Path Loss in Clutter (NLOS) Conditions

We investigated above situations in communication links where the LOS conditions occur between two terminal antennas, the transmitter and receiver. Now we consider radio propagation above the terrain in the situation where both antennas are placed above the ground surface in non-line-of-sight (NLOS) conditions. Here a new effect of diffraction phenomena arises from various kinds of obstacles, such as trees or hills, placed on the terrain. The diffraction phenomenon is based on the Huygens' principle [1–12]. Let us briefly describe the diffraction from obstructions, using the Huygens' principle and replacing each obstruction by a *knife edge* [1–3,5–8].

Propagation over a single knife edge. If there is some obstacle that we may model as a simple knife edge (denoted as OO′, see Fig. 2.14) which lies between the receiver and the transmitter, the phase difference $\Delta\Phi$ between the direct ray from the source (at point O), denoted TOR, and that diffracted from the point O′, denoted TO′R, can be obtained in the standard manner by use of a simple

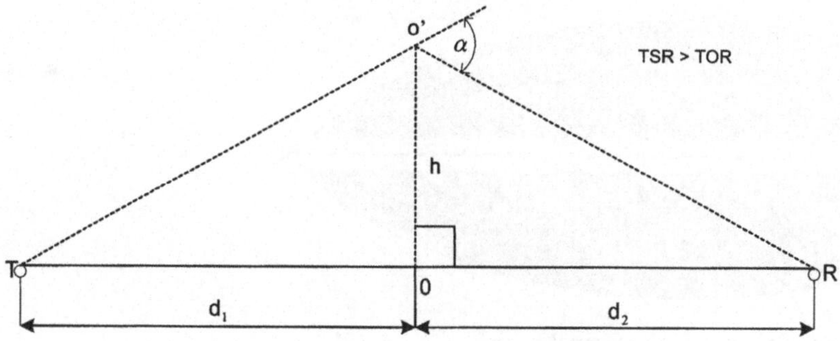

Figure 2.14 Schematical presentation of simple knife-edge model.

presentation of the path difference, Δd, and the phase difference, $\Delta\Phi$, between these rays assuming that the height of the obstacle is much smaller than the characteristic ranges between the antennas and the obstacle ($h \ll d_1, d_2$) [1–3,8]:

$$\Delta d \approx \frac{h^2}{2}\frac{d_1+d_2}{d_1 d_2}$$
$$\Delta\Phi = \frac{2\pi}{\lambda}\Delta d = \frac{2\pi}{\lambda}\frac{h^2}{2}\frac{d_1+d_2}{d_1 d_2} \tag{2.45}$$

Fresnel–Kirchhoff diffraction parameter. If we now introduce the Fresnel–Kirchhoff diffraction parameter v according to [1–3,8]

$$v = h\sqrt{\frac{2(d_1+d_2)}{\lambda d_1 d_2}} \tag{2.46}$$

the phase difference may be rewritten in terms of this parameter, i.e.,

$$\Delta\Phi = \frac{\pi}{2}v^2 \tag{2.47}$$

To estimate the effect of diffraction around obstructions we need a quantitative measure of the required clearance over any terrain obstruction, and as was shown in [1,6,8], this may be obtained analytically in terms of Fresnel-zone ellipsoids drawn around both ends of the radio link, receiver and transmitter (Fig. 2.15). The reader can find a full discussion of Fresnel ellipsoids in [1–3,6,8]. Here we will only repeat that the cross-sectional radius of any ellipsoid with number n from the family at a distance d_1 and $d_2 = d - d_1$ can be presented as a function of the parameters n, d_1, and d_2 as

$$r_n \equiv h_n = \left(\frac{n\lambda d_1 d_2}{d_1+d_2}\right)^{1/2} \tag{2.48}$$

From Eq. (2.46) one can obtain the physical meaning of the Fresnel–Kirchhoff diffraction parameter:

$$v_n = h_n\left[\frac{2(d_1+d_2)}{\lambda d_1 d_2}\right]^{1/2} = \left[\frac{2(d_1+d_2)}{\lambda d_1 d_2}\frac{n\lambda d_1 d_2}{d_1+d_2}\right]^{1/2} = (2n)^{1/2} \tag{2.49}$$

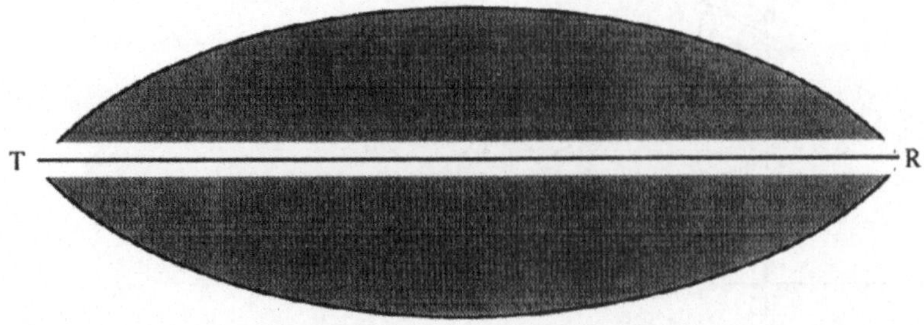

Figure 2.15 Geometrical presentation of the Fresnel zones.

Thus the diffraction parameter v increases with the number n of ellipsoids. All the above formulas are correct for $h_n \ll d_1, d_2$, i.e., far from both antennas. The volume enclosed by the ellipsoid defined by $n = 1$ is known as a *first Fresnel zone*. The volume between this ellipsoid and that defined by $n = 2$ is the *second Fresnel zone*. The contributions to the total field at the receiving point from successive Fresnel zones tend to be in phase opposition and therefore interfere destructively rather than constructively. If an obstruction OO′ is placed at the middle of radio path TO′R (i.e., TO′ = O′R, see Fig. 2.14), then if the height of obstruction h increases from $h = r_1$ (corresponding to the *first* Fresnel zone) to $h = r_2$ (defining the limit of the *second* Fresnel zone), then to $h = r_3$ (i.e., to the *third* Fresnel zone), etc., then the field at the receiver R would oscillate. The amplitude of oscillations would essentially decrease since a smaller amount of wave energy penetrates into the outer zone.

Diffraction losses. When there is a single obstacle between the transmitter and receiver, which can be modeled by a single "knife edge," losses of the wave energy take place. Such losses in the literature are called *diffraction losses*. They can be obtained analytically by use of so-called Fresnel complex integrals by the use of Huygens' principle [1–8]:

$$E = E_0 \frac{1+j}{2} \int_v^\infty \exp\left(-j\frac{\pi}{2}t^2\right) dt \tag{2.50}$$

The integral in the right side of Eq. (2.50) is the complex integral with parameter of integration v defined by Eq. (2.46) for the height of the obstruction under consideration. We note that if the path TR between the transmitter and receiver (line-of-sight path) is actually obstructed by some obstacle modeled by a knife edge, as is shown in Fig. 2.16a, then the height h and the diffraction parameter v are positive [it follows from Eq. (2.46)]. If the knife edge lies below the line-of-sight path (line TR in Fig. 2.16b), so that there is no interruption between T and R, then h and, hence, v are negative [see again, Eq. (2.46)]. As is known, the Fresnel integral in Eq. (2.50) can be presented in the standard manner using the following integral presentations

$$\int_v^\infty \cos\left(-\frac{\pi}{2}t^2\right) dt = \frac{1}{2} - \int_0^v \cos\left(-\frac{\pi}{2}t^2\right) dt = \frac{1}{2} - C(v) \tag{2.51a}$$

and

$$\int_v^\infty \sin\left(-\frac{\pi}{2}t^2\right) dt = \frac{1}{2} - \int_0^v \sin\left(-\frac{\pi}{2}t^2\right) dt = \frac{1}{2} - S(v) \tag{2.51b}$$

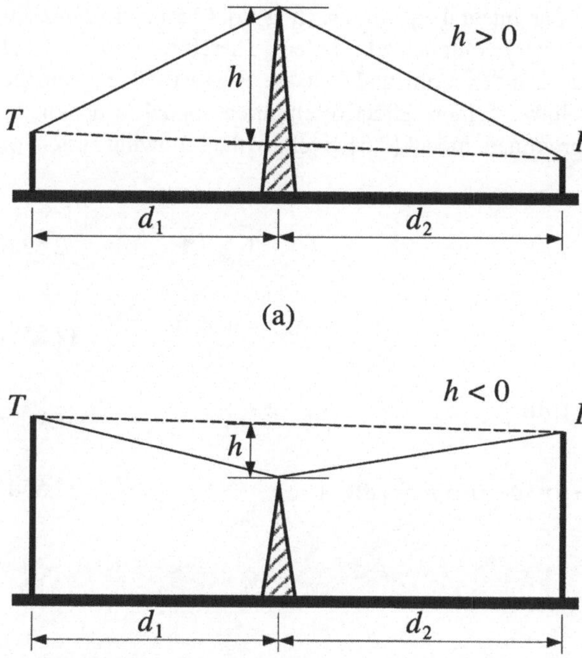

Figure 2.16 Two variants of antenna locations relative to knife edge: (a) above and (b) below the direct visibility line TR [according to Refs. 1–3,8].

At the same time, as follows from the classical theory of plane wave propagation [1,2], the total wave field E_{total} after diffraction at the edge of some arbitrary obstruction can be presented as

$$E_{\text{total}} = E_i \cdot \hat{D} \cdot \exp\,(j\Delta\Phi) \tag{2.52}$$

where E_i is the incident wave from the transmitter located in free space, $\hat{D}$ is the diffraction coefficient or matrix [1,2,5,6], and $\Delta\Phi$ is the phase difference between the diffracted and direct waves mentioned above. Taking into account Eqs. (2.50) and (2.51), the total field according to Eq. (2.52) can be rewritten as [1–3,8]

$$E = E_0 \frac{1+j}{2}\left[\left(\frac{1}{2} \pm C(v)\right) - j\left(\frac{1}{2} \pm S(v)\right)\right] \tag{2.53}$$

The main goal of strict diffraction theory is to obtain parameters D and $\Delta\Phi$ by use of Fresnel integrals. Comparing now Eqs. (2.52) and (2.53), it is easy to obtain the diffraction coefficient and the phase difference $\Delta\Phi$ through the Fresnel integrals [1–3,8]:

$$\hat{D} = \frac{S + (1/2)}{\sqrt{2}\,\sin(\Delta\Phi + (\pi/2))} \tag{2.54a}$$

$$\Delta\Phi = \tan^{-1}\left[\frac{(S + (1/2))}{(C + (1/2))}\right] - \frac{\pi}{4} \tag{2.54b}$$

However, to obtain an exact solution by use of an integral equation such as Eq. (2.50), which is connected with the complex Fresnel integral, is a very complicated problem. Therefore, *empirical* and *semiempirical* models, which are based on numerous experimental data, are usually used to obtain the diffraction losses in NLOS communication links. A more effective empirical model to obtain the knife-edge diffraction losses is the Lee's approximate model [2,3] given by the following system of empirical equations:

$$L(v) = L_{\Gamma}^{(0)} = 0 \text{ dB} \qquad v \le -1 \tag{2.55a}$$

$$L(v) = L_{\Gamma}^{(1)} = 20 \log (0.5 - 0.62v) \text{ dB} \qquad -0.8 < v < 0 \tag{2.55b}$$

$$L(v) = L_{\Gamma}^{(2)} = 20 \log [0.5 \exp (-0.95v)] \text{dB} \qquad 0 < v < 1 \tag{2.55c}$$

$$L(v) = L_{\Gamma}^{(3)} = 20 \log \{0.4 - [0.1184 - (0.38 - 0.1v)^2]^{1/2}\} \text{dB} \quad 1 < v < 2.4 \tag{2.55d}$$

$$L(v) = L_{\Gamma}^{(4)} = 20 \log \frac{0.225}{v} \text{dB} \qquad v > 2.4 \tag{2.55e}$$

Results of calculations of such a model are shown in Fig. 2.17 according to Refs. 1–3, 8 with the corresponding notations according to Eqs. (2.55a)–(2.55e). Other empirical models, which give solution for diffraction losses after two, three and more obstructions are based on Lee's model (see detailed discussions in Refs. 1 and 2). All formulas above can be used mostly for open and rural communication links with LOS and NLOS conditions. More complicated situations with communication between terminal antennas are observed in mixed built-up environments with or without vegetation.

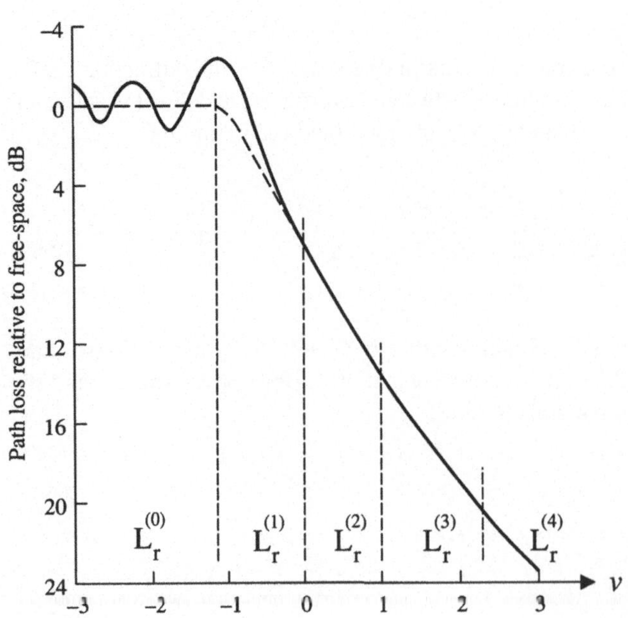

Figure 2.17 Computations of path loss according to Lee's knife-edge empirical model.

2.3.4. Path Loss Models in Land Communication Links with Regular Built-up Terrain

Here we consider several urban propagation environments and we start with the simplest case of wireless communication in the urban areas, when both antennas are placed above the flat ground surface in conditions of direct visibility (LOS conditions), but below the rooftops' level. Here we refer to the multislit street waveguide model [19–23], which was found to be in good agreement with experimental data of wave propagation in urban areas with a regular straight crossing streets. Then we will present the 2D diffraction model of propagation along the rows of buildings with different location of both terminal antennas relative to building rooftops.

Street-Multislit-Waveguide Model

The street is seen as a planar multislit waveguide. One waveguide plane is placed at the street side $z = 0$, and the second one at $z = a$ (see Fig. 2.18), so a denotes the street width. The screen (building) L_n and slit (gap) l_n lengths are distributed according to the Poisson law with the average values of $\langle L\rangle = L$ and $\langle l\rangle = l$, respectively [20–23]:

$$f(L_n) = L^{-1}\exp\left(-\frac{L_n}{L}\right) \quad f(l_n) = l^{-1}\exp\left(-\frac{l_n}{l}\right) \tag{2.56}$$

Following this model [1], we consider the resulting reflected and diffracted fields as a sum of the fields reaching the observer from the virtual image sources Π_n^+ (for the reflections from plate $z = a$) and Π_n^- (for the reflections from plate $z = 0$), (see Fig. 2.18), which finally gives us the approximate

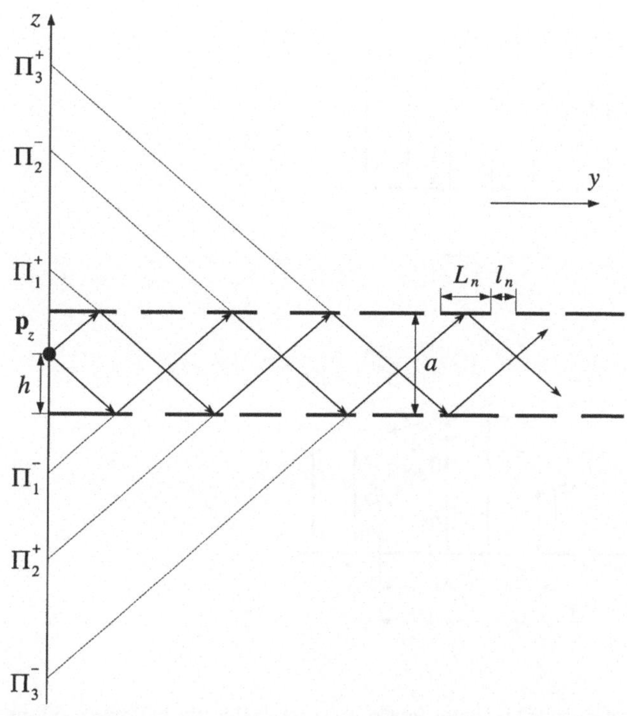

Figure 2.18 Two-dimensional geometry of street waveguide model according to [20–23].

expression for the path loss at a large range from the source ($r \gg a$) [20–23]

$$L \approx 32.1 + 20\log f_0 - 20\log_{10}\frac{(1-\chi)^2}{(1+\chi)^2} + 17.8\log r$$
$$+ 8.6\left(|\ln\chi|\frac{\pi n}{a}\frac{r}{\rho_n^{(0)}a}\right) \tag{2.57}$$

Here $\chi = L/(L+l)$ is the parameter of breakness, $\rho_n^{(0)} = \sqrt{k^2 - (n\pi/a)^2}$, n is a number of waveguide modes (number of reflections), which, as was shown in Refs. 1 and 20–23, must be less than $n = 2$ at the distances more than 50 m along the street (this range can be changed with changes of the street width a). Usually, a main mode with $n = 1$ propagates along the street waveguide.

Two-Dimensional Model of Straight Rows of Buildings

In obstructive ("clutter") conditions the receiver or transmitter antennas (or both) are placed in the shadow zones, when there are many nontransparent buildings surrounding them. In this case the diffraction from the roofs and corners of buildings plays a significant role and the total field depends not only on the reflected, but mostly on the diffracted waves [6,24,25]. Let us consider, according to [6,24,25], that an elevated antenna (base station) radiates a field that propagates in an environment with regularly distributed nontransparent buildings with various heights h_i and different separation distances d_i ($i = 1, 2, 3, \ldots$) between them. The height of the base station antenna, H, can be greater or smaller than the height of the first (near the antenna) building, h_1 (see Fig. 2.19a and b, respectively).

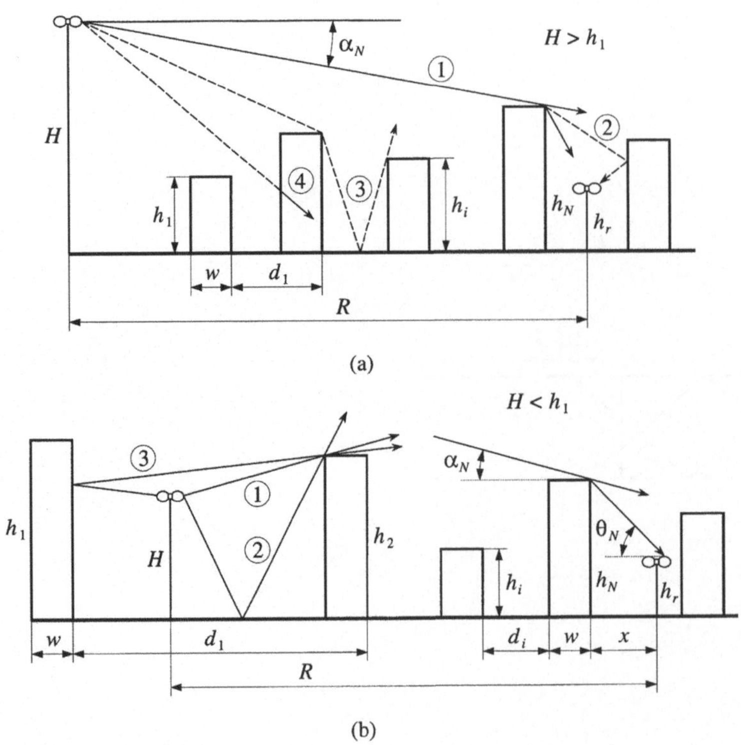

Figure 2.19 Two variants of the base station antenna locations: (a) above and (b) below the building rooftops in 2D-multiple-diffraction model according to [6].

The propagation over the rooftops involves diffraction past a series of buildings with dimensions larger than wavelength λ, i.e., h_i, $d_i \gg \lambda$. At each building a portion of the field will be diffracted toward the ground. The field reaching street level results from diffraction of the waves incident on the rooftops in the vicinity of the receiving antenna [6,24,25].

Treating the base station as a transmitter and assuming that the receiver is at street level, we can obtain the path loss in dB as the sum of the free space path loss

$$L_0 = -10 \log \frac{1}{(4\pi R)^2} \tag{2.58}$$

and excess loss L_{ex}. The last can be presented as the sum of two parts [6,24,25]:

1. The diffraction of the fields at the rooftops before the receiver down to the street level is

$$L_{e1} = -10 \log \left\{ \frac{G_1(\theta_N)}{\pi k r} \left[\frac{1}{\theta_N} - \frac{1}{(2\pi + \theta_N)} \right]^2 \right\} \tag{2.59}$$

   where $G_1(\theta_N)$ is the gain of the receiving antenna pattern in the direction θ_N as shown in Fig. 2.19a; simple geometrical constructions give $\theta_N = \tan^{-1}[(h_N - h_r)/x]$ and $r = [(h_N - h_r)^2 + x^2]^{1/2}$, where x is the distance between the receiver and the building closest to the receiver; h_r is the receiver antenna height.
2. The reduction of the field at the rooftop before the receiver as a result of propagation past the previous rows of buildings

$$L_{e2} = -10 \log(G_2 W^2) \tag{2.60}$$

   where G_2 is the gain in the direction of the highest building edge visible from the base station antenna. To determine parameter W, let us consider two typical cases occur in the urban scene.

In the case when the base antenna is *higher* than the first building ($H > h_1$, see Fig. 2.19a) parameter W can be presented for small angle $\alpha_N = \tan^{-1}\{[H - h_N]/R\}$ and for $x \ll R$ [6,24,25] as

$$W = \frac{(d_N - w)/[R - (d_N - w)]}{\sqrt{2\pi k[(h_N - H)^2 + (d_N - w)^2]^{1/2}}} \times \left\{ \frac{1}{\tan^{-1}[(h_N - H)/(d_N - w)]} + \frac{1}{2\pi + \tan^{-1}[(h_N - H)/(d_N - w)]} \right\} \tag{2.61}$$

In the case when the base antenna is *lower* than the first building ($H < h_1$, see Fig. 2.19b), according to [6,24,25] we have

$$W = \left(2.35 \sqrt{\frac{d_N - w}{\lambda}} \tan^{-1}\left(\frac{H}{d_N - w} \right) \right)^{0.9} \tag{2.62}$$

Using above formulas, one can easily predict path loss effects in built-up areas with regularly distributed rows of straight crossing streets.

2.3.5. Path Loss Models in Land Communication Links with Irregular Built-up Terrain

Below we will describe propagation in built-up areas, when both terminals, the transmitter and receiver, are located in LOS and/or NLOS conditions at the street level, but with the assumption that buildings are randomly distributed over irregular terrain, as a main case of city topography, and will present some more realistic and more specific models that describe the propagation phenomena within the urban communication channel and predict the loss characteristics within it. We will start with empirical and semiempirical models, which are mostly used in urban communication link design, then we will present stochastic model on how to obtain path loss in built-up areas with array of buildings randomly distributed on the rough ground surface.

Okumura's Empirical Model

Based on numerous measurements carried out in and around Tokyo, Okumura proposed an empirical method of predicting the average power within the communication channel "mobile-base station" [26]. The method is based on a series of curves describing the average attenuation $A_{Ru}(f, d)$ relative to free space for quasismooth terrain in an urban environment. We present the average path loss, L_{50}, according to [26], as

$$L_{50} = L_{FS} + A_{Ru}(f, d) + H_{Tu}(h_T, d) + H_{Ru}(h_R, d) \tag{2.63}$$

Here as above, L_{FS} is the path loss in free space. The first correction factor in Eq. (2.63), $A_{Ru}(f, d)$, is expressed in Fig. 2.20 versus frequencies from 100 MHz to 1 GHz and distance from the transmitter (denoted by T) in the range 1–100 km. The reference transmitter antenna height is $h_T = 200$ m, and the reference moving vehicle antenna (denoted by R) height is $h_R = 3$ m. The second correction factor in (2.63), $H_{Tu}(h_T, d)$, is the base station antenna gain factor presented in Fig. 2.21 for the same reference heights of both antennas, $h_T = 200$ m and $h_R = 3$ m. The third correction factor in Eq. (2.63),

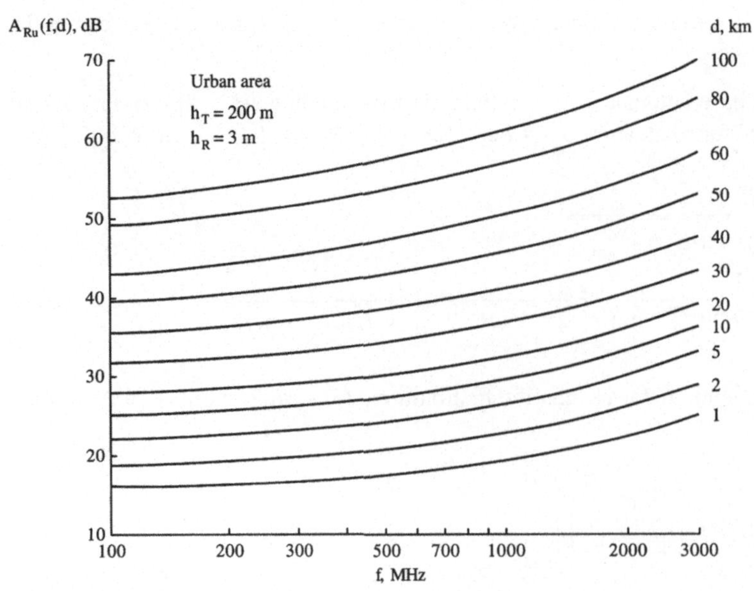

Figure 2.20 The correction factor $A_{Ru}(f, d)$ versus frequencies from 100 MHz to 1 GHz and ranges from 1 km to 100 km.

$H_{Ru}(h_R, d)$, is the moving vehicle antenna height gain that is shown in Fig. 2.22. Here once more, the reference antenna heights are $h_T = 200\,\text{m}$ and $h_R = 3\,\text{m}$. All corrections in Figs. 2.21 and 2.22 are changed in the positive or negative directions as the antenna height differ becoming greater or smaller than $h_T = 200\,\text{m}$ and $h_R = 3\,\text{m}$.

As was mentioned in [1,2], the Okumura approach is probably the most widely quoted of the available models. It takes into account not only urban, suburban and rural environments, but also describes the effects of different kind of terrain. All phenomena and effects can be computed well in practice. However it is rather cumbersome to implement this model with all correction factors in a

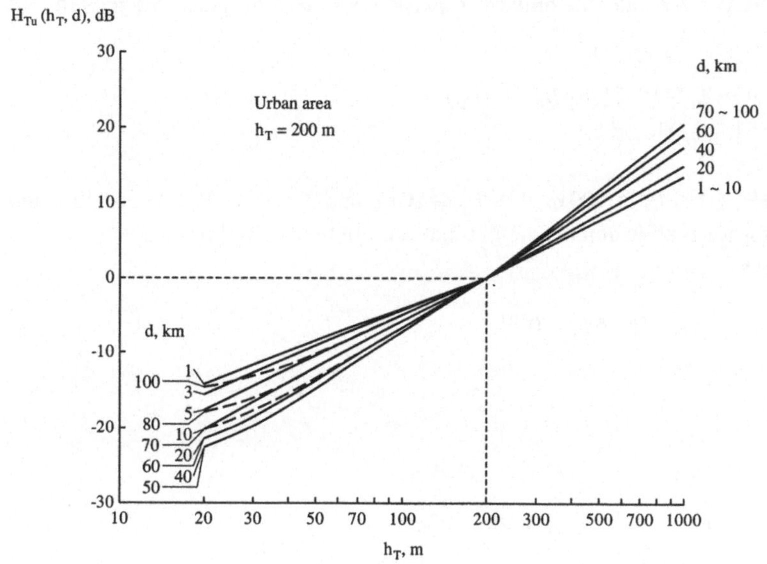

Figure 2.21 The base station antenna gain factor $H_{Tu}(h_T, d)$.

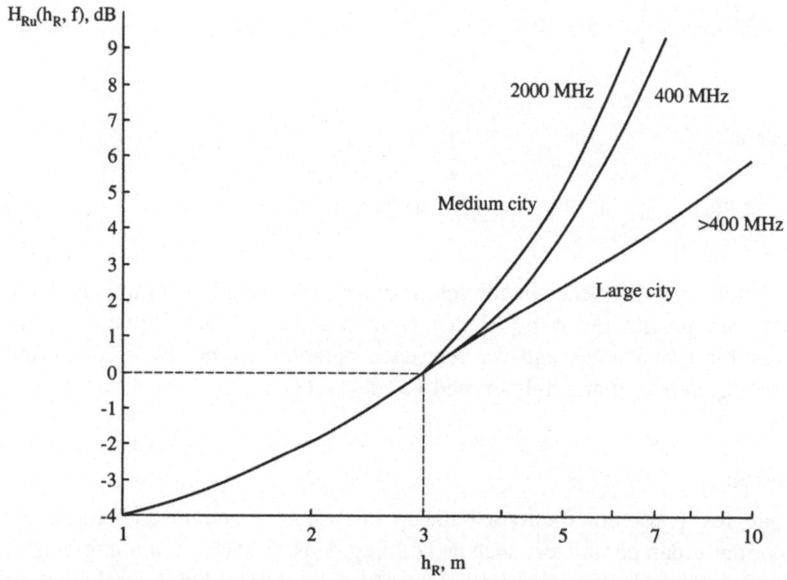

Figure 2.22 The vehicle antenna gain factor $H_{Ru}(h_R, d)$.

computer, because the data is available in graphical form. Thus, for computer implementation data has to be entered in the computer memory in *point-to-point* form and interpolation routines have to be written for intermediate computations.

Hata Model

In an attempt to make the Okumura technique suitable for computer implementation and easy to apply, Hata [27] developed an empirical model to describe the graphical information given by Okumura and presented in Figs. 2.20–2.22. His analytical expressions for average path loss, L_{50}, for urban, suburban and rural areas are applicable only over quasismooth terrain. The average path loss is given in dB as

$$\begin{aligned} L_{50} &= 69.55 + 26.16 \log f_0 - 13.82 \log h_T - a(h_R) \\ &\quad + (44.9 - 6.55 \log h_T) \log d \end{aligned} \tag{2.64}$$

where $150 \leq f_0 \leq 1500\,\text{MHz}$, $30 \leq h_T \leq 200\,\text{m}$, $1 \leq h_R \leq 10\,\text{m}$, and $1 \leq d \leq 20\,\text{km}$. The function $a(h_R)$ is the correlation factor for mobile antenna height that is computed as follows [27]:

For medium-size cities,

$$a(h_R) = (1.1 \log f_0 - 0.7) h_R - (1.56 f_0 - 0.8) \tag{2.65a}$$

For a large city,

$$a(h_R) = \begin{cases} 8.29(\log 1.54\, h_R)^2 - 1.1 & f_0 \leq 200\ \text{MHz} \\ 3.2(\log 11.75\, h_R)^2 - 4.97 & f_0 \geq 400\ \text{MHz} \end{cases} \tag{2.65b}$$

For suburban areas,

$$L_{50} = L_{50}(\text{urban}) - 2\left(\log \frac{f_0}{28}\right)^2 - 5.4\,\text{dB} \tag{2.66}$$

For open and rural areas,

$$L_{50} = L_{50}(\text{urban}) - 4.78(\log f_0)^2 + 18.33 \log f_0 - 40.94\ \text{dB} \tag{2.67}$$

The last formula also account for the difference in correction function for small, medium, and large cities. A comparison between results given by Hata's formulations and data obtained from Okumura's original curves for urban areas and for reference antenna heights $h_T = 200\,\text{m}$ and $h_R = 3\,\text{m}$ reveals negligible differences that, rarely exceed 1–2 dB [1,11].

Walfisch–Ikegami Model

This model gives a good path loss prediction for dense built-up areas such as medium and large cities [1,28]. It is based on important urban parameters such as building density, average building height, and street width. In this model antenna height is generally lower than the average buildings' height, so that the waves are guided along the street.

For *LOS conditions*, the path loss formula has the same form as the free-space formula changing only constants before log d, the distance between terminals d:

$$L_{50}(\text{LOS}) = 42.6 + 20\log f_0 + 26\log d \tag{2.68}$$

As for *NLOS conditions*, the semiempirical path loss formula is [1,28]:

$$L_{50}(\text{NLOS}) = 32.4 + 20\log f_0 + 20\log d + L_{\text{RD}} + L_{\text{MD}} \tag{2.69}$$

where L_{RD} represents rooftop diffraction loss, and L_{MD} represents multiple diffraction loss due to surrounding buildings. The rooftop diffraction loss is characterized as

$$L_{\text{RD}} = -16.9 - 10\log \Delta a + 10\log f_0 + 20\log \Delta h_R + L(0) \tag{2.70}$$

where Δa is the distance between the vehicle and the building, h_R is the mobile vehicle antenna height, $L(0)$ is the loss due to elevation angle, and $\Delta h_R = h_{\text{roof}} - h_R$.

The multiple-diffraction component is characterized by following equation [1,28]:

$$L_{\text{MD}} + K_0 + K_a + K_d \log d + K_f \log f_0 - 9\log a \tag{2.71}$$

where

$$\begin{aligned}
K_0 &= -18\log(1+\Delta h_T)\\
K_a &= \begin{cases} 54 - 0.8\Delta h_T & d \geq 0.5\text{ km} \\ 54 - 1.3\Delta h_T & d < 0.5\text{ km} \end{cases}\\
K_d &= 18 - 15\left(\frac{\Delta h_T}{h_{\text{roof}}}\right)\\
K_f &= \begin{cases} -4 + 0.7\left(\dfrac{f_0}{925} - 1\right) & \text{for suburban} \\ -4 + 0.7\left(\dfrac{f_0}{925} - 1\right) & \text{for urban} \end{cases}
\end{aligned}$$

a is the street width, h_T is the base station antenna height, h_{roof} is the average height of small buildings ($h_{\text{roof}} < h_T$), $\Delta h_T = h_T - h_{\text{roof}}$. In Walfisch–Ikegama model it was initially assumed that the base station antenna height is lower than a tall building but higher than the small buildings surrounding it.

Comparison between both empirical models, the Hata model and the Walfisch–Ikegama model for a dense urban area, shows that both of them have approximately the same polynomial signal power decay versus distance from both terminals with the parameter of attenuation $2.5 \leq \gamma \leq 4$.

Statistical Model of Path Loss in Outdoor Communication Links

We will present now a statistical approach, which is based on the stochastic models described in [1,29–31] for different kinds of the terrain, that is, on the knowledge of the terrain parameters and features.

To estimate the path loss, we, first of all, need information about the terrain features introduced and defined in Refs. 1 and 29–31:

Terrain elevation data, i.e., digital terrain map, consisting of ground heights as grid points $h_q(x,y)$.

A clutter map, that is, the ground cover of artificial and natural obstructions as a distribution of grid points, $h_0(x,y)$, for built-up areas this is the buildings' overlay profile; the average length or width of obstructions, $\langle L \rangle$ or $\langle d \rangle$; the average height of obstructions in the test area, $\bar{h}$; the obstructions density per km^2, ν.

The effective antenna height, that is, the antenna height plus a ground or obstruction height, if the antenna is assembled on a concrete obstruction: z_1 and z_2 for the transmitter and receiver, respectively.

As the result, there is a digital map (cover) with actual heights of obstructions can be performed according to buildings' overlay profile and topographic map of the built-up terrain. Using now all parameters of built-up terrain and both antennas, transmitter and receiver, the three-dimensional digital map can be analyzed. In the general case of rough terrain with randomly distributed obstacles (see Fig. 2.23), in obstructive conditions between both antennas, the following parameters in addition to those presented above must be used: the typical correlation scales, ℓ_v and ℓ_h, of the complex reflection coefficient from the obstacles with absolute value Γ and the type of building material dominant in the tested area, defining the reflecting properties of obstacles. The geometrical parameters of the built-up terrain allow us to obtain the density of building contours at the ground level, $\gamma_0 = 2\langle L \rangle \nu / \pi$, and then the clearance conditions between receiver and transmitter, e.g., the average horizontal distance of the line of sight $\langle \rho \rangle$ as $\langle \rho \rangle = \gamma_0^{-1}$.

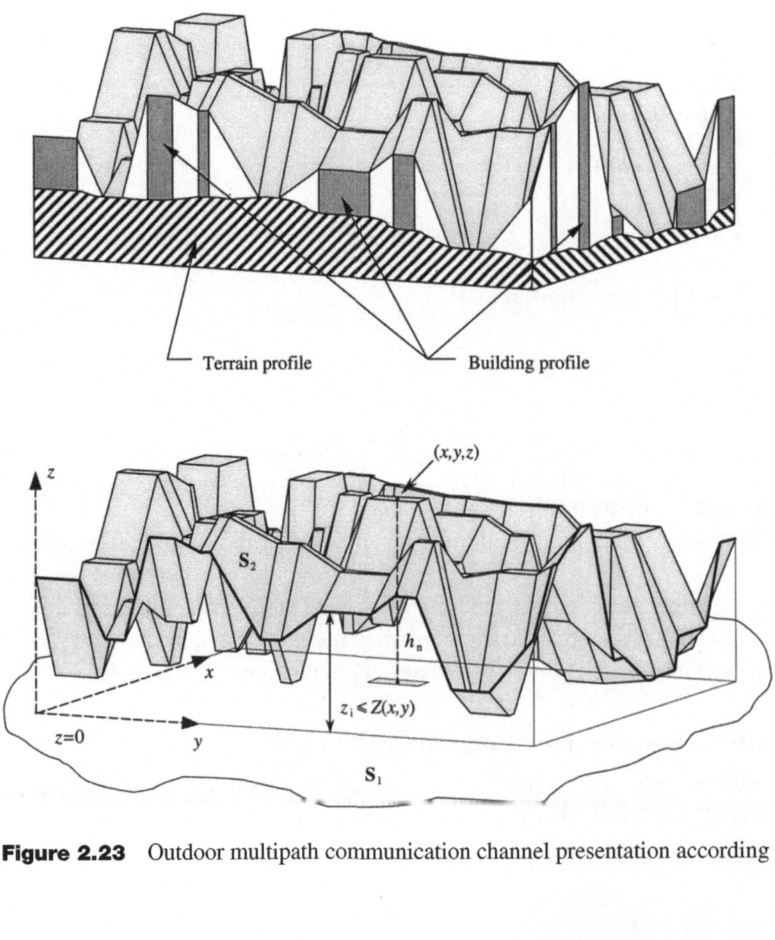

Figure 2.23 Outdoor multipath communication channel presentation according to [31].

Path Loss ($\bar{L}$)

The various factors obtained above are then used for the computer program based on the three-dimensional parametric model for three types of irregular terrain.

In the case of *forested environments* the following formulas are used according to Ref. 31.

1. For description of the incoherent part of the total average field intensity created by multipath field components due to multiple scattering from trees:

$$\langle I_{\text{inc}} \rangle \approx \frac{\gamma_0 \Gamma \exp(-\gamma_0 d)}{(4\pi)^2} \left[\frac{\Gamma^3}{4(8)^3} \frac{1}{d} + \frac{\Gamma}{32} \left(\frac{\pi}{2\gamma_0} \right)^{1/2} \frac{1}{d^{3/2}} + \frac{1}{2\gamma_0} \frac{1}{d^2} \right] \qquad (2.72)$$

2. For the coherent part $\langle I_{\text{co}} \rangle$ of the total field intensity created by the waves coming from the source and specularly reflected from the ground surface,

$$\langle I_{\text{co}} \rangle = \frac{1}{(4\pi)^2} \frac{\exp(-\gamma_0 d)}{d^2} \left[2 \sin \frac{k z_1 z_2}{d} \right]^2 \qquad (2.73)$$

Here d is the distance between the terminal antennas; all other parameters of the terrain and the terminal antennas are defined above.

In the case of *mixed residential areas* the following formulas are used according to Refs. 30 and 31.

1. For description of the incoherent part of the total average field intensity created by multipath components due to independent (single) scattering and diffraction from each obstacle:

$$\langle I_{\text{inc}} \rangle = \frac{\Gamma \lambda \ell_h}{\lambda^2 + (2\pi \ell_h \gamma_0)^2} \frac{\lambda \ell_v}{\lambda^2 + [2\pi \ell_v \gamma_0 (\bar{h} - z_1)]^2} \times \frac{[(\lambda d / 4\pi^3)^2 + (z_2 - \bar{h})^2]^{1/2}}{8\pi d^3} \qquad (2.74)$$

2. For the coherent part $\langle I_{\text{co}} \rangle$ of the total field intensity created by the waves coming from the source and specularly reflected from the ground surface,

$$\langle I_{\text{co}} \rangle = \exp\left(-\gamma_0 d \frac{\bar{h} - z_1}{z_2 - z_1} \right) \left[\frac{\sin(k z_1 z_2 / d)}{2\pi d} \right]^2 \qquad (2.75)$$

In the case of *built-up* (*urban and suburban*) *areas* the following formulas are used according to Refs. 29–31:

1. The incoherent part of the total field intensity due to single scattering and diffraction from buildings' corners and rooftops:

$$\langle I_{\text{incl}} \rangle = \frac{\Gamma \lambda l_v}{8\pi \left[\lambda^2 + [2\pi \ell_v \gamma_0 (\bar{h} - z_1)^2] \right] d^3} \left[\frac{\lambda d}{4\pi^3} + (z_2 - \bar{h})^2 \right]^{1/2} \qquad (2.76a)$$

2. The incoherent part of the total field intensity due to double scattering and diffraction from buildings' corners and rooftops:

$$\langle I_{\text{inc2}} \rangle = \frac{\Gamma^2 \lambda^2 l_v [(\lambda d / 4\pi^3) + (z_2 - \bar{h})^2]}{24\pi^2 [\lambda^2 + [2\pi l_v \gamma_0 (\bar{h} - z_1)]^2]^2 d^3} \tag{2.76b}$$

The coherent part of the total field intensity is described by the same expression, as Eq. (2.75). The total average intensity of the receiving signal for all three types of the terrain is determined by the following formulas:

$$\langle I_{\text{total}} \rangle = \langle I_{\text{co}} \rangle + \langle I_{\text{inc}} \rangle \tag{2.77a}$$

for rural forested and residential areas and

$$\langle I_{\text{total}} \rangle = \langle I_{\text{co}} \rangle + \langle I_{\text{incl}} \rangle + \langle I_{\text{inc2}} \rangle \tag{2.77b}$$

for urban and suburban areas.

The corresponding mean path loss in decibels (dB), taking into account the free space propagation, can be defined as [1,29–31]

$$\bar{L} = -10 \log \left(\lambda^2 \langle I_{\text{total}} \rangle \right) \tag{2.78}$$

All these formulas are used to design a link budget for different land wireless communication links, which also needs knowledge about fading phenomena characteristics, slow and fast, for different kinds of environment. Let us briefly describe fading characteristics in a multipath communication system.

2.4. FADING PHENOMENA IN WIRELESS OUTDOOR COMMUNICATION LINKS

As was mentioned above, most wireless communication systems operate in built-up areas where there is no direct line-of-sight (LOS) radio path between the terminals, the transmitter and the receiver, and where due to natural and artificially made obstructions (hills, trees, buildings, towers, etc.), there occur multidiffraction, multireflection, and multiscattering effects (see Fig. 2.3), which cause not only additional losses (with respect to those obtained in LOS above-the-terrain conditions) but also the multipath fading of the signal strength observed at the receiver, which can be separated into fully independent phenomena, the *slow* and the *fast* fading (see definitions in Sec. 2.1) Below we, first of all, will determine the main parameters of the multipath communication channel, the relations between channel parameters and those of the actual signal passing through it. Then, some general statistical descriptions of the multipath outdoor communication links, which are based on well-known stochastic laws, will be introduced.

2.4.1. Parameters of the Multipath Communication Links

In order to compare different multipath communication links and develop some general understanding of how to design such systems, some specific parameters, which grossly quantify the multipath channel, are used. First of all, we will describe the small-scale fast variations of mobile radio signal, which, as was shown in Refs. 4–8 and 31, directly relate to the impulse response of the channel. Next is a wideband (pulse) channel characterization and contains all information necessary to analyze and to simulate any type of radio transmission through the channel. Because of a time-varying impulse response, due to receiver/transmitter or obstructions motion in space, one can finally represent a total received signal as a sum of amplitudes and time delays of the multipath components arrived at the receiver at any instant of time. If through measurements one can obtain information about the signal power delay profile, one can finally determine the main parameters of the multipath communication channel.

Delay Spread Parameters

First important parameters for wideband channels, which can be determined from a signal power delay profile, are *mean excess delay, rms delay spread* and *excess delay spread* for the concrete threshold level X (in dB) of the channel (see Fig. 2.24). The *mean excess delay* is the first moment of the power delay profile of the pulse signal and is defined, using multipath signal presentation introduced in Sec. 2.1 by the formula Eq. (2.15), as

$$\langle\tau\rangle = \frac{\sum_{i=0}^{N-1} A_i^2 \tau_i}{\sum_{i=0}^{N-1} A_i^2} = \frac{\sum_{i=0}^{N-1} P(\tau_i)\tau_i}{\sum_{i=0}^{N-1} P(\tau_i)} \tag{2.79}$$

The *rms delay spread* is the square root of the second central moment of the power delay profile and is defined as

$$\sigma_\tau = \sqrt{\langle\tau^2\rangle - \langle\tau\rangle^2} \tag{2.80}$$

where

$$\langle\tau^2\rangle = \frac{\sum_{i=0}^{N-1} A_i^2 \tau_i^2}{\sum_{i=0}^{N-1} A_i^2} = \frac{\sum_{i=0}^{N-1} P(\tau_i)\tau_i^2}{\sum_{i=0}^{N-1} P(\tau_i)} \tag{2.81}$$

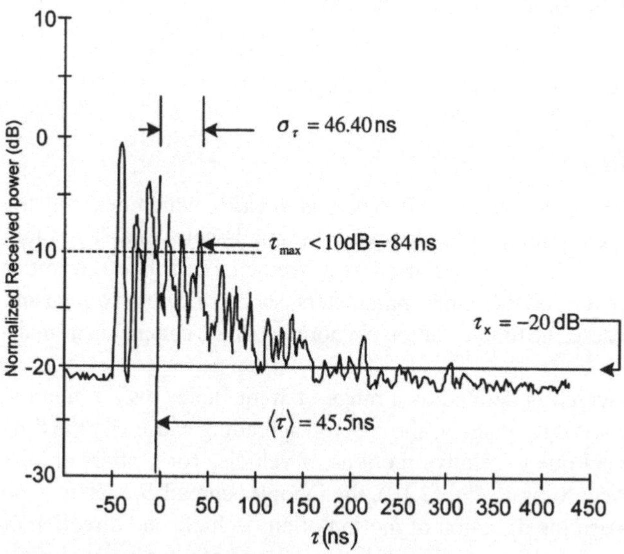

Figure 2.24 Example of the computation of the maximum excess delay for multipath components within 10 dB of the maximum according to [8].

Usually [4–8,31], an additional parameter, the *maximum excess delay*, is introduced as the time delay during which multipath energy falls to the threshold level X (dB) below the maximum, that is,

$$\tau_{max} = \tau_X - \tau_0 \tag{2.82}$$

Here, as above, τ_0 is the time of the first arriving signal at the receiver, τ_X is the maximum delay at which a multipath component is within X dB of the strongest arriving multipath signal (which does not necessarily arrive at τ_0). Figure 2.24 from Ref. 8 illustrates the computation of the maximum excess delay for multipath components within 10 dB of the maximum threshold. The value of τ_X is also called the *excess delay spread* of the power delay profile and in any case must be specified with a threshold of the ratio of the noise floor and the maximum received component.

Coherence Bandwidth

As shown in Refs. 4–8 and 31, the power delay profile in the time domain and the power spectral response in the frequency domain is related through the Fourier transform. Therefore, for the multipath channel full description, the *time delay* parameters in the time domain and the *coherence bandwidth* in the frequency domain are used simultaneously. The coherence bandwidth is the statistical measure of the frequency range over which the channel is considered to be "flat." In other words, this is the frequency range over which two frequency signals are strongly amplitude correlated. Depending on the degree of amplitude correlation of two frequency separated signals, there are different definitions of this parameter.

The *first definition*: *the coherence bandwidth*, B_c, is a bandwidth over which the frequency correlation function is above 0.9 or 90%, and it equals [8]

$$B_c \approx 0.02\sigma_\tau^{-1} \tag{2.83}$$

The *second definition*: the *coherence bandwidth*, B_c, is a bandwidth over which the frequency correlation function is above 0.5 or 50%, and it equals [8]

$$B_c \approx 0.2\sigma_\tau^{-1} \tag{2.84}$$

Doppler Spread and Coherence Time

Above we considered two parameters, *delay spread* and *coherence bandwidth*, which describe the time dispersive nature of the multipath communication channel in a small-scale area. To obtain information about the time varying nature of the channel caused by movements of either transmitter or receiver or obstructions scatters located around them, new parameters, such as *Doppler spread* and *coherence time*, are usually introduced to describe time variation phenomena of the channel in a small-scale region.

Doppler spread B_D is a measure, which is defined as a range of frequencies over which the received Doppler spectrum is essentially nonzero. It shows the spectral spreading caused by the time rate of change of the mobile radio channel due to relative motions of vehicles (or scatters around them) with respect to the base station. According to Eq. (2.10), the Doppler spread B_D depends on Doppler shift f_D and on the angle θ between the direction of motion of any vehicle and direction of arrival of the reflected and/or scattered waves. If we deal with the complex base-band signal presentation Eq. (2.13), then we can introduce some criterion: if the baseband signal bandwidth is greater than the Doppler spread B_D, the effects of Doppler shift are negligible at the receiver.

Coherence time T_c is the time domain dual of *Doppler spread* and it is used to characterize the time varying nature of the frequency dispersive properties of the channel in time coordinates. There is a simple relationship between these two channel characteristics, that is [see also Eq. (2.10) and all notation there]:

$$T_c \approx \frac{1}{f_{D\max}} \approx \frac{\lambda}{v} \tag{2.85a}$$

We can also define the *coherence time* more strictly, according to Refs. 4–8 and 31, as "the time duration over which two multipath components of receiving signal have a strong potential for amplitude correlation." If so, one can, as above for coherence bandwidth, define the coherence time as the time over which the correlation function of two various signals in the time domain is above 0.5 (or 50%). Then according to Refs. 8 and 11 we get

$$T_c \approx \frac{9}{16\pi f_m} = \frac{9\lambda}{16\pi v} = 0.18\frac{\lambda}{v} \tag{2.85b}$$

As was shown in Ref. 12, this definition can be improved for modern digital communication channels by means of combination of Eqs. (2.85a) and (2.85b) as the geometric mean of them, that is,

$$T_c \approx \frac{0.423}{f_m} = 0.423\frac{\lambda}{v} \tag{2.85c}$$

The definition of coherence time implies that two signals, arriving at the receiver with a time separation greater than T_c, are affected differently by the channel.

2.4.2. Types of Fading

It is clear from channel parameters definitions that the type of signal fading within the mobile radio channel depends on the nature of the transmitting signal with respect to the characteristics of the channel. In other words, depending on the relation between the signal parameters, such as *bandwidth* B_S and *symbol period* T_S, and the corresponding channel parameters, such as *coherence bandwidth* B_c and *rms delay spread* σ_τ (or *Doppler spread* B_D and *coherence time* T_c), different transmitted signals will undergo different types of fading. As was shown by Rappaport [8], there are *four possible effects* due to the time and frequency dispersion mechanisms in a mobile radio channel, which are manifested depending on the balance of the above-mentioned parameters of the signal and of the channel. The multipath time delay spread leads to *time dispersion* and *frequency selective fading*, whereas Doppler frequency spread leads to *frequency dispersion* and *time selective fading*. Separation between these four types of small-scale fading for impulse response of multipath radio channel is explained in Table 2.1 according to Ref. 8:

A. Fading due to multipath time delay spread. Time dispersion due to multipath phenomena causes small-scale fading, either *flat* or *frequency selective*:

A.1. The small-scale fading is characterized as *flat* if the mobile channel has a constant-gain and linear-phase impulse response over a bandwidth, which is *greater* than the bandwidth of the transmitted signal. Moreover, the signal bandwidth in the time domain exceeds the signal delay spread, i.e., $T_S \gg \sigma_\tau$ (see the top rows of the last column of Table 2.1). As it can be seen, a flat fading channel can be defined as *narrowband* channel, since the bandwidth of the applied signal is *narrow* with respect to the channel flat fading bandwidth in

Table 2.1 Types of Fading according to [8]

General type of fading	Type of fading	Type of channel
		Narrowband
(A) *Small-scale fading*	(A.1): *Flat fading*	(A.1.1): $B_S \ll B_c$ (A.1.2): $T_S \gg \sigma_\tau$
(Based on multipath time Delay Spread)		*Wideband*
	(A.2): *Frequency selective fading*	(A.2.1): $B_S > B_c$ (A.2.2): $T_S < \sigma_\tau$
		Narrowband
(B) *Small-scale fading*	(B.1): *Fast fading*	(B.1.1): $B_S < B_D$ (B.1.2): $T_S > T_c$
(Based on Doppler frequency spread)		*Wideband*
	(B.2): *Slow fading*	(B.2.1): $B_S \gg B_D$ (B.2.2): $T_S \ll T_c$

the frequency domain, that is, $B_S \ll B_c$. At the same time, the flat fading channel is *amplitude-varying* channel, since there is a deep fading of the transmitted signal which occurs within such a channel.

A.2. The small-scale fading is characterized as a *frequency selective*, if the mobile channel has a constant-gain and linear-phase impulse response over a bandwidth which is *smaller* than the bandwidth of the transmitted signal in the frequency domain, as well as its impulse response has a multiple delay spread greater than the bandwidth of the transmitted signal waveform, i.e., $T_S \ll \sigma_T$. These conditions are presented at the last column of Table 2.1. As can be seen, a frequency-selective fading channel can be defined as *wideband* channel, since the bandwidth of the spectrum $S(f)$ of the transmitted signal is *greater* than the channel frequency-selective fading bandwidth in the frequency domain (the coherence bandwidth), that is, $B_S > B_c$.

B. Fading due to Doppler spread. Depending on how rapidly the transmitted baseband signal changes with respect to the rate of change of the channel, a channel may be classified as a *fast fading* or *slow fading* channel.

B.1. The channel, in which the channel impulse response changes rapidly within the pulse (symbol) duration, is called a *fast fading* channel. In other words, in such a channel its coherence time is smaller than the symbol period of the transmitted signal. At the same time the Doppler spread bandwidth of the channel in the frequency domain is greater than the bandwidth of the transmitted signal (see the last column in Table 2.1). That is, $B_S < B_D$ and $T_c < T_S$. These effects cause frequency dispersion (also called *time-selective fading* [8]) due to Doppler spreading, which leads to signal distortion.

B.2. The channel, in which the channel impulse response changes at a rate *slower* than the transmitted baseband signal $u(t)$, is called a *slow fading* channel. In this case the channel may be assumed to be static over one or several bandwidth intervals. In the time domain, this implies that the reciprocal bandwidth of signal is much smaller than the coherence time of the channel and in the frequency domain the Doppler spread of the channel is less than the bandwidth of the baseband signal, that is, $B_S \gg B_D$ and $T_S \ll T_c$. Both these conditions are presented at the bottom rows of the last column in Table 2.1. It is important to note that velocity of the moving vehicle or moving obstructions within the channel, as well as the baseband signal determine whether a signal undergoes fast fading or slow fading. All situations within the wireless communication channel, described above, are summarized in Fig. 2.25 [8].

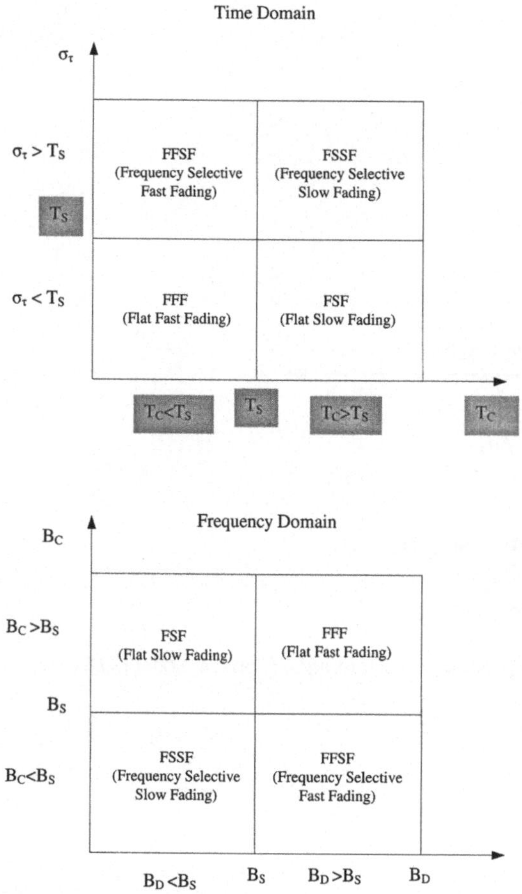

Figure 2.25 Types of fading experienced in the multipath outdoor communication link [8].

2.4.3. Mathematical Modeling of Fast Fading

Now we will discuss the question of the existence of a suitable statistical model for satisfactory description of multipath fast fading channels. Several multipath models have been proposed to describe the observed random signal envelope and phase in a mobile channel. The earliest 2D models were developed in Refs. 32–34 and were based on the random interference of direct (incident) waves and waves scattered from the flat sides of buildings, screens randomly distributed above the terrain. Because such a model is widely used for the description of wireless short-scale fading communication channels, we will present this model to the reader. The model assumes a fixed transmitter with a vertically polarized omnidirectional antenna and a moving receiver also with omnidirectional antenna. The signal at the receiver is assumed to comprise N horizontally traveling plane waves with each wave with number i having equal average amplitude A_i and with statistically independent angles of arrival (α_i) (azimuth angles) and phase angle (ϕ_i) distributions. The assumption of equal average amplitude of each ith wave is based on the absence of an LOS component with respect to scattered components arriving at the receiver. Moreover, phase angles distribution is assumed to be uniform in the interval [0, 2π], that is, the angle distribution function is equal $P(\phi_i) = (2\pi)^{-1}$. A typical ith wave arriving at an angle α_i to the x axis is shown in Fig. 2.26. The receiver moves with

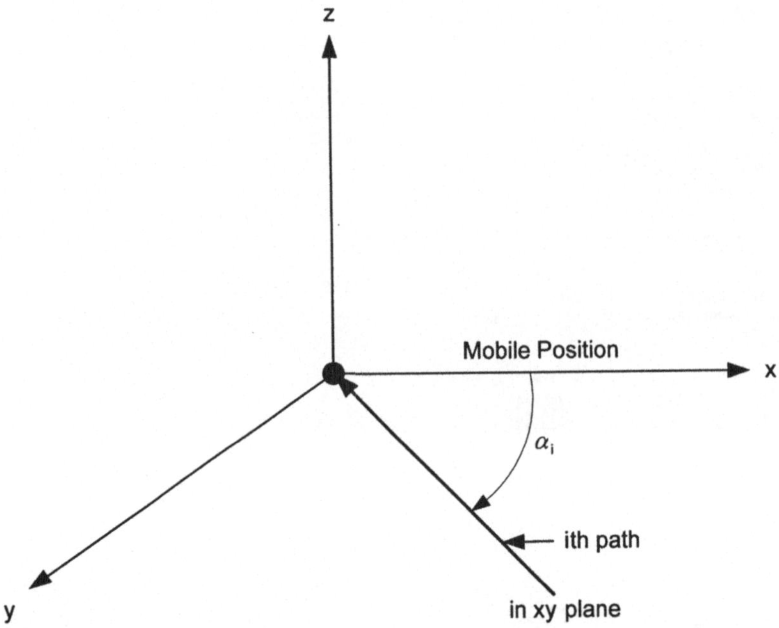

Figure 2.26 Graphical presentation of the multipath phenomena according to Clarke's model [32].

a velocity v in the x direction, so the Doppler shift in z axis, according to Eq. (2.10), can be now rewritten as

$$f_D = \frac{v}{\lambda}\cos\alpha_i \tag{2.86}$$

The mean square value of the amplitude A_i of such uniformly distributed individual waves is constant

$$E\{A_i^2\} \equiv \langle A_i^2 \rangle = \frac{E_0}{N} \tag{2.87}$$

because $N =$ constant and the real amplitude of local average field E_0 is also assumed to be a constant.

Let us consider the vertically polarized plane electromagnetic waves arriving at the moving receiver, which usually have one E-field component (E_z) and two H-field components (H_x and H_y, see [4–8]). Without any loss of generality of the problem, because for each field component the same technique is used, let us consider only the E-field component and present it at the receiving point as [32–34].

$$E_z = E_0 \sum_{i=1}^{N} A_i \cos(\omega_c t + \theta_i) \tag{2.88}$$

where $\omega_c = 2\pi f_c$, f_c is the carrier frequency, $\theta_i = \omega_i t + \phi_i$ is the random phase of the ith arriving component of total signal, and $\omega_i = 2\pi f_i$ represents the Doppler shift experienced by the ith individual wave. The amplitudes of all three electromagnetic field components are normalized such that the ensemble average of the amplitude A_i is given by $\sum_{i=1}^{N}\langle A_i^2 \rangle = 1$.

Since the Doppler shift is small with respect to the carrier frequency, all field components may be modeled as narrow band random processes and approximated as gaussian random variables, if $N \to \infty$, with a uniform phase distribution in the interval $[0, 2\pi]$. If so, the E-field component can be expressed in the following form:

$$E_z = C(t)\cos(\omega_c t) - S(t)\sin(\omega_c t) \tag{2.89}$$

where $C(t)$ and $S(t)$ are the in phase and quadrature components that would be detected by a suitable receiver [32–34]:

$$\begin{aligned} C(t) &= \sum_{i=1}^{N} A_i \cos(\omega_i t + \theta_i) \\ S(t) &= \sum_{i=1}^{N} A_i \sin(\omega_i t + \theta_i) \end{aligned} \tag{2.90}$$

According to the assumptions above, both components $C(t)$ and $S(t)$ are independent Gaussian random processes. They are uncorrelated zero-mean gaussian random variables, that is,

$$\langle S \rangle = \langle C \rangle = \langle E_z \rangle = 0 \tag{2.91}$$

with an equal variance σ^2 (the mean signal power) given by

$$\sigma^2 \equiv \langle |E_z|^2 \rangle = \langle S^2 \rangle = \langle C^2 \rangle = \frac{E_0^2}{2} \tag{2.92}$$

The envelope of the received E-field component can be presented as

$$|E(t)| = \sqrt{S^2(t) + C^2(t)} = r(t) \tag{2.93}$$

Since components $C(t)$ and $S(t)$ are independent gaussian random variables that satisfy Eqs. (2.91)–(2.93), the random received signal envelope r has a Rayleigh distribution [4–8,12] (below, we will talk about the probability density (PDF) and cumulative distribution (CDF) functions of the signal envelope $r(t)$). Using such a definition of the signal envelope, we can now describe two mainly used descriptions of wireless multipath communication links, the Rayleigh and the Rician, according to Refs. 4–8 and 12.

Rayleigh Multipath Fast Fading Statistics

In wireless communication channels, stationary or mobile, the Rayleigh distribution is commonly used to describe the signal's spatial or temporal (i.e., small-scale or fast) fading. As was shown above, a Rayleigh distribution can be obtained mathematically as the limit envelope of the sum of two quadrature gaussian signals, $C(t)$ and $S(t)$. Again, if the phase of multipath components is uniformly distributed over the range of $[0, 2\pi]$, then we deal with a zero-mean Rayleigh distribution of random variable r, the PDF of which can be presented in the following form [2–5,8,9]:

$$\text{PDF}(r) = \frac{r}{\sigma_r^2}\exp\left(-\frac{r^2}{2\sigma_r^2}\right) \quad \text{for } r \geq 0 \tag{2.94}$$

Here the variance σ_r^2 or average power of the received signal envelope for the Rayleigh distribution can be determined as $\sigma_r^2 \equiv E[r^2] - E^2[r]$, where $E[r]$ is an expected value usually used in statistics [2]. The PDF distribution Eq. (2.94) completely describes the random received signal envelope $r(t)$ defined in Clarke's model by the Eq. (2.93). In the formula Eq. (2.94) the maximum value of $\text{PDF}(r) = \exp(-0.5)/\sigma_r = 0.6065/\sigma_r$ corresponds to random variable $r = \sigma_r$.

One can also operate with the so-called *mean value*, the *rms value* and the *median value* of random variable x. The definition of these parameters follows from the Rayleigh CDF presentation, which describes the probability of the event that the envelope of received signal strength (voltage) does not exceed a specified value R [2–5,8,9]:

$$\text{CDF}(R) = \Pr(r \le R) = \int_0^R \text{PDF}(r)\, dr = 1 - \exp\left(-\frac{R^2}{2\sigma_r^2}\right) \tag{2.95}$$

The mean value of the Rayleigh distributed signal strength (voltage), r_{mean} (in the literature it is also denoted as an expected value $E[x]$ [2]), can be obtained from the following conditions:

$$r_{\text{mean}} \equiv E[r] = \int_0^\infty r\, \text{PDF}(r)\, dr = \sigma_r \sqrt{\frac{\pi}{2}} \approx 1.253\sigma_r \tag{2.96a}$$

If so, the *rms value* of the signal envelope is defined as the square root of the mean square, that is,

$$\text{rms} = \sqrt{2}\, \sigma_r \approx 1.414\sigma_r \tag{2.96b}$$

The *median value* of Rayleigh distributed signal strength envelope is defined from the following conditions [2]:

$$\frac{1}{2} = \int_0^{r_{\text{median}}} \text{PDF}(r)\, dr$$

from which follows that

$$r_{\text{median}} = 1.177\sigma_r \tag{2.96c}$$

As follows from Eqs. (2.96a)–(2.96c), the difference between the *mean* and the *median* values is $\sim 0.076\sigma_r$ and their PDF for a Rayleigh fading signal envelope differ by only 0.55 dB. The differences between the rms value and two other values are higher.

Rician Multipath Fading Statistics

As was mentioned in Sec. 2.1, in a wireless communication link, multipath components arrive at the receiver due to multiple reflection, diffraction and scattering from various obstruction around the two terminals, the transmitter and the receiver. Also, a line-of-sight (LOS) component, which describes signal loss along the path of direct visibility (called the *dominant path* [4–9,12]) between both antennas, is often found at the receiver. The PDF of such a received signal is usually said to be *Rician*. To estimate the contribution of each component, dominant (or LOS) and multipath, for the resulting signal at the receiver, the Rician parameter K is usually introduced, as a ratio between these components, i.e.,

$$K = \frac{\text{LOS} - \text{component power}}{\text{multipath} - \text{component power}} \tag{2.97}$$

The Rician PDF distribution of the signal strength or voltage envelope r can be defined as [4–9,12]:

$$\text{PDF}(r) = \frac{r}{\sigma_r^2}\exp\left(-\frac{r^2 + A^2}{2\sigma_r^2}\right) I_0\left(\frac{Ar}{\sigma_r^2}\right) \qquad \text{for } A > 0, \quad r \geq 0 \tag{2.98}$$

where A denotes the peak strength or voltage of the dominant component envelope and $I_0(\cdot)$ is the modified Bessel function of the first kind and zero order. According to the definition Eq. (2.98), we can now rewrite the parameter K, which was defined above as the ratio between the *dominant* and the *multipath* component power. It is given by

$$K = \frac{A^2}{2\sigma_r^2} \tag{2.99a}$$

or in terms of dB

$$K = 10\log\frac{A^2}{2\sigma_r^2}\ \text{dB} \tag{2.99b}$$

Using Eqs. (2.99a) in (2.98), we can rewrite Eq. (2.98) as a function only of K:

$$\text{PDF}(x) = \frac{r}{\sigma_r^2}\exp\left(-\frac{r^2}{2\sigma_r^2}\right)\exp(-K) I_0\left(\frac{r}{\sigma_r}\sqrt{2K}\right) \tag{2.100}$$

from which for $K = 0$ and $\exp(-K) = 1$ follows the worst-case Rayleigh PDF Eq. (2.94) when there is no dominant signal component. Conversely, in a situation of good clearance between two terminals with no multipath components, that is $K \to \infty$, the Rician fading approaches a gaussian one yielding a "Dirac-delta shaped" PDF described by the formula Eq. (2.101) (see below). Hence, the Rician distribution's PDF approaches the Rayleigh PDF and the gaussian PDF, if the Rician K factor approaches zero and infinity, respectively. These features of the Rician PDF and CDF (in dB) can be seen from illustrations presented in Figs. 2.27 and 2.28, respectively.

2.4.4. Mathematical Modeling of Slow Fading

Gaussian Fading Statistics

A very interesting situation within the wireless communication link is that, when *good clearance* between two terminals or *slow* fading at the receiver occurs, we find a tendency to gaussian (also called *normal*) distribution of the random received signal strength or voltage r [4–8,12,31] with the following PDF:

$$\text{PDF}(r) = \frac{1}{\sigma_L\sqrt{2\pi}}\exp\left[-\frac{(r - \bar{r})^2}{2\sigma_L}\right] \tag{2.101}$$

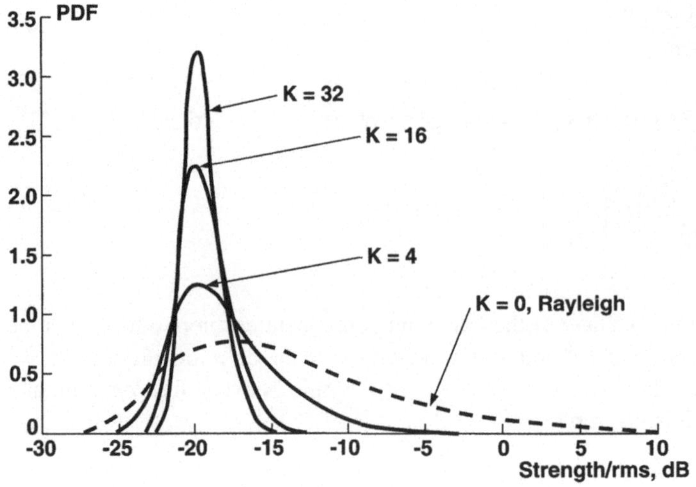

Figure 2.27 The Rician PDF distribution in the logarithmic scale for different parameters $K = 0, 4, 16$, and 32 [11].

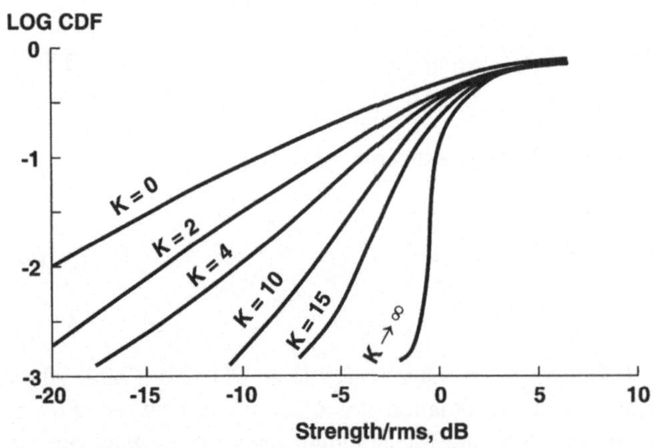

Figure 2.28 The Rician LOG CDF distribution for different parameters K: $K = 0$ corresponds to the Rayleigh distribution; $K \to \infty$ corresponds to the gaussian distribution [11].

Here $\bar{r} \equiv \langle r \rangle$ is the mean value of the random signal level, σ_L is the value of the received signal strength or voltage envelope and $\sigma_L^2 = \langle r^2 - \bar{r}^2 \rangle$ is the variance or time-average power ($\langle w \rangle$ is a sign of averaging of variable w) of the received signal envelope. This PDF can be obtained only as a result of the random interference of a large number of signals with randomly distributed amplitudes (strength or voltage) and phase. If the phase of the interfering signals is uniformly distributed over the range of $[0, 2\pi]$, then one can talk about a zero-mean gaussian distribution of random variable r. In this case we define the PDF of such a process by Eq. (2.101) with $\bar{r} = 0$ and $\sigma_L^2 = \langle r^2 \rangle$, and CDF as [10–12]

$$\mathrm{CDF}(R) = \Pr(r \leq R) = \int_0^R \mathrm{PDF}(r)\, dr$$
$$= \frac{1}{\sigma_L \sqrt{2\pi}} \int_{-\infty}^{X} \exp\left[-\frac{(r-\bar{r})^2}{2\sigma_L^2}\right] = \frac{1}{2} + \frac{1}{2}\operatorname{erf}\left(\frac{R-\bar{r}}{\sqrt{2}\sigma_L}\right) \tag{2.102}$$

where the error function is defined by

$$\text{erf}(w) = \frac{2}{\sqrt{\pi}} \int_0^w \exp(-y^2)\, dy \tag{2.103}$$

From Eq. (2.101) it is clearly seen that the value $\bar{r} = 0$ corresponds to the maximum of the PDF, which equals $\text{PDF}(0) = 1/\sigma_L\sqrt{2\pi}$. For $r = \sigma_L$, it follows from Eq. (2.101) that $\text{PDF}(\sigma_L) = 1/\sigma_L\sqrt{2\pi e}$, where $e \approx 2.71$. Using probability distribution functions Eqs. (2.101) and (2.102) and obtaining during measurements some information about the variance $\sigma_L^2 = \langle r^2 \rangle$ of received signals, we can easily predict the slow fading in corresponding communication links.

We must note here that *in decibels* slow fading, described in voltage by normal or gaussian distributions Eqs. (2.101) and (2.102), is usually described by the log-normal distribution [4–9,12]. In the propagation channels with log-normal *slow fading* or *shadowing*, the effects of fading can be represented by the local path loss $L(r)$ at an arbitrary local point x of radio path randomly distributed inside the propagation channel [8,12]:

$$L(x) = \bar{L}(x) + R_\sigma \text{ dB} \tag{2.104}$$

where R_σ is a gaussian distributed random variable (in dB) with mean value $\bar{r} = 0$ and with standard deviation $\sigma_L = \sqrt{(r - \bar{r})^2}$ (also in dB) and $\bar{L}(x)$ is the average large-scale path loss for arbitrary point r between both terminal antennas, which can be presented as follows [8]:

$$\bar{L}(x) = \bar{L}(x_0) + 10n \log \frac{x}{x_0} \text{ dB} \tag{2.105}$$

Here x_0 is the reference distance, close to the transmitter, which is determined from concrete measurements in the urban scene, $\bar{L}(x_0)$ is the average path loss at the distance x_0 from the transmitter, and n is the path loss exponent, which indicates the rate of signal power attenuation with distance.

As follows from Eqs. (2.104) and (2.105), the *log-normal shadowing*, as a slow fading phenomenon, implies that measured signal levels at a concrete distance between the transmitter and the receiver have a normal, or gaussian, distribution about the distance-dependent mean path loss from Eq. (2.105), where the measured signal levels have values in dB units. The standard deviation σ_L of the gaussian distribution, which describes the shadowing effect also has units in dB. The probability density function of the shadowing component R_σ in Eq. (2.104), as a zero-mean gaussian variable with standard deviation σ_L, can be presented at the same manner than Eq. (2.101), that is, [8]

$$\text{PDF}(R_\sigma) = \frac{1}{\sigma_L\sqrt{2\pi}} \exp\left(-\frac{R_\sigma^2}{2\sigma_L^2}\right) \tag{2.106}$$

The probability that the shadowing increases in Eq. (2.105) the average path loss $\bar{L}(r)$ by at least Z dB can be presented using the *complimentary cumulative distribution* function $\text{CCDF}(R) = 1 - \text{CDF}(R)$, which we denote as $Q(Z/\sigma_L)$,

$$Q\left(\frac{Z}{\sigma_L}\right) \equiv \text{CCDF}\left(\frac{Z}{\sigma_L}\right) = 1 - \text{CDF}\left(\frac{Z}{\sigma_L}\right) \equiv \Pr(R_\sigma > Z) \tag{2.107}$$

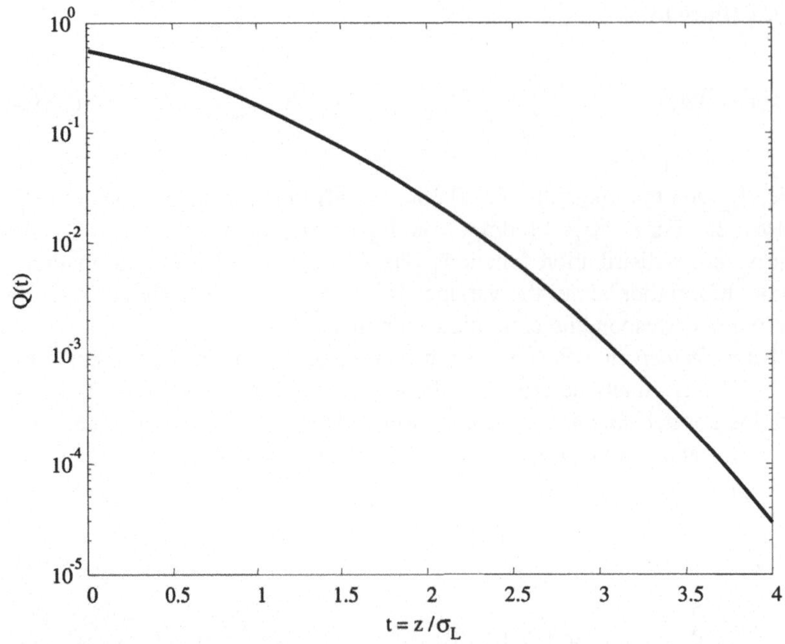

Figure 2.29 The complementary cumulative normal distribution function $Q(Z/\sigma_L)$ versus normalized parameter $w = Z/\sigma_L$ [8].

where CDF(Z/σ_L) is described in the same manner, as in Eq. (2.102), through the error function (erf) defined by Eq. (2.103), that is,

$$\mathrm{CDF}\left(\frac{Z}{\sigma_L}\right) \equiv \Pr(R_\sigma < Z) = \int_0^Z \mathrm{PDF}(R_\sigma)\, dR_\sigma \tag{2.108}$$

If so, we finally have for the complementary cumulative normal distribution function $Q(Z/\sigma_L)$ the following expression by putting in Eq. (2.107) the normalized variable $w = Z/\sigma_L$ [8]:

$$Q(w) = \frac{1}{\sqrt{2\pi}} \int_{r=w}^{\infty} \exp\left(-\frac{r^2}{2}\right) dx = \frac{1}{2}\,\mathrm{erf}\left(\frac{w}{\sqrt{2}}\right) \tag{2.109}$$

This function is plotted in Fig. 2.29 and can be used to evaluate the shadowing margin needed for any location variability in accordance with Eq. (2.109).

2.5. LINK BUDGET DESIGN IN WIRELESS OUTDOOR COMMUNICATION SYSTEMS

Link budget is the main parameter of wireless communication systems, both indoor and outdoor. Because the subject of our chapter is the outdoor communication, below we will deal with the land communication links, taking into account all essential parts of the communication system, as is shown in Fig. 2.1, that is, the total path loss within the communication channel, including fading propagation phenomena, the losses inside the terminal antennas, the transmitter and receiver, and also the thermal noise (adaptive) inside the electronic channels. Some characteristics, such as thermal noise, gains of antennas and antenna losses, and link average losses, (called above the *average path loss*) are simple to evaluate using knowledge obtained in Secs. 2.1–2.3. A more complicated question is how to obtain information about the multiplicative noise, that is, about the long-term (or slow) and short-term (or fast) fading. Let us briefly discuss this subject and give some simple examples on how to estimate such propagation characteristics within the outdoor communication link.

2.5.1. Link Budget Accounting Shadowing Effects (Slow Fading)

To take into account the slow fading or the shadow effects within the communication link, the corresponding graph is plotted in Fig. 2.30 for link budget design. The later takes into account both LOS and NLOS (clutter) conditions, and by the slow fading component caused by shadowing, that is,

$$L_{\text{total}} = L_{\text{LOS}} + L_{\text{NLOS}} + R_{\sigma} \tag{2.110}$$

An example of such a link budget design is shown in Fig. 2.30 (according to Ref. 5) for a wireless communication system providing 90% successful communications at the fringe of radio coverage. In other words, the communication system was examined for the case where 90% of locations at the boundaries of the tested area had acceptable radio coverage. Within the tested area, a greater percentage of vehicle antenna location has acceptable coverage, so here the total path loss will be less. We can illustrate this effect by using the results presented in Fig. 2.30. In fact, if the maximum acceptable pass loss is 120 dB, the probability that $L_{\text{total}} > 120$ dB is [5]

$$\begin{aligned} \Pr(L_{\text{total}} > 120) &= \Pr(L_{\text{LOS}} + L_{\text{NLOS}} + R_{\sigma} > 120) \\ &= \Pr(R_{\sigma} > 120 - L_{\text{LOS}} - L_{\text{NLOS}}) \\ &\equiv Q\left(\frac{120 - L_{\text{LOS}} - L_{\text{NLOS}}}{\sigma_L}\right) \end{aligned} \tag{2.111}$$

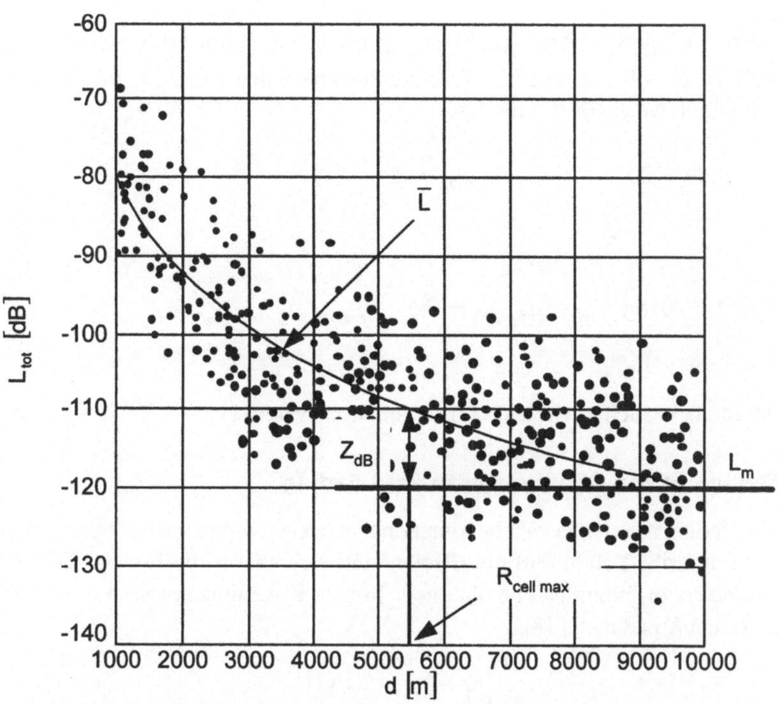

Figure 2.30 Effect of shadowing margin on tested site range.

This probability was denoted as an outage probability P_{out}. According to Eq. (2.107) the fraction of locations covered by the transmitter at a range r is simply [5]

$$\text{Coverage fraction} = P_{cf}(d) = 1 - P_{out} \tag{2.112}$$

We must note here, that the outage probability Eq. (2.112) does not take into account the effects of signal-to-interference ratio, which we consider below, and here is purely caused by shadow effects. In general terms Eq. (2.112) can be expressed as [5]

$$P_{cf}(d) = 1 - Q\left(\frac{L_m - \bar{L}(d)}{\sigma_L}\right) = 1 - Q\left(\frac{Z}{\sigma_L}\right) \tag{2.113}$$

where L_m is the maximum acceptable path loss and $\bar{L}(d)$ is the median path loss within the actual communication system, evaluated at a distance d; $Z = L_m - \bar{L}(d) \equiv L_{SF}$ is the fade margin chosen for such a system (see all notations in Fig. 2.30).

Example. Find a distance d between the terminal antennas within a wireless land communication link, which operates at frequency of 1 GHz and provides 80% of successful communications at the fringe of coverage. Let us assume that propagation LOS effects are described by a free space model Eq. (2.32) for isotropic antennas and the clutter factor of the rough terrain (hills and trees) is $L_{NLOS} = 38.5$ dB, with shadowing of location variability $\sigma_L = 8$ dB. The maximum acceptable path loss is $L_m = 150$ dB.

Solution. According to link budget Eq. (2.110) and free-space propagation model Eq. (2.32) for isotropic antennas:

$$L_{total} = 32.44 + 20\log d_{km} + 20\log f_{MHz} + L_{NLOS} + Z$$

To find Z we take value $t = Z/\sigma_L$ from Fig. 2.26, taking into account that the probability of shadow is $Q(t) = 100\% - 80\% = 20\%$ or $Q = 0.2$. From Fig. 2.26 this occurs when $t = 0.75$, that is, $Z = t\sigma_L = 0.75(8) = 6$ dB, from which we easily obtain that

$$20\log d_{km} = L_{total} - 32.44 - 20\log f_{MHz} - L_{NLOS} - Z$$

or

$$\log d_{km} = \frac{L_{total} - 32.44 - 20\log f_{MHz} - L_{NLOS} - Z}{20} \approx 0.65$$

So the distance between antennas in such a wireless communication system is $d = 10^{0.65} = 4.5$ km.

2.5.2. Link Budget Design of the Channels with Fast Fading

As was mentioned above, fast fading is a stochastic phenomenon that occurs within the multipath communication links and can be described commonly by Rician statistics, that is, by Rician PDF and CDF, from which main parameters of fading can be obtained. In fact, if we now rewrite Eq. (2.110) through parameter rms and Rician K parameter [8],

$$\text{PDF}(r) = 2\frac{r}{(\text{rms})^2}\exp\left(-\frac{r^2}{(\text{rms})^2}\right)\exp(-K)I_0\left(2\frac{r}{\text{rms}}\sqrt{K}\right) \tag{2.114}$$

we can present the fast fade margin, L_{FF}, as [9]

$$L_{\text{FF}} = 10\log\sigma_{\text{FF}} = 10\log\left[\int_0^{\infty} r^2\text{PDF}(r)\,dr - \left(\int_0^{\infty} r\text{PDF}(r)\,dr\right)^2\right]\text{dB}$$

Using derivations carried out in Ref. 9, we finally get the expression

$$\sigma_{\text{FF}} = [2(\text{rms})^2 e^{-K}]\left[\frac{1}{2}e^K\int_0^{\infty} y^3 e^{-y^2} I_0(2y\sqrt{K})\,dy - \left(\int_0^{\infty} y^2 e^{-y^2} I_0(2y\sqrt{K})\,dy\right)^2\right]^{1/2} \tag{2.115}$$

Then, the total path loss in the multipath channel with the fast fading only, can be easily obtained by the knowledge of average path loss, consisting the LOS effects and NLOS effects, according to the commonly used models described in Sec. 2.3, and the fade margin, i.e.,

$$L_{\text{total}} = \bar{L} + L_{\text{FF}} \tag{2.116a}$$

In the *common case* of a communication link with multipath and shadowing effects of radio propagation, the corresponding link budget equation is [9]

$$L_{\text{total}} = \bar{L} + L_{\text{FF}} + L_{\text{SF}} \tag{2.116b}$$

Through the corresponding knowledge of the signal-to-noise ratio, the antenna gains and the sensitivity of wireless system, i.e., the maximum acceptable path loss (see above all definitions), we can finally design a full budget of a wireless outdoor communication system.

Example. Let us consider a communication system designed in the built-up area with, as discussed above, a probability of shadowing $Q(t) = 0.2$ with shadowing of location variability $\sigma_L = 8$ dB, and with a probability of fast fading CDF $(\sigma_{\text{FF}}/\text{rms}) = 0.1$ described by the Rician statistics with the standard deviation about the mean $\sigma_r = 5$ dB and with Rician parameter $K = 15$.

Find the link budget of such a wireless outdoor communication system, if the average path loss inside the system is 105 dB, the antenna gains are $G_T = G_R = -5$ dB, and signal-to-noise ratio is 10 dB. Estimate the efficiency of its performance if the maximum acceptable path loss of the system $L_m = 120$ dB.

Solution. Taking into account Eq. (2.32) for mean path loss in free space for isotropic antennas and the effects of slow and fast fading described by Eq. (2.116b), we will rewrite the link budget of the system as

$$L_{\text{total}} = \bar{L} + L_{\text{FF}} + L_{\text{SF}} + \frac{S}{N} + G_T + G_R$$

1. For $Q(t) = 0.2$ and $\sigma_L = 8$ dB, we get $t = 0.75$ (see example above) and then the slow fade margin is $L_{\text{SF}} \equiv Z = 8 \times 0.75 = 6$ dB.
2. Taking into account that CDF$(\sigma_{\text{FF}}/\text{rms}) = 0.1$ and that rms $\approx 1.4\sigma_r = 7$ dB, we have from Fig. 2.28 (check curve for $K = 15$), that $\sigma_{\text{FF dB}} - (rms)_{\text{dB}} \equiv L_{\text{FF}} - 7 = -3$ dB, from which we get $L_{\text{FF}} = 4$ dB.
3. The total path loss of the system is

$$L_{\text{total}} = 105 + 6 + 4 + 10 - 5 - 5 = 115\,\text{dB}$$

Because the maximum acceptable path loss is 120 dB, than is, higher than the real loss within the system, the sensitivity of the system is enough to obtain information from any vehicle or subscriber located within the area of service.

2.6. CELLULAR CONCEPT FOR WIRELESS SYSTEMS

Usually the actual design of modern wireless systems relates to the so-called *cellular concept* of wireless communications in built-up areas [1–8,28,35,36], which allows the designers to decrease natural background noises within the propagation channels and to exclude deep fading affecting the signal at the input of the receiver.

Let us ask a question: What is the "cellular principle" and how may we construct each "cell" in a completed cellular system? The simplest "radio cell" one can construct uses a base station at the center of such a cell, which determines the coverage area from its antenna. This coverage area is defined by the range where a stable signal from this station can be received. Figure 2.31 illustrates the distribution of such cells. It is seen that there exist regions of overlap with neighboring "radio cells," where stable reception from neighboring base stations can be obtained. From this scheme it also follows that different frequencies should be used in these cells which surround the tested central "cell." On the other hand, the same frequencies can be used for the cells farthest from the central one. This is the so-called *cells repeating or reuse of operating frequencies* principle. At the same time, the reuse of the same radio channels and frequencies within the neighboring cells is limited by preplanned *cochannel interference*. Moreover, in the process of cellular systems design in various built-up areas, it is very important to predict the influence of propagation phenomena within the

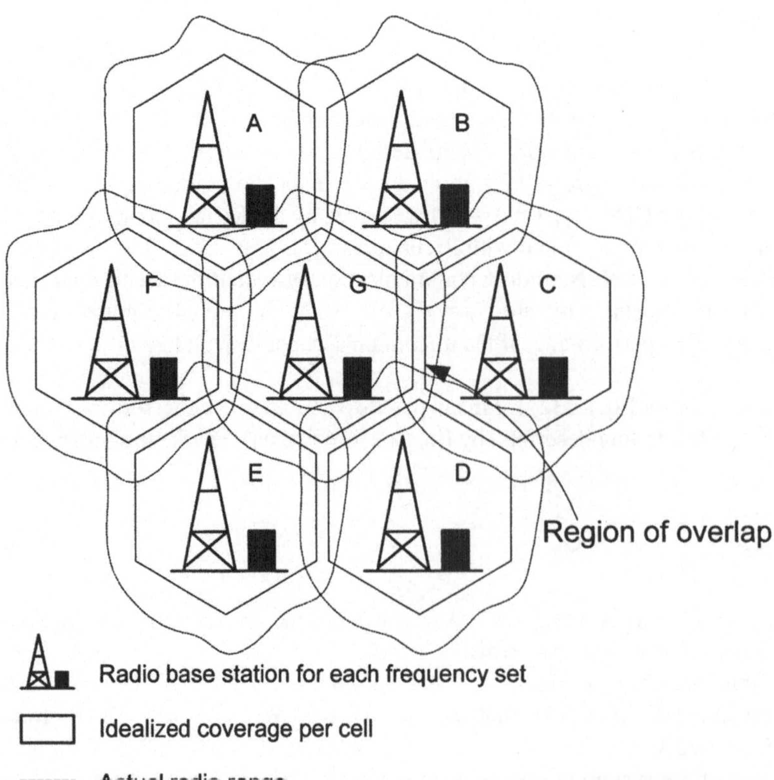

Figure 2.31 The concept of cell distribution and cellular map pattern according to [35,36].

corresponding communication channels on variations of the main parameters of the cellular system, on the construction and splitting of cellular maps. All these questions will be discussed below.

2.6.1. Main Characteristics of a Cell

The main question is why is it useful to use such a cell structure with a lot of base stations, as shown in Fig. 2.31, instead of using a more powerful antenna which will cover a large area and will service enough subscribers? Let us present a simple example. For a flat terrain, R can be defined as the radius of a circle surrounding the base station on the topographical map of selected area. Then the area that is covered by the base station antenna is approximately πR^2(in km^2). If, for example, radius $R = 2\,\text{km}$, the single cell coverage area is $S = 6.28\,\text{km}^2$, which provides service to about 200 subscribers of a wireless personal communication channel [5]. For $R = 20\,\text{km}$, $S = 628\,\text{km}^2$ and the number of subscribers grows to 20,000. But to service 120,000 subscribers, the cell should be designed with a radius $R = 25\,\text{km}$, with an area $S = 1960\,\text{km}^2$ [5]. As follows from the above estimations, to use only a single antenna (or single cell) for stable wireless communication in urban conditions with the complicated multipath propagation phenomena, caused by the multireflections, multidiffraction, multiscattering, etc., is in practice quite unrealistic.

This is why, the concept of cellular wireless communication has been introduced with numerous cells of a small radius, which provide a sufficient signal-to-noise ratio and a low level of fading, slow and fast, of the received signals within the communication channel. As an example, a characteristic cell layout plan for London, U.K. is presented in Fig. 2.32 according to Ref. 35, at an early stage of its implementation. As follows from this figure, the early strategy of cell communications design is based on the following principles:

With an increase of the number of subscribers, the dimensions of the cells become smaller (usually this was done for centers of cities, where the number of users is bigger and building density is higher).

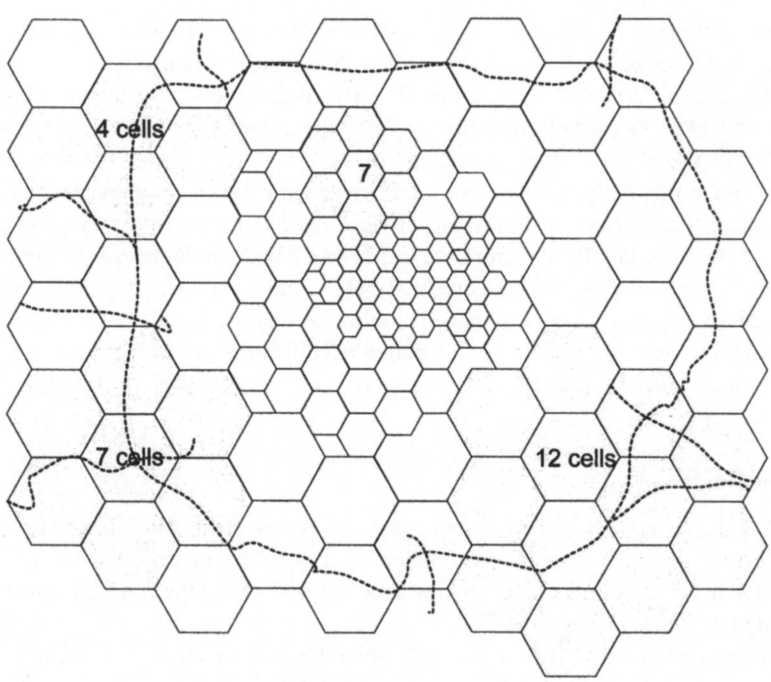

Figure 2.32 A typical city cellular map, where cluster sizes of 4, 7, and 12 are also indicated [35].

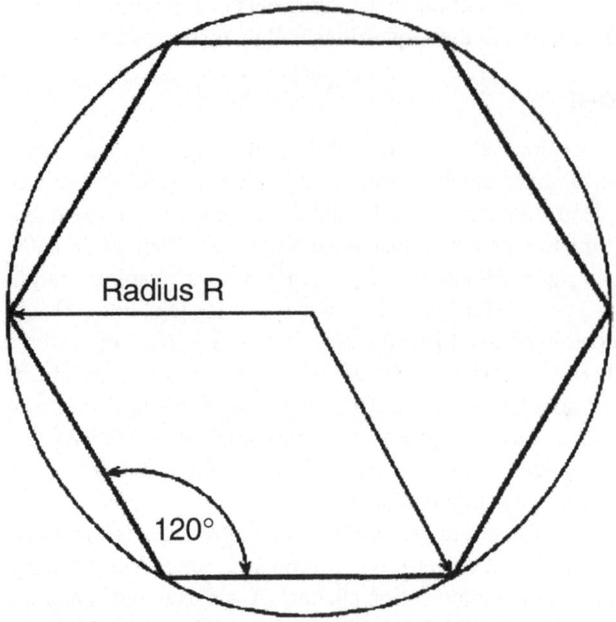

Figure 2.33 Circle-shaped and regular hexagon-shaped cells presentation according to [35].

Cells are arranged in *clusters*. Only clusters with a hexagonal shape are possible; the designed cluster sizes of 4, 7, and 12 cells are shown in Fig. 2.32;

Cells are split. The installation of additional base stations within each cell depends on the degree of cell density in each cluster and on the coverage effect of each base station antenna.

Moreover, this proposed strategy of cells design has shown [5] that each base station antenna in such cells, with an effective power of 100 mW, covers a cell with radius $R = 1$ mil (~1.6 km). At the same time, to cover the area of one cell with radius $R = 10$ mil (~16 km), the transmitting antenna requires a power of 100 W, i.e., 1000 times higher. Thus the antenna power problem has been successfully solved by the use of a cell splitting strategy.

However the question regarding the regions of overlap of coverage between neighboring cells is not solved yet (as can be clearly seen from Fig. 2.31). The circle-shaped cell was therefore replaced by a regular hexagon-shaped cell. It is clearly seen from Fig. 2.33, where both circle-shaped and regular hexagon-shaped cells are presented, that the hexagon-shaped cell is more geometrically attractive than the circle-shaped cell. Moreover, in the hexagon-shaped multiple cells structure (plan), the hexagonal cells are closely covered by each other. Thus, each hexagonal cell can be packed into clusters "side to side" with neighboring cells. The size of such a hexagonal cell can be defined by use of its radius R and the angle of 120° (see Fig. 2.33).

2.6.2. Cell Design Strategy

Now we will describe the main characteristics of a cell and will show how to create the cell structure. The real distance from the center of a cell, where the base station is located ("based cell"), to the center of the "repeat cell," which is denoted in Fig. 2.34 by the same letter, is called the *reuse distance*, D; the cell size is determined by its *radius* R.

The cluster size is designated by the letter N and is determined by the equation [1–5,35,36]

$$N = i^2 + ij + j^2 \tag{2.117}$$

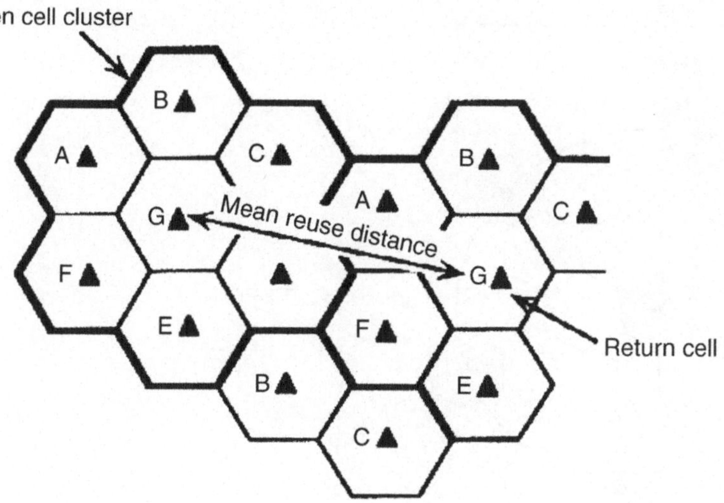

Figure 2.34 The popular 7-cell cluster arrangement. D is the reuse distance, and R is the cell's radius according to [35].

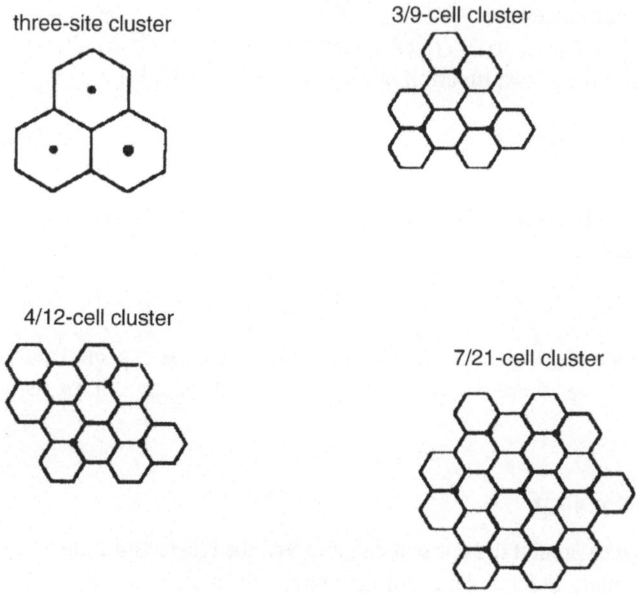

Figure 2.35 Different variants of sectored clusters according to [35].

where $i, j = 0, 1, 2, \ldots$, etc. As follows from Eq. (2.117), only the cluster sizes 3, 4, 7, 9, 12, etc., are possible. However each cluster can be divided into 3 clusters each consisting of 3 cells. It is called a 3/9-cell cluster (see Fig. 2.35). Other variants of sectored clusters are presented in Fig. 2.35. As can be seen, each sector has one base station antenna (or radio port).

We do not enter into this subject deeply, but will only remark that it is necessary to divide clusters into sub-clusters because of the necessity to use the same repeating frequencies in different cells. In fact, if we focus on the popular 7-cell cluster arrangement, which is depicted in Fig. 2.34, we first notice that the allocation of frequencies into seven sets is required. In Fig. 2.34, the mean reuse distance, is

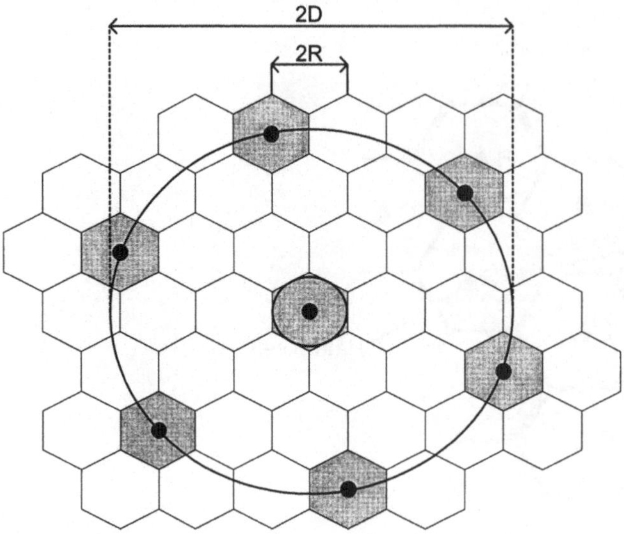

Figure 2.36 Directional frequency reuse plan according to [35].

illustrated, in which the cells (say, denoted by $G \leftrightarrow G$) use the *same frequency set*. This is a simple way to use the repeat frequency set in the other clusters.

Between D and the cell radius R (see Fig. 2.36) there exists a relationship, which is called the *reuse ratio*. This parameter for a hexagonal cell is a function of cluster size, [1–5,35,36], i.e.,

$$\frac{D}{R} = \sqrt{3N} \tag{2.118}$$

Example. For a 7-cell cluster of 2-mile-radius cells, the repeat cell centers, which operate with the same frequency set would be separated by

$$D = R\sqrt{3N} = 2\sqrt{21} \approx 9.2 \text{ mil}$$

Within other cells in a cluster, interference inside the communication channel can be expected at the same frequencies. Hence, for a 7-cell cluster there could be up to six immediate interferers, as is shown in Fig. 2.36.

2.6.3. Cochannel Interference Concept

Now we will discuss the question of how to predict the optimal cell size and the cluster splitting using the law of signal decay, described by many independent radio propagation models constructed to predict the propagation effects within various wireless communication channels. This question is related closely to another major problem of cochannel interference caused by frequent reuse of channels within the cellular communication system. To illustrate the cochannel interference concept, let us consider a pair of cells with radius R, separated by a reuse distance D, as shown in Fig. 2.36. Since the cochannel site is located far from the transmitter ($D \gg R$), which is located within the initial cell, its signal at the servicing site will suffer multipath attenuation. We consider here the situation in the built-up environments where both antennas are lower than the surrounding buildings' rooftops. To predict the degree of cochannel interference in such a situation with moving subscribers within the cellular system, a new parameter, "carrier-to-interference ratio," C/I, is introduced in the literature [2–9]. This parameter in turn depends on frequency planning and antenna engineering. As pointed out in Ref. 9, a cochannel interferer has the same nominal frequency as the desired frequency. It arises from the multiple use of the same frequency. Thus, referring to the part of cellular map depicted in Fig. 2.34, we find that cochannel sites are located in the second cluster. For omnidirectional antennas located inside each site, the theoretical cochannel interference in dB is given by [1–4,8]:

$$\frac{C}{I} = 10\log\left[\frac{1}{j}\left(\frac{D}{R}\right)^{\gamma}\right] \tag{2.119}$$

where j is the number of cochannel interferers ($j = 1, 2, \ldots, 6$), γ is the path loss slope constant, which determines the signal decay in various propagation environments. For a typical seven-cell cluster ($N = 7$) with one cell as basic (with the transmitter inside it) and with six other interferers ($j = 6$) as the cochannel sites in first tier (see Fig. 2.36), this parameter depends on conditions of wave propagation within the urban communication channel. To understand this fact, let us present, according to Ref. 1, a simple propagation model for the regular urban environment with $\gamma = 4$. In this case one can rewrite Eq. (2.119) as

$$\frac{C}{I} = 10\log\left[\frac{1}{6}\left(\frac{D}{R}\right)^{4}\right] \tag{2.120}$$

In this case, according to Eq. (2.120), $D/R = \sqrt{3N} = 4.58$, and $C/I = 18.6\,\text{dB}$.

In the general case, by introducing Eq. (2.118) in Eq. (2.120) we have that [1–4,8]

$$\frac{C}{I} = 10\log\left[\frac{1}{6}(3N)^2\right] = 10\log(1.5N^2) \tag{2.121}$$

i.e., C/I is also a function of cluster size N and is increased with increase of cells' number in each cluster or with decrease of cell radius R.

According to the propagation situation in urban scene the servicing and cochannel sites can lie both inside and outside the break point range r_B (see Fig. 2.37). If both of them are within this range, as follows from Fig. 2.38a, the cochannel interference parameter can be described instead [Eq. (2.119)] by the C/I-ratio prediction equation (in dB) as [1–4,8]

$$\frac{C}{I} = 10\log\left[\frac{1}{6}\left(\frac{D}{R}\right)^{2}\right] \tag{2.122}$$

For cell sites located beyond the break point range (see Fig. 2.38b) this equation can be modified taking into account the multipath phenomenon and obstructions which change the signal decay law from D^{-2} to $D^{-\gamma}$, $\gamma = 2 + \Delta\gamma$, $\Delta\gamma \geq 1$. Hence, we finally have instead of Eq. (2.118) [1]:

$$\frac{C}{I} = 10\log\left[\frac{1}{6}\left(\frac{D^{(2+\Delta\gamma)}}{R^2}\right)\right] \tag{2.123}$$

We can now rewrite Eq. (2.123) versus number of cells in cluster, N, and of radius of the individual cell, R, by use of Eq. (2.118):

$$\frac{C}{I} = 10\log\left[\frac{N}{2}(3N)^{\Delta\gamma/2}R^{\Delta\gamma}\right] \tag{2.124}$$

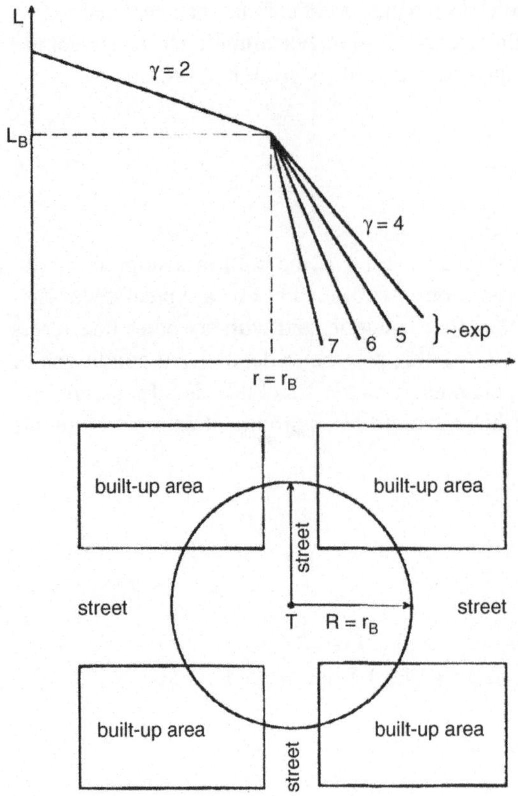

Figure 2.37 Cochannel interference evaluation scheme.

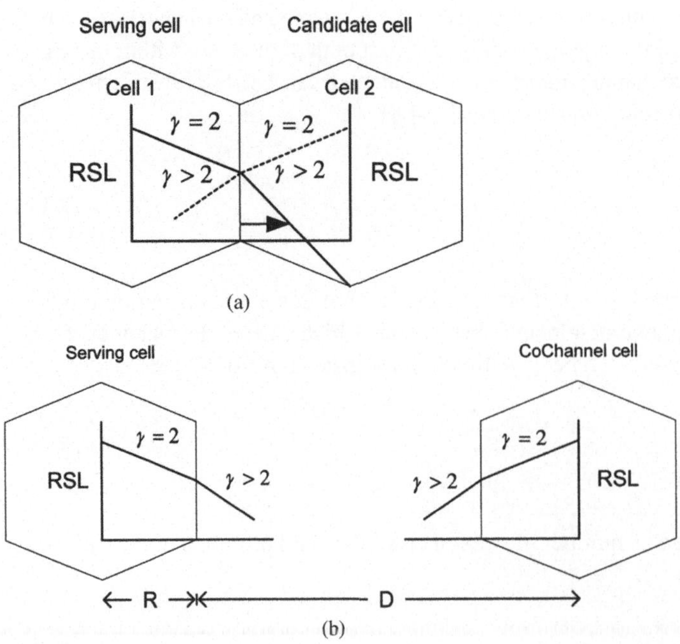

Figure 2.38 A cell in an urban area with grid-plan streets (a) and (b) according to [1,35,36].

Let us examine this equation for two typical cases in the urban scene described above in Sec. 2.3.

City with Regularly Planned Streets

In this case for a typical straight wide avenue, for which according to street-multislit-waveguide model $\Delta\gamma = 2(\gamma = 4)$ (see Sec. 2.3) and

$$\frac{C}{I} = 10\log\left[\frac{3}{2}N^2R^2\right] \tag{2.125}$$

For the case of narrow streets (more realistic case in urban scene) one can put in Eq. (2.125) $\Delta\gamma = 3 - 7(\gamma = 5 - 9)$, which is close to the exponential signal decay that follows from the street waveguide model. In this case, the cell size R can be approximately described, using the multislit street waveguide model (see Sec. 2.3), as [1]

$$R_{\text{cell}} \equiv r_B = \frac{4h_T h_R}{\lambda}\frac{(1+\chi)}{(1-\chi)}\left(1+\frac{h_b}{a}+\frac{h_T h_R}{a^2}\right) \tag{2.126}$$

where all parameters in Eq. (2.126) are described in Sec. 2.3.

City with Nonregularly Planned Streets

For the case of propagation over irregular built-up terrain, as follows from the probabilistic approach, described in Sec. 2.3, $\Delta\gamma = 1$ and the C/I-ratio prediction equation is as follows [1]:

$$\frac{C}{I} = 10\log\left[\frac{N}{2}(3N)^{1/2}R\right] \tag{2.127}$$

As follows from this approach, the average distance of the direct visibility $\bar{\rho}$ between two arbitrary points, the source and the observer, is described by the following formula:

$$\bar{\rho} = (\gamma_0)^{-1}\text{km} \tag{2.128}$$

where all parameters are presented in Sec. 2.3.

As follows from Eqs. (2.125)–(2.128), the C/I ratio strongly depends on conditions of wave propagation within the urban communication channels (on path loss slope parameter $\gamma = 2 + \Delta\gamma$, $\Delta\gamma \geq 1$) and on the cellular map splitting strategy (on parameters N and R).

Let us now introduce the important cellular parameters and present them in Table 2.2. Here in column (A) the reuse ratio D/R is presented; number of channels per cell is presented in column (B). The data presented in column (C) is obtained by use of the standard presentation of formula Eq. (2.121), that is, $C_i = C/I = 1.5N^2$. To obtain the number of subscribers per cell, described by column (D) in Table 2.2, we need additional information about the urban area and additional formulations, such as [35]

The urban area of operation and servicing: A, km^2
The number of citizens in the operating urban area: P (per thousands)
The mean radius of the cell: R, km
The number of channels in one cell: n_c

Table 2.2 Important Cellular Parameters in Column (a) the Reuse Ratio D/R; in Column (b) Number of Channels Per Cell; in Column (c) $C_i = C/I = 1.5 \cdot N^2$; in Column (d) the Number of Subscribers Per Cell [35]

Cluster size (N)	(A) Reuse ratio (D/R)	(B) Number of channels per cell ($279/N$)	(C) Cochannel ($C_i = C/I = 1.5\,N^2$) (dB)	(D) Number of subscribers per cell ($\tilde{n}$)
3	3	93	11	2583
4	3.5	69	14	1840
7	4.6	39	18	937
9	5.2	31	21	707
12	6	23	23	483
21	7.9	14	28	245

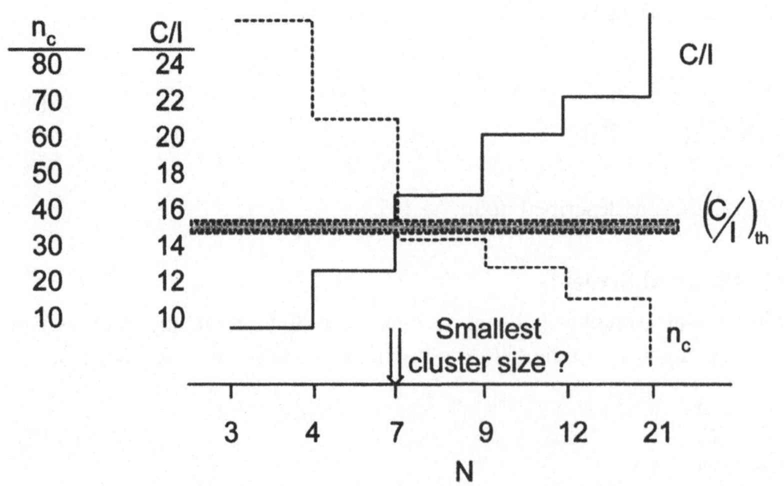

Figure 2.39 Number of channels per cell n_c versus cells number N for various C/I ratio.

Thus, in Fig. 2.39 according to Ref. 35 the dependence of n_c versus cells' number N for various C/I ratio is shown. If, for example, 30 subscribers use the same channel in the considered cell, then the number of subscribers in this cell equals

$$\tilde{n} = 30\, n_c \approx 10\,\pi n_c \tag{2.129}$$

For regularly distributed cells over the built-up terrain, the number of cells in the urban area concerned equals

$$K = \frac{A}{\pi R^2} \tag{2.130}$$

Then the total number of subscribers in the urban area considered equals

$$\tilde{N} = \tilde{n}K = \frac{10\, n_c A}{R^2} \tag{2.131}$$

The parameter $\tilde{N}$ calculated by use of this formula is presented in column (D) in Table 2.2. From Eq. (2.131) we can estimate as a percentage the number of subscribers from the population located in the urban area. In fact,

$$\tilde{N}(\%) = \frac{10\, n_c A}{R^2 P(1000)} 100\% = \frac{A n_c}{P R^2} (\%) \tag{2.132}$$

Example. The number of citizens is 600,000 located within an area with radius $R_a = 8\,\text{km}$. The cell size is $R = 2\,\text{km}$, the number of channels in each cell is $n_c = 40$. Find the number of subscribers (in %) for effective servicing by wireless communication system.

Solution.
First step. We calculate a city area: $A = \pi R_a^2 \approx 200\,\text{km}^2$.
Second step. We calculate the number of citizen per thousands: $P = 600{,}000/1000 = 600$.
Third step. We calculate, using Eq. (2.132) in first tier, $\tilde{N}(\%) = 200(40)/600(4) \approx 3.3\%$.

The result of this example shows that for cities with a high density of population (A/P is small), it is very hard to plan the wireless service by use of a simple propagation model. If we try to increase (up to the maximum) the number of channels by splitting the operating radio frequency band, the cell size (radius R) remains critically limited by the conditions of radio wave propagation in the urban area.

We note that the formula Eq. (2.132) can be rewritten by introducing into it new parameters as the frequency band of the total cellular service system, ΔF, and the frequency band of each channel, Δf_c. In this case the number of radio channels in each cell equals [35,36]

$$n_c = \frac{\Delta F}{\Delta f_c N} \tag{2.133}$$

Then the number of subscribers, which can effectively communicate by using existing cellular cervicing system equals, as a percentage,

$$\tilde{N}(\%) = \frac{A \Delta F}{\Delta f_c P N R^2} \tag{2.134}$$

Equations (2.132) and (2.134) show that to increase the efficiency of the cellular communication system in various urban environments, an effective frequency splitting strategy over the channels within each cell is required. Moreover, by a decrease of the cell size and the cluster size (or number N) one can also increase the efficiency of the cellular system. The latter depends on the strategy of cellular map construction and splitting. However, the experience of cellular systems designers shows that it is very difficult to decrease the number N of cells in each cluster (see Fig. 2.39 according to Ref. 35). Apparently, as follows from this picture, number $N = 7$ is the smallest size of cluster constructed, because for $N < 7$ the acceptable C/I level of 16 dB cannot be reached. Initially the parameter N was selected as $N = 12$ by the TACS cellular system constructed in England (more detailed information is presented in Refs. 2–9). However, while analyzing the C/I ratio and its optimization, the optimal number $N = 7$ was found. In fact, as follows from Fig. 2.39, for 300 working radio channels with 21 channels required to control the total cellular system, we obtain for $N = 12$ and $N = 7$, respectively, $n_c = 39$ and $n_c = 23$ communication channels in each cell. This result follows from Table 2.2 and Fig. 2.39, where value 23 from first column in table lies between 20 and 30 (the level corresponding to $N = 7$) and value 39 from this column lies between 30 and 40 (the level corresponding to $N = 12$).

REFERENCES

1. Blaunstein, N. *Radio Propagation in Cellular Networks*; Artech Houses: Boston–London, 1999, p. 386.
2. Parsons, L.D. *The Mobile Radio Propagation Channels*; Pentech Press: New York– Toronto, 1992, p. 313.
3. Lee, W.Y.C. *Mobile Communication Design Fundamentals*; McGraw Hill: New York, 1993, p. 365.
4. Yacoub, M.D. *Foundations of Mobile Radio Engineering*; CRC Press: NY, 1993, p. 290.
5. Saunders, S.R. *Antennas and Propagation for Wireless Communication Systems*; Wiley: New York, 1999, p. 409.
6. Bertoni, H.L. *Radio Propagation for Modern Wireless Systems*; Prentice Hall PTR: New Jersey, 2000, p. 258.
7. Feuerstein, M.L.; Rappaport, T.S. *Wireless Personal Communication*; Artech House: Boston–London, 1992, p. 315.
8. Rappaport, T.S. *Wireless Communications*; Prentice Hall PTR: New York, 1996, p. 641.
9. Blaunstein, N.; Jorgen Bach Andersen, *Multipath Phenomena in Cellular Networks*; Artech Houses: Boston–London, 2002, p. 296.
10. Jakes, W.C. *Microwave Mobile Communications*; Wiley: New York, 1974, p. 642.
11. Steele, R. *Mobile Radio Communication*; IEEE Press: New York, 1992, p. 779.
12. Proakis, J.G. *Digital Communications*; McGraw Hill: New York, 1995.
13. Balanis, C.A. *Advanced Engineering Electromagnetics*; Wiley: New York, 1997, p. 941.
14. Kraus, J.D. *Antennas*, 2nd Ed.; McGraw-Hill: New York, 1988.
15. Siwiak, K. *Radiowave Propagation and Antennas for Personal Communications*, 2nd Ed.; Artech House: Boston–London, 1998.
16. Vaughan, R.; Bach Andersen, J. *Channels, Propagation, and Antennas for Mobile Communications*; IEEE: London, 2002.
17. Milstein, L.B.; Schilling, D.L.; Pickholtz, R.L. On the feasibility of a CDMA overlay for personal communications networks. IEEE Select. Areas Commun. **May 1992**, *10*(4), 665–668.
18. Rustako, A.J., Jr.; Amitay, N.; Owens, M.J.; Roman, R.S. Radio propagation at microwave frequencies for line-of-sight microcellular mobile and personal communications. IEEE Trans. Veh. Technol. **Feb. 1991**, *40*(2), 203–210.
19. Tan, S.Y.; Tan, H.S. UTD propagation model in an urban street scene for microcellular communications. IEEE Trans. Electromag. Compat. **1993**, *35*(4), 423–428.
20. Blaunstein, N., Levin, M. VHF/UHF wave attenuation in a city with regularly spaced buildings. Radio Sci. **1996**, *31*(2), 313–323.
21. Blaunstein, N.; Levin, M. Propagation loss prediction in the urban environment with rectangular grid-plan streets. Radio Sci. **1997**, *32*(2), 453–467.
22. Blaunstein, N.; Giladi, R.; Levin, M. Los characteristics' prediction in urban and suburban environments. IEEE Trans. on Vehic. Tech. **1998**, *47*(1), 11–21.
23. Blaunstein, N. Average field attenuation in the nonregular impedance street waveguide. IEEE Trans. Anten. Propagat. **1998**, *46*(12), 1782–1789.
24. Xia, H.H.; Bertoni, H.L.; Maciel, L.R.; Honcharenko, W. Radio propagation characteristics for line-of-sight microcellular and personal communications. IEEE Trans. Anten. Propag. **Oct. 1993**, *41*(10), 1439–1447.
25. Bertoni, H.L.; Honcharenko, W.; Maciel, L.R.; Xia, H.H. UHF propagation prediction for wireless personal communications. Proc. IEEE. **Sept. 1994**, *82*(9), 1333–1359.
26. Okumura, Y.; Ohmori, E.; Kawano, T.; Fukuda, K. Field strength and its variability in the VHF and UHF land mobile radio service. Review Elec. Commun. Lab. **1968**, *16*, 825–843.
27. Hata, M. Empirical formula for propagation loss in land mobile radio services. IEEE Trans. Veh. Technol. **1980**, *VT-29*, 317–325.
28. Saleh Faruque, *Cellular Mobile Systems Engineering*; Artech House: Boston–London, 1994.
29. Ponomarev, G.A.; Kulikov, A.N.; Telpukhovsky, E.D. *Propagation of Ultra-Short Waves in Urban Environments*; Tomsk: Rasko, Russia, 1991.
30. Blaunstein, N. Prediction of cellular characteristics for various urban environments. IEEE Anten. Propagat. Magazine. **1999**, *41*(6), 135–145.

31. Blaunstein, N.; Katz, D.; Censor, D.; Freedman, A.; Matityahu, I.; Gur-Arie, I. Prediction of loss characteristics in built-up areas with various buildings' overlay profiles. IEEE Anten. Propagat. Magazine. **2001**, *43*(6) 181–191.
32. Clarke, R.H. A statistical theory of mobile-radio reception. Bell Systems Tech. J. **1968**, *47*, 957–1000.
33. Aulin, T. A modified model for the fading signal at a mobile radio channel. IEEE Trans. Veh. Technol. **1979**, *28*(3), 182–203.
34. Suzuki, H. A statistical model for urban propagation. IEEE Trans. Communication **1977**, *25*, 673–680.
35. Mehrotra, A. *Cellular Radio Performance Engineering*; Artech House: Boston–London, 1994, p. 249.
36. Linnartz, J.P. *Narrowband Land-Mobile Radio Networks*; Artech House: Boston–London, 1993, p. 335.

3
Satellite Communication Systems

Matthew N. O. Sadiku
Prairie View A&M University
Prairie View, Texas

Satellite-based communication has become a major facet of the telecommunication industry for two major reasons. First, it provides a means of broadcasting information to a large number of people simultaneously. Thus, satellite communication systems are an important ingredient in the implementation of a global communication infrastructure. Second, satellite communication provides a means of reaching isolated places on earth, where terrestrial telecommunications infrastructure does not exist or teledensity is low.

Satellite communication was first deployed in the 1960s and has its roots in military applications. Since the launch of the Early Bird satellite (first commercial communication satellite also known as Intelsat I) by NASA in 1965 proved the effectiveness of satellite communication, satellites have played an important role in both domestic and international communications networks. They have brought voice, video, and data communications to areas of the world that are not accessible with terrestrial lines. By extending communications to the remotest parts of the world, virtually everyone can be part of the global economy.

Satellite communications is not a replacement of the existing terrestrial systems but rather an extension of the wireless system. However, satellite communication has the following merits over terrestrial communications:

Coverage: Satellites can cover a much large geographical area than the traditional ground-based systems. They have the unique ability to cover the globe.

High bandwidth: A Ka-band (27–40 GHz) system can deliver a throughput of gigabits per second rate.

Low cost: A satellite communications system is relatively inexpensive because there are no cable-laying costs and one satellite covers a large area.

Wireless communication: Users can enjoy untethered mobile communication anywhere within the satellite coverage area.

Simple topology: Satellite networks have simpler topology, which results in more manageable network performance.

Broadcast/multicast: Satellite are naturally attractive for broadcast/multicast applications.

Maintenance: A typical satellite is designed to be unattended, requiring only minimal attention by customer personnel.

Immunity: A satellite system will not suffer from disasters such as floods, fire, and earthquakes and will, therefore, be available as an emergency service should terrestrial services be knocked out.

Of course, satellites systems do have some disadvantages. These are weighed with their advantages in Table 3.1. Some of the services provided by satellites include fixed satellite service (FSS),

Table 3.1 Advantages and Disadvantages of Satellite Communication [1]

Advantages	Disadvantages
Wide-area coverage	Propagation delay
Easy access to remote sites	Dependency on a remote facility
Costs independent of distance	Less control over transmission
Low error rates	Attenuation due to atmospheric particles (e.g., rain) can be severe at high frequencies
Adaptable to changing network patterns	Continual time-of-use charges
No right-of-way necessary, earth stations located at premises	Reduced transmission during solar equinox

mobile satellite service (MSS), broadcasting satellite service (BSS), navigational satellite service, and meteorological satellite service.

This chapter explores the integration of satellites with terrestrial networks to meet the demands of highly mobile communities. After looking at the fundamentals of satellite communications, we will discuss the orbital and propagation characteristics and the various applications of satellite-based communications systems.

3.1. FUNDAMENTALS

A satellite communication system may be viewed as consisting of two parts: the space and ground segments. The space segment consists of the satellites and all their on-board tracking and control systems. The earth segment comprises the earth terminals, their associated equipment, and the links to terrestrial networks [2].

3.1.1. Types of Satellites

There were only 150 satellites in orbit by September 1997. The number was expected to be roughly 1700 by the year 2002. With this increasing trend in the number of satellites, there is a need to categorize them. According to the height of their orbit and "footprint" or coverage on the earth's surface, they are classified as follows [3].

Geostationary Earth Orbit (GEO) Satellites

They are launched into a geostationary or geosynchronous orbit, which is 35,786 km above the equator. (Raising a satellite to such an altitude, however, required a rocket, so that the achievement of a GEO satellite did not take place until 1963.) A satellite is said to be in geostationary orbit when the space satellite is matched to the rotation of the earth at the equator. A GEO satellite can cover nearly one-third of the earth's surface, i.e., it takes three GEO satellites to provide global coverage. Due to their large coverage, GEO satellites are ideal for broadcasting and international communications. [GEO is sometimes referred to as *high earth orbit* (HEO).] Examples of GEO satellite constellations are Spaceway designed by Boeing Satellite Systems and Astrolink by Lockheed Martin. Another example is Thuraya, designed by Boeing Satellite Systems to provide mobile satellite services to the Middle East and surrounding areas.

There are at least three major objections to GEO satellites [4]. First, there is a relatively long propagation delay (or latency) between the instant a signal is transmitted and when it returns to earth (about 240 ms). This is caused by speed-of-light transmission delay and signal processing delay. This may not be a problem if the signal is going only one way. However, for signals such as data and voice, which go in both directions, the delay can cause problems. GEO satellites, therefore, are less attractive for voice communication. Second, there is lack of coverage at far northern and southern latitudes. This is unavoidable because a GEO satellite is below the horizon and may not provide coverage at latitudes as close to the equator as 45°. Unfortunately, many of the European capitals, including London, Paris, Berlin, Warsaw, and Moscow, are north of this latitude. Third, both the mobile unit and the satellite of a GEO system require a high transmit power. In spite of these objections, the majority of satellites in operation today are GEO satellites but that may change in the near future.

Middle Earth Orbit (MEO) Satellites

They orbit the earth at 5,000 to 12,000 km. GEO satellites do not provide good coverage for places far north and satellites in inclined elliptical orbits are an alternative. Although the lower orbit reduces propagation delay to only 60 to 140 ms round trip, it takes 12 MEO satellites to cover most of the planet. MEO systems represent a compromise between LEO (see below) and GEO systems, balancing the advantages and disadvantages of each. [MEO is sometimes referred to as *intermediate circular orbit* (ICO).]

Low Earth Orbit (LEO) Satellites

They circle the earth at 500 to 3000 km. For example, the Echo satellite circled the earth every 90 min. To provide global coverage may require as many as 200 LEO satellites. Latency in a LEO system is comparable with terrestrial fiber optics, usually less than 30 ms round trip. LEO satellites are suitable for personal communication systems (PCS). However, LEO systems have a shorter life space of 5–8 years (compared with 12–15 years for GEO systems) due to the increased amount of radiation in low earth orbit. The LEO systems have been grouped as Little LEO and Big LEO. The Little LEOs have less capacity and are limited to nonvoice services such as data and message transmission. An example is OrbComm designed by Orbital Corporation, which consists of 36 satellites, each weighing 85 lb. The Big LEOs have larger capacity and voice transmission capability. An example is Loral and Qualcomm's Globalstar which will operate in the L-band frequencies and employ 48 satellites organized in eight planes of six satellites each.

The arrangement of the three basic types of satellites is shown in Fig. 3.1. The evolution from GEO to MEO and LEO satellites has resulted in a variety of global satellite systems. The convenience of GEO was weighed against the practical difficulty involved with it and the inherent technical advantages of LEO, such as lower delay and higher angles of elevation. While it has been conceded that GEO is in many respects theoretically preferable, LEO or MEO systems would be preferred for many applications. Although a constellation (a group of satellites) is required instead

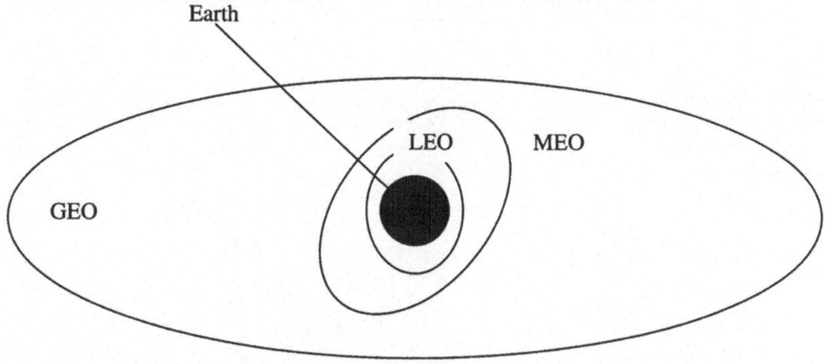

Figure 3.1 The three common types of satellites: GEO, MEO, and LEO.

of only one for hemispheric coverage, the loss of individual satellites would cause only gradual degradation of the system rather than a catastrophic failure. A comparison of the three satellite types is given in Table 3.2.

3.1.2. Frequency Bands

Every nation has the right to access the satellite orbit and no nation has a permanent right or priority to use any particular orbit location. Without a means for the nations to coordinate the use of satellite frequency bands, the satellite services of one nation could interfere with those of another, thereby creating a chaotic situation in which neither country's signals could be received clearly.

To facilitate satellite communications and eliminate interference between different systems, international organizations govern the use of satellite frequency. The International Telecommunication Union (ITU) is responsible for allocating frequencies to satellite services. Since the spectrum is a limited resource, the ITU has reassigned the same parts of the spectrum to many nations and for many purposes throughout the world.

The frequency spectrum allocations for satellite services are given in Table 3.3. Notice that the assigned segment is the 1–40 GHz frequency range, which is the microwave portion of the spectrum. As microwaves, the signals between the satellite and the earth stations travel along line-of-sight paths and experience free-space loss that increases as the square of the distance.

Satellite services are classified into 17 categories [6]: fixed, intersatellite, mobile, land mobile, maritime mobile, aeronautical mobile, broadcasting, earth exploration, space research, meteorological,

Table 3.2 Comparison of the GEO, MEO, and LEO [5]

Type	Altitude	Coverage	Advantages	Disadvantages
LEO	300–1000 km	Spot	Low path loss High data rate Low delay Low launch cost Less fuel	Coverage is less Need many satellites Short orbital life High Doppler Highly complex
MEO	1000–10,000 km	Region	Moderate path loss Moderate launch cost Less fuel	Multiple satellites Moderate coverage Highly complex
GEO	36,000 km	Earth	Global coverage Need few satellites Long orbital life Low Doppler Less complex	High path loss Long delay Low data rate High launch cost Fuel for station keeping

Table 3.3 Satellite Frequency Allocations

Frequency band	Range (GHz)
L	1–2
S	2–4
C	4–8
X	8–12
Ku	12–18
K	18–27
Ka	27–40

Table 3.4 Typical Uplink and Downlink Satellite Frequencies

Uplink frequencies (GHz)	Downlink frequencies (GHz)
5.925–6.426	3.700–4.200
7.900–8.401	7.250–7.750
14.00–14.51	11.70–12.20
27.50–31.0	17.70–20.20

space operation, amateur, radiodetermination, radionavigation, maritime radionavigation, and standard frequency and time signal. The Ku band is presently used for broadcasting services, and also for certain fixed satellite services. The C band is exclusively used for fixed satellite services, and no broadcasting is allowed. The L band is employed by mobile satellite services and navigation systems.

A satellite band is divided into separation portions: one for earth-to-space links (the uplink) and one for space-to-earth links (the downlink). Like a terrestrial microwave relay, a satellite must use separate frequencies for sending to the satellite (the uplink) and receiving from the satellite (the downlink); otherwise, the powerful signal transmitted by the satellite would interfere with the weak incoming signal. Table 3.4 provides the general frequency assignments for uplink and downlink satellite frequencies. We notice from the table that the uplink frequency bands are slightly higher than the corresponding downlink frequency band. This is to take advantage of the fact that it is easier to generate higher frequency RF power within a ground station than it is onboard a satellite. In order to direct the uplink transmission to a specific satellite, the uplink radio beams are highly focused. In the same way, the downlink transmission is focused on a particular *footprint* or area of coverage.

All satellite systems are constrained to operate in designed frequency bands depending on the kind of earth station used and service provided. The satellite industry, particularly in the United States, is subject to several regulatory requirements, domestically and internationally, depending upon which radio services and frequency bands are proposed to be used on the satellite. In the United States, the Federal Communications Commission (FCC) is the independent regulatory agency that ensures that the limited orbital or spectrum resource allocated to space radiocommunications services is used efficiently. After receiving an application for a U.S. domestic satellite, FCC initiates the advance publication process for a U.S. satellite. This is to ensure the availability of an orbit position when the satellite is authorized. FCC does not guarantee international recognition and protection of satellite systems unless the authorized satellite operator complies with all coordination requirements and completes the necessary coordination of its satellites with all other administrations whose satellites are affected [7].

3.1.3. Basic Satellite Components

Every satellite communication involves the transmission of information from a ground station to the satellite (the uplink), followed by a retransmission of the information from the satellite back to the earth (the downlink). Hence the satellite system must typically have a receiver antenna, a receiver, a transmitter antenna, a transmitter, some mechanism for connecting the uplink with the downlink, and a power source to run the electronic system. These components are illustrated in Fig. 3.2 and explained as follows [8,9]:

Transmitters: The amount of power required by a satellite transmitter to send out depends on whether it is GEO or LEO satellite. The GEO satellite is about 100 times farther away than the LEO satellite. Thus, a GEO would need 10,000 times as much power as a LEO satellite. Fortunately, other parameters can be adjusted to reduce this amount of power.

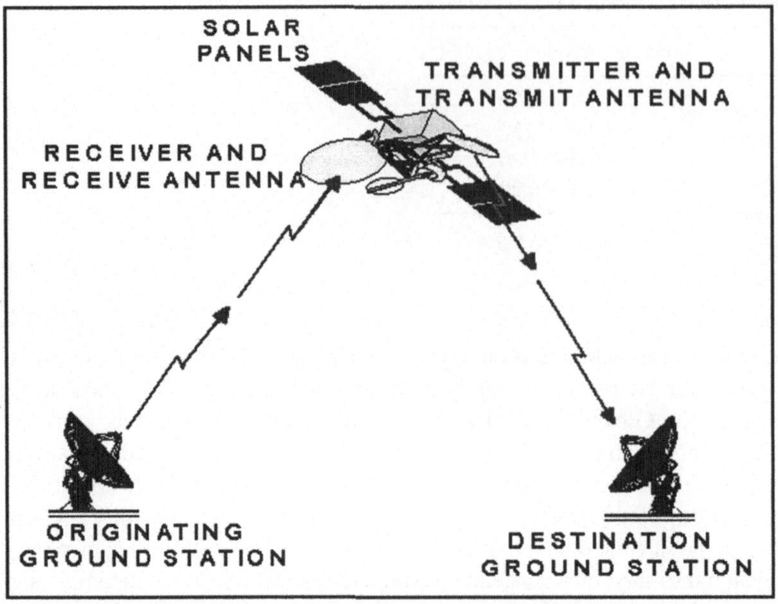

Figure 3.2 Basic components of a communication satellite link (with permission of Regis Leonard, NASA Lewis Research Center).

Antennas: The antennas dominate the appearance of a communication satellite. Antenna design is one of the more difficult and challenging parts of a communication satellite project. The antenna geometry is constrained physically by the design and the satellite topology. A major difference between GEO and LEO satellites is their antennas. Since all the receivers are located in the coverage area, which is relatively small, a properly designed antenna can focus most of the transmitter power within that area. The easiest way to achieve this is to simply make the antenna larger. This is one of the ways the GEO satellite makes up for the apparently larger transmitter power it requires.

Power generation: The satellite must generate all of its own power. The power is often generated by large solar cells, which convert sunlight into electricity. Since there is a limit to how large the solar panel can be, there is also a practical limit to the amount of power that can be generated. Satellites must also be prepared for periods of eclipse, when the earth is between the sun and the satellite. This necessitates having batteries on board that can supply power during eclipse and recharge later.

Transponders: These are the communication devices each satellite must carry. A transponder is a piece of equipment that receives a weak signal at one frequency, amplifies it, and changes its frequency to another for transmission to another earth station. The block diagram of a typical transponder is shown in Fig. 3.3. For example, a GEO satellite may have 24 transponders with each assigned a pair of frequencies (uplink and downlink frequencies).

Ground Stations: The ground (or earth) stations form the ground segment of the satellite communication system. The ground station is responsible for interacting and communicating with the satellites. Most ground or earth stations simply transmit and receive signals with a fixed antenna. At least one ground station must perform the task of controlling and monitoring the satellite. In a transmitting ground station, the information signal (voice, video, or data) is processed, amplified, and transmitted. In a receiving ground station, the reverse process takes place. In the past, ground stations were massive, expensive, and owned by common carriers and the military. Now, earth stations are

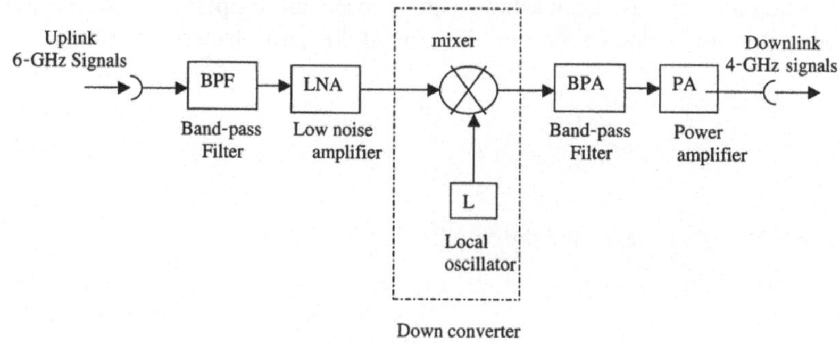

Figure 3.3 A simplified block diagram of a typical transponder.

small, less expensive, and owned or leased by private organizations. The antenna is a vital component of the ground station, and its size varies considerably from a 1-m diameter parabolic reflector used for TV programs at home to a 64-m diameter reflector used in the deep space network.

Telemetry and Control: This is partly implemented by the ground stations and partly by the satellite. Telemetry, tracking, and command (TTC) are used to monitor and control the satellite while in orbit. Telemetry is the means by which a measurement made is transmitted to an observer at a distance. Tracking is collecting data to monitor the movement of an object. Command is the process of establishing and maintaining control. Tracking and commands are used by the terrestrial control station to determine the position of the satellite and predict its future location. The satellite contains telemetry instrumentation that continuously gathers information and transmits it to the ground station.

3.2. ORBITAL CHARACTERISTICS

Since a satellite is a spacecraft that orbits the earth, an intuitive question to ask is "What keeps objects in orbit?" The answer to the question is found in the orbital mechanical laws governing satellite motion. Satellite orbits are essentially elliptical and obey the same laws of Johannes Kepler that govern the motion of planets around the sun. The three Kepler's laws are stated as follows [10]:

First law: The orbit of each planet follows an elliptical path in space with the sun serving as the focus.

Second law: The line linking a planet with the sun sweeps out equal areas in equal time.

Third law: The square of the period of a planet is proportional to the cube of its mean distance from the sun.

Besides these laws, Newton's law of gravitation states that any two bodies attract each other with a force proportional to the product of their masses and inversely proportional to the square of the distance between them, i.e.,

$$\mathbf{F} = -\frac{GMm}{r^2}\mathbf{a}_r \tag{3.1}$$

where M is the mass of one body (earth), m is the mass of the other body (satellite), $\mathbf{F}$ is the force on m due to M, r is the distance between the two bodies, $\mathbf{a}_r = \mathbf{r}/r$ is a unit vector along the displacement vector $\mathbf{r}$, and $G = 6.672 \times 10^{-11}\,\text{Nm/kg}^2$ is the universal gravitational constant. If M is the mass of the earth, the product $GM = \mu = 3.99 \times 10^{14}\,\text{m}^3/\text{s}^2$ is known as Kepler's constant.

Kepler's laws in conjunction with Newton's laws can be used to completely describe the motion of the planets around the sun or that of the satellite around the earth. Newton's second law can be written as

$$\mathbf{F} = m\frac{d^2r}{dt^2}\mathbf{a}_r \tag{3.2}$$

Equating this with the force between the earth and the satellite in Eq. (3.1) gives

$$\frac{d^2r}{dt^2}\mathbf{a}_r = -\frac{\mu}{r^2}\mathbf{a}_r \tag{3.3}$$

or

$$\ddot{\mathbf{r}} + \frac{\mu}{r^3}\mathbf{r} = 0 \tag{3.4}$$

where $\ddot{\mathbf{r}}$ is the vector acceleration. The solution to the vector second-order differential Eq. (3.4) is not simple but it can be shown that the resulting trajectory is in the form of an ellipse given by [11,12]

$$r = \frac{p}{1 + e\cos\theta} \tag{3.5}$$

where r is the distance between the geocenter and any point on the trajectory, p is a geometric constant, $e(0 \le e < \mathbf{1})$ is the eccentricity of the ellipse, and θ (known as the *true anomaly*) is the polar angle between r and the point on the ellipse nearest to the focus. These orbital parameters are illustrated in Fig. 3.4. The eccentricity e is given by

$$e = \sqrt{1 - \left(\frac{a}{b}\right)^2} \tag{3.6}$$

The point on the orbit where the satellite is closest to the earth is known as the *perigee*, while the point where the satellite is farthest from the earth is known as the *apogee*. The fact that the orbit is

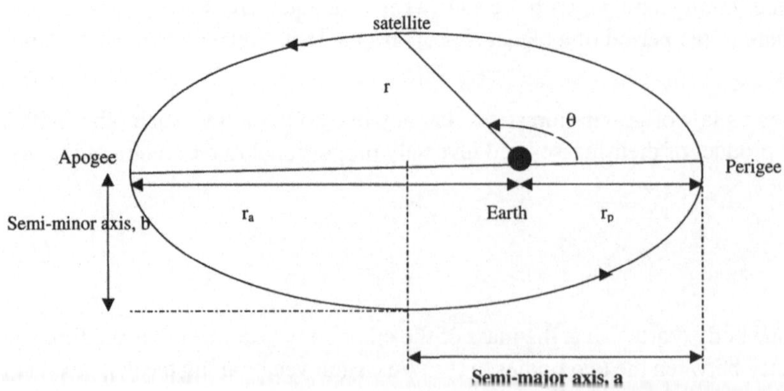

Figure 3.4 Orbital parameters.

an ellipse confirms Kepler's first law. If a and b are the semimajor and semiminor axes (see Fig. 3.4), then

$$b = a\sqrt{(1 - e^2)} \tag{3.7a}$$

$$p = a(1 - e^2) \tag{3.7b}$$

Thus, the distance between a satellite and the geocenter is given by

$$r = \frac{a(1 - e^2)}{1 + e\cos\theta} \tag{3.8}$$

Note that the orbit becomes circular orbit when $e = 0$.

The apogee height and perigee height are often required. From the geometry of the ellipse, the magnitudes of the radius vectors at apogee and perigee can be obtained as

$$r_a = a(1 + e) \tag{3.9}$$

$$r_p = a(1 - e) \tag{3.10}$$

To find the apogee and perigee heights, the radius of the earth must be subtracted from the radii lengths.

The period T of a satellite is related to its semimajor axis a using Kepler's third law as

$$T = 2\pi\sqrt{\frac{a^3}{\mu}} \tag{3.11}$$

For a circular orbit to have a period equal to that of the earth's rotation (a sidereal day 23 h, 56 min, 4.09 s), an altitude of 35,803 km is required. In this equatorial plane, the satellite is "geostationary."

The velocity of a satellite in an elliptic orbit is obtained as

$$v^2 = \mu\left(\frac{2}{r} - \frac{1}{a}\right) \tag{3.12}$$

For a synchronous orbit ($T = 24$ h), $r = a = 42{,}230$ km, and $v = 3074$ m/s or 11,070 km/h. The closer the satellite is to the earth, the stronger is the effect of gravity, which constantly pulls it toward the earth, and so the greater must be the speed of the satellite to avoid falling to the earth.

A constellation is a group of satellites. The total number N of satellites in a constellation depends on the earth central angle γ and is given by [13]

$$N \approx \frac{4\sqrt{3}}{9}\left(\frac{\pi}{\gamma}\right)^2 \tag{3.13}$$

3.3. PROPAGATION CHARACTERISTICS

There are two major effects space has on satellite communications. First, the space environment, with radiation, rain, and space debris, is harsh on satellites. The satellite payload, which is responsible for the satellite communication functions, is expected to be simple and robust. Traditional satellites, specially GEOs, serve as "bent pipes" and act as repeaters between communication points on the ground, as shown in Fig. 3.5a. There is no onboard processing (OBP). However, new satellites allow OBP, including decoding/recoding, demodulation/remodulation, transponder, beam switching, and routing [14], as in Fig. 3.5b where a network of satellites is connected by intersatellite links (ISL).

The second effect is that of wave propagation. Attenuation due to atmospheric particles (rain, ice, dust, snow, fog, etc.) is not significant at L, S, and C bands. Above 10 GHz, the main propagation effects are [15,16]

Tropospheric propagation effects: attenuation by rain and clouds, scintillation, and depolarization

Effects of the environment on mobile terminals: shadowing, blockage, and multipath caused by objects in the surrounding of the terminal antenna

The troposphere can produce significant signal degradation at the Ku, Ka, and V bands, particularly at lower elevation angles. Most satellite systems are expected to operate at an elevation angle above roughly 20°. Rain constitutes the most fundamental obstacle encountered in the design of satellite communication systems at frequencies above 10 GHz. The resultant loss of signal power makes for unreliable transmission. Based on empirical data, the specific attenuation per unit α (dB/km) is related to rain intensity or rain rate R (in mm/h) as

$$\alpha = aR^b \tag{3.14}$$

where a and b are frequency dependent coefficients. The approximate expressions for $a(f)$ and $b(f)$ given in Table 3.5 are suitable for engineering purposes. The total attenuation loss in **dB** is given by

$$A = \alpha L_{eq} \tag{3.15}$$

Figure 3.5 Satellite configuration types: (a) bent pipe and (b) onboard processing (OBP) switching and routing.

Table 3.5 Attenuation Coefficients [17]

Frequency f (GHz)	a	b
8.5–25	$4.21 \times 10^{-5}(f)^{2.42}$	$1.41(f)^{-0.0779}$
25–54	$4.21 \times 10^{-5}(f)^{2.42}$	$2.63(f)^{-0.272}$
54–100	$4.09 \times 10^{-2}(f)^{0.699}$	$2.63(f)^{-0.272}$

where L_{eq} is the equivalent length, which is determined by the height of the freezing level and it depends on the rain rate R and the elevation angle θ. It is given by

$$L_{eq} = [7.413 \times 10^{-3} R^{0.766} + (0.232 - 1.803 \times 10^{-4} R) \sin \theta]^{-1} \tag{3.16}$$

The rain rate R is given by Rice-Holmberg model. The percent of an average year for which the rain rate exceeds R at a medium location is given by the Rice-Holmberg distribution

$$P(R) = ae^{-0.03R} + be^{-0.258R} + ce^{-1.63R} \tag{3.17}$$

where

$$a = \frac{M\beta}{2922} \qquad b = M(1-\beta)/438.3 \qquad c = 1.86\beta \tag{3.18}$$

M is the total mean yearly rainfall in millimeters and β is the ratio of thunderstorm rain accumulation to total accumulation. Attenuation due to other hydrometeors such as oxygen, water vapor, and fog is discussed in Refs. 18 and 19.

To determine the amount of power received on the ground due to satellite transmission, we consider the power density

$$\Psi = \frac{P_t}{S} \tag{3.19}$$

where P_t is the power transmitted and S is the terrestrial area covered by the satellite. The value of P_t is a major requirement of the spacecraft. The coverage area is given by

$$S = 2\pi R^2(1 - \cos \gamma) \tag{3.20}$$

where $R = 6378$ km is the radius of the earth. S is usually divided into a cellular pattern of spot beams, thereby enabling frequency reuse. The effective area of the receiving antenna is a measure of the ability of the antenna to extract energy from the passing electromagnetic wave and is given by

$$A_e = G_r \frac{\lambda^2}{4\pi} \tag{3.21}$$

where G_r is the gain of the receiving antenna and λ is the wavelength. The power received is the product of the power density and the effective area. Thus,

$$P_r = \Psi A_e = \frac{G_r \lambda^2}{4\pi S} P_t \tag{3.22}$$

This is known as the *Friis equation* relating the power received by one antenna to the power transmitted by the other. We first notice from this equation that for a given transmitted power P_t, the received power P_r is maximized by minimizing the coverage area S. Second, mobile terminals prefer having nondirectional antennas, thereby making their gain G_r fixed. Therefore, to maximize P_r encourages using as long a wave-length as possible, i.e., as low a frequency as practicable within regulatory and technical constraints.

The path loss accounts for the phenomenon, which occurs when the received signal becomes weaker as the distance between the satellite and the earth increases. In free space, the strength of the radiated signal diminishes as the square of the distance it travels, so the received power density is inversely proportional to the square of the distance. The path "free-space" loss (in dB) is given by

$$L_p = 92.45 + 20\log_{10} f + 20\log_{10} r \tag{3.23}$$

where f is the frequency (in GHz) and r is the distance (in km).

The noise density N_o is given by

$$N_o = kT_o \tag{3.24}$$

where $k = 1.38 \times 10^{-23}$ Ws/K is Boltzmann's constant and T_o is the equivalent system temperature, which is defined to include antenna noise and thermal noise generated at the receiver. Shannon's classical capacity theorem for the maximum error-free transmission rate in bits per second (bps) over a noisy power-limited and bandwidth-limited channel is

$$C = B\log_2\left(1 + \frac{P_r}{N_o}\right) \tag{3.25}$$

where B is the bandwidth and C is the channel capacity.

3.4. APPLICATIONS

Satellite communication services are uniquely suited for many applications involving wide area coverage. Satellites provide the key ingredient in the development of broad-band communications and information processing infrastructure. Here, we consider five major applications of satellite communications: the use of very small aperture terminals (VSATs) for business applications; fixed satellite service (FSS), which interconnects fixed points, and mobile satellite (MSAT) service (MSS), which employs satellite to extend cellular network to mobile vehicles; satellite radio, which continuously provide entertainment to listeners; and satellite-based Internet, which enables IP-over-satellite connectivity.

3.4.1. VSAT Networks

A very small aperture terminal (VSAT) is a dish antenna that receives signals from a satellite. (The dish antenna has a diameter that is typically in the range of 1.2 m to 2.8 m but the trend is toward smaller dishes, not more than 1.5 m in diameter.) A VSAT may also be regarded as a complete earth station that can be installed on the user's premises and provide communication services in conjunction with a larger (typically 6–9 m) earth station acting as a network management center (NMC), as illustrated in Fig. 3.6.

VSAT technology brings features and benefits of satellite communications down to an economical and usable form. VSAT networks have become mainstream networking solutions for long-distance, low-density voice and data communications because they are affordable to both small and large companies. Other benefits and advantages of VSAT technology include lower operating costs, ease of installation and maintenance, ability to manage multiple protocols, and ability to bring locations where the cost of leased lines is very high into the communication loop.

Satellite links can support interactive data applications through two types of architectures [3,20]: mesh topology (also called *point-to-point connectivity*) and star topology (also known as

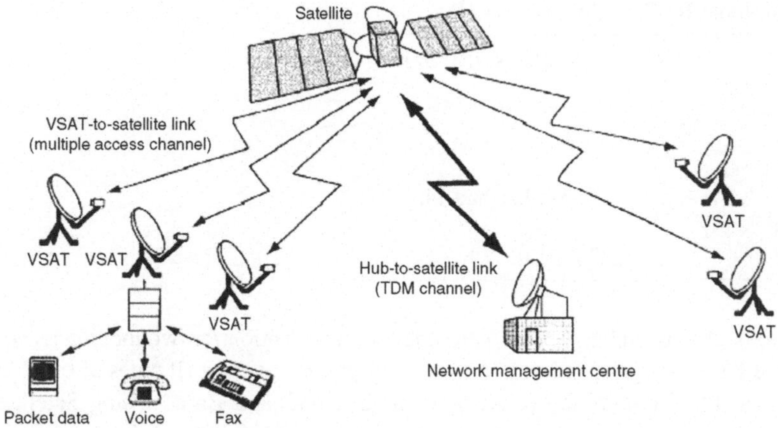

Figure 3.6 A typical VSAT network.

point-to-multipoint connectivity). Single-hop communications between remote VSATs can be achieved by full-mesh connectivity. Although the mesh and star configurations have different technical requirements, it is possible to integrate the two if necessary.

The star network employs a hub station. The hub consists of an RF terminal, a set of baseband equipment and network equipment. A VSAT network can provide transmission rates of up to 64 kbps. As common with star networks, all communication must past through the hub. That is, all communication is between a remote node and the hub; no direct node-to-node information transfer is allowed in this topology. This type of network is highly coordinated and can be very efficient. The point-to-multipoint architecture is very common in modern satellite data networks and is responsible for the success of the current VSAT.

A mesh network is more versatile than star network because it allows any-to-any communications. Also, the star network can provide transmission rates up to 64 kbps per remote terminal, whereas the mesh network can have its data rates increased to 2 Mbps or more. Mesh topology was used by the first satellite networks to be implemented. With time, there was a decline in the use of this topology but it remains an effective means of transferring information with least delay. Mesh topology applies to either temporary connections or dedicated links to connect two earth stations. All full-duplex point-to-point connectivities are possible and provided, as typical of a mesh configuration. If there are N nodes, the number of connections is equal to the permutation of $N(N - 1)/2$. Mesh networks are implemented at C and Ku bands. The transmission rate ranges from about 64 kbps to 2.048 Mbps (E1 speed). Users have implemented 45 Mbps.

Several types of VSAT networks are now in operation, both domestically and internationally. There were over 1000 VSATs in operation at the beginning of 1992. Today, there are over 100,000 two-way Ku band VSATs installed in the United States and over 300,000 worldwide. Almost all of these VSATs are designed primarily to provide data for private corporate networks, and almost all two-way data networks with more than 20 earth stations are based on some variation of an ALOHA protocol for access [21,22]. The price of a VSAT started around $20,000 and dropped to around $6,000 in 1996.

3.4.2. Fixed Satellite Service

Several commercial satellite applications are through earth stations at fixed locations on the ground. The international designation for such an arrangement is *fixed satellite service* (FSS). The FSS is to provide communication service between two or more fixed points on earth, as opposed to mobile

Table 3.6 Frequency Allocations for FSS (Below ~30 GHz)

Downlinks (in GHz)	Uplinks (in GHz)
3.4–4.2 and 4.5–4.8	5.725–7.075
7.25–7.75	7.9–8.4
10.8–11.7	
11.7–12.2 (Region 2 only)	12.75–13.25 and 14.0–14.5
12.6–12.75 (Region 1 only)	
17.7–21.2	27.5–31.0

satellite services (MSS) (to be discussed later), which provides communication for two moving terminals. Although ITU defined FSS as a space radiocommunication service covering all types of satellite transmissions between given fixed points, the borderline between FSS and Broadcasting Satellite Service (BSS) for satellite television is becoming more and more blurred [23]. FSS applies to systems which interconnect fixed points such as international telephone exchanges. It involves GEO satellites providing 24 hour per day service.

Table 3.6 shows the WARC (World Administrative Radio Conference) frequency allocations for FSS. The table only gives a general idea and is by no way comprehensive. The FSS shares frequency bands with terrestrial networks in the 6/4 GHz and 14/12 GHz bands. Thus, it is possible that a terrestrial network could affect a satellite on the uplink or that a terrestrial network may be affected by the downlink from a satellite.

As exemplified by Intelsat, FSS has been the most successful part of commercial satellite communications. Early applications were point-to-point telephony and major trunking uses. Current applications of the FSS can be classified according to frequency (from about 3 MHz to above 30 GHz), the lowest frequency being the HF band. They include high-frequency (HF) service, private fixed services, auxiliary broadcasting (AUXBC) services, cable relay service (CARS), and federal government fixed services.

Although the telecommunications industry as a whole is growing rapidly, the FSS industry is not. The market trend is toward the replacement of long-haul microwave system with fiber. Fiber provides much greater capacity than microwaves.

3.4.3. Mobile Satellite Service

There is the need for global cellular service in all geographical regions of the world. The terrestrial cellular systems serve urban areas well; they are not economical for rural or remote areas where the population or teledensity is low. Mobile satellite (MSAT) systems can complement the existing terrestrial cellular network by extending communication coverage from urban to rural areas. Mobile satellite services (MSS) are not limited to land coverage but include marine and aeronautical services [6,24]. Thus, the coverage of mobile satellite is based on geographical and not on population coverage as in terrestrial cellular system and could be global.

MSAT or satellite-based PCS/PCN is being developed in the light of the terrestrial constraints. The low cost of installation makes satellite-based PCS simple and practical. The American Mobile Satellite Corporation (AMSC) along with Telesat Mobile of Canada are designing a geosynchronous MSAT to provide PCS to North America. The concept of MSAT is illustrated in Fig. 3.7.

Satellite communication among mobile earth stations is different from the cellular communication. First, the cells move very rapidly over the earth, and the mobile units, for all practical purposes, appear stationary—a kind of inverted cellular telephone system. Second, due to different designs, use of a handheld is limited to the geographical coverage of a specific satellite constellation and roaming of handheld equipment between different satellite systems will not be allowed. With personal communication systems (PCS), there will be a mix of broad types of cell sizes: the picocell

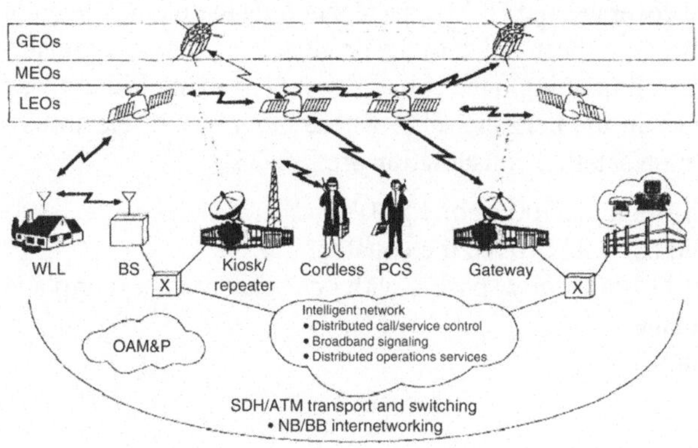

Figure 3.7 MSAT concept.

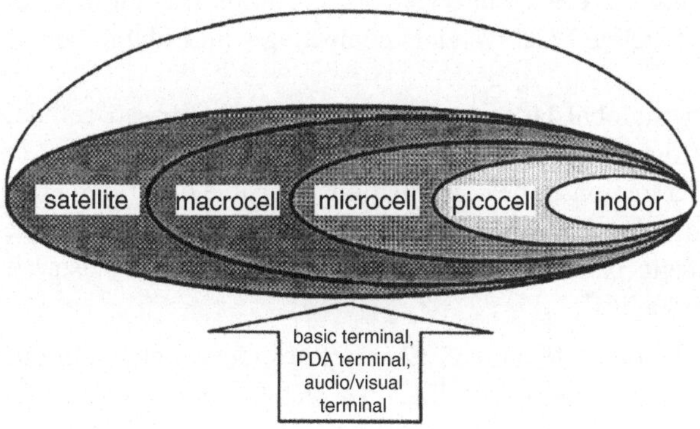

Figure 3.8 Various cell sizes.

for low-power indoor applications, the microcell for lower-power outdoor pedestrian application; macrocell for high-power vehicular applications; and supermacro cell with satellites, as shown in Fig. 3.8. For example, a mirocell of a PCS has a radius of 1 to 300 m.

There are two types of constellation design approaches to satellite-based PCS. One approach is to provide coverage using three GEO satellites at approximately 36,000 km above the equator. The other approach involves using the LEO and MEO satellites at approximately 500 to 1500 km above the earth's surface. Thus, MSS are identified as either GEO or nongeostationary orbit (NGSO) satellites [25].

The main purpose of MSAT or MSS is to provide data and/or voice services into a fixed or portable personal terminal, close to the size of today's terrestrial cellular phones, by means of interconnection via satellite. LEO and MEO satellites have been proposed as an efficient way to communicate with these handheld devices. The signals from the handheld devices are retransmitted via a satellite to a gateway (a fixed earth station) which routes the signals through the public switched telephone network (PSTN) to its final destination or to another handheld device.

Satellite systems designed for personal communications include the Iridium, Globalstar, and ICO systems [26–29]. All are global system covering everywhere on earth. Each of these is characterized by two key elements: a constellation of non-geosynchronous satellites (LEO or MEO) arranged in multiple planes and a handheld terminal (handset) for accessing PCS.

Iridium (www.iridium.com), which began in 1990, is the first mobile satellite telephone network to offer voice and data services to and from handheld telephones anywhere in the world. It uses a network of inter-satellite switches for global coverage and GSM-type technology to link mobile units to the satellite network. Several modifications have been made to the original idea, including reducing the number of satellites from 77 to 66 by eliminating one orbital plane. (The name Iridium was based on the fact that Iridium is the element in the periodic table whose atom has 77 electrons.) Some of the key features of the current Iridium satellite constellation are [30–33]:

Number of (LEO) satellites: 66 (each weighing 700 kg or 1500 lb)
Number of orbital planes: 6 (separated by 31.6° around the equator)
Number of active satellites per plane: 11 (uniformly spaced, with one spare satellite per plane at 130 km lower in the orbital plane)
Altitude of orbits: 780 km (or 421.5 nmi)
Inclination: 86.4°
Period of revolution: 100 min
Design life: 8 y

In spite of some problems expected of a complex system, Iridium is already at work. Its 66 LEO satellites were fully commercial as of November 1, 1998. But on August 13, 1999, Iridium filed for bankruptcy and was later bought by Iridium Satellite LCC. Vendors competing with Iridium include Aries, Ellipso, Globalstar, and ICO.

The second system is Globalstar (www.globalstar.com), which is a satellite-based cellular telephone system that allows users to talk from anyplace in the world. It serves as an extension of terrestrial systems world-wide except for polar regions. The constellation is capable of serving up to 30 million subscribers. Globalstar is being developed by the limited partnership of Loral Aerospace Corporation and Qualcomm with ten strategic partners. A functional overview of Globalstar is presented in Fig. 3.9. The key elements are [34–36]:

Space segment: It comprises a constellation of 48 active LEO satellites located at an altitude of 1414 km and equally divided in 8 planes (6 satellites per plane). The satellite orbits

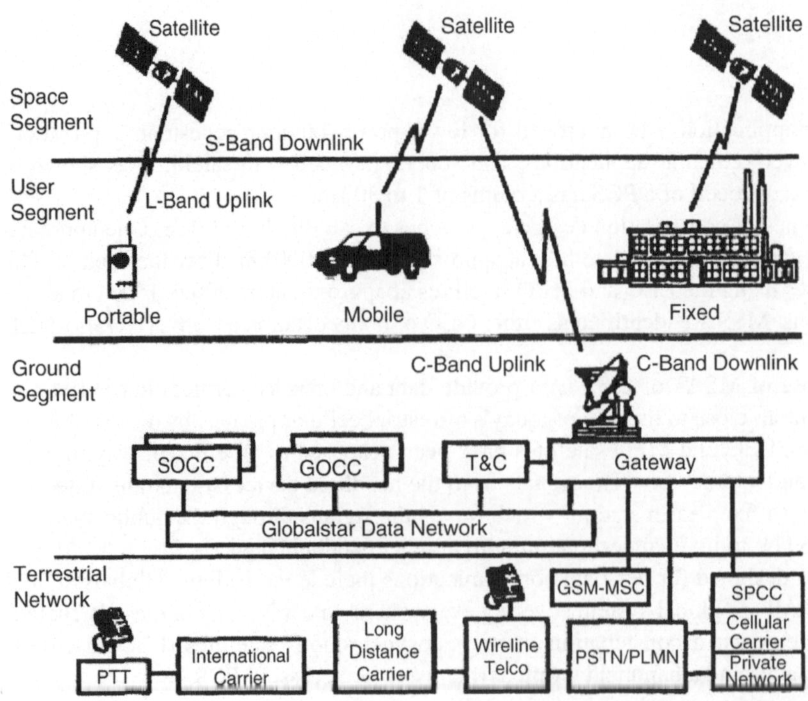

Figure 3.9 Globalstar system architecture.

are circular and are inclined at 52° with respect to the equator. Each satellite illuminates the earth at 1.6 GHz L band and 2.5 GHz S band with 16 fixed beams with service links, assignable over 13 FDM channels.

User segment: This includes mobile and fixed users.

Ground segment: This consists of gateways (large ground station), ground operations control center (GOCC), satellite operations control center (SOCO), and Globalstar data network (GDN). The gateway enables communications to and from handheld user terminals (UTs), relayed via satellite, with Public Switched Telephone Network (PSTN). A gateway with a single radio channel transmits on a single frequency.

The Globalstar satellites employ "bent pipe" transponders with the feeder link at C band. Each satellite weighs about 704 lb and has a capacity of 2800 full-duplex circuits. It covers the earth with only 16 spots beams.

Since Globalstar plans to serve the military with commercial subscriptions, it employs signal encryption for protection from unauthorized calling party. Unlike Iridium, which offers a global service, Globalstar's business plan calls for franchising its use to partners in different countries.

The third system is the ICO system (originally called Inmarsat-P), which was built by Hughes Space and Communications (now Boeing Satellite Systems). ICO constellation is made of [37–39]:

Ten operational MEO satellites with five in each of the two inclined circular orbits at an altitude of 10,355 km.

One spare satellite in each plane, making 12 total launched.

Each satellite employs 163 spot beams.

Each satellite will carry an integrated C- and S-band payload.

Twelve satellite access nodes (SANs) located globally.

The inclination of the orbits is 45°—making it the lowest of the systems described. Although this reduces the coverage at high latitudes, it allows for the smallest number of satellites. The ICO system (www.ico.com) is designed to provide the following services:

Global paging

Personal navigation

Personal voice, data, and fax

The three constellations are compared in Table 3.7.

Table 3.7 Characteristics of Satellite PCS Systems

Parameter	Iridium	Globalstar	ICO
Company	Motorola	Loral/Qualcomm	ICO Global
No. of satellites	66	48	10
No. of orbit planes	6	8	2
Altitude (km)	780	1414	10,355
Weight (lb)	1100	704	6050
Bandwidth (MHz)	5.15	11.35	30
Frequency up/down (GHz)	30/20	5.1/6.9	14/12
Spot beams/satellite	48	16	163
Carrier bit rate (kps)	50	2.4	36
Multiple access	TDMA/FDMA	CDMA/FDMA	TDMA/FDMA
Cost to build ($ billion)	4.7	2.5	4.6
Service start date	1998	1999	2003

3.4.4. Satellite Radio

Satellite radio is broadcasting from satellite. With satellite radio, one can drive from Washington DC to Los Angeles, CA without changing the radio station and without static interference. Satellite eliminates localization, which is the major weakness of conventional radio. It transforms radio from a local medium into a national one. Satellite radio will permanently change radio just as cable changed television. It is regarded as radio beyond AM, beyond FM, or radio to the power of *X*.

Figure 3.10 displays a typical architecture of satellite radio. Satellite radio is based on digital radio, which produces a better sound from radio than analog radio. Digital radio systems are used extensively in communication networks. Digital radio offers CD quality sound, efficient use of the spectrum, more programming choice, new services, and robust reception even under the most challenging condition.

Satellite radio is both a new product and a service. As a product, it is a new electronic device that receives the satellite signal. As a service, it will provide consumers with 100 national radio stations, most of which will be brand-new, comprising various music, news, sports, and comedy stations.

Satellite radio service is being provided by DC two companies: XM Satellite Radio (also known as XM Radio), based in Washington, DC, and Sirius Satellite Radio, based in New York. The two companies obtained FCC licenses to operate digital audio radio service (DARC) system coast-to-coast throughout continental U.S.A. To avoid competition with terrestrial radio broadcasters, both satellite broadcasters will carry advertisement of nationally branded products.

XM Satellite Radio is made possible by two satellites, officially named "Rock and Roll," placed in geostationary orbit, one at 85 degrees West longitude and the other at 115 degrees West longitude. Rock and Roll are Boeing 702 satellites, built by Boeing Satellite Systems. The satellites will be positioned above the United States. In September, 2001, XM Satellite Radio started to broadcast. Subscribers pay as little as $9.95 per month after they purchase an AM/FM/XM radio.

Sirius Satellite Radio, on the other hand, does not use GEO satellites. Rather, it is flying three satellites which are equally spaced in an elliptical 47,000 × 24,500-km orbit that takes 24 h to complete. This ensures that each satellite spends about 16 h a day over the continental U.S.A., with at least one satellite over the country at any time. It also means that Sirius will be higher in the sky than

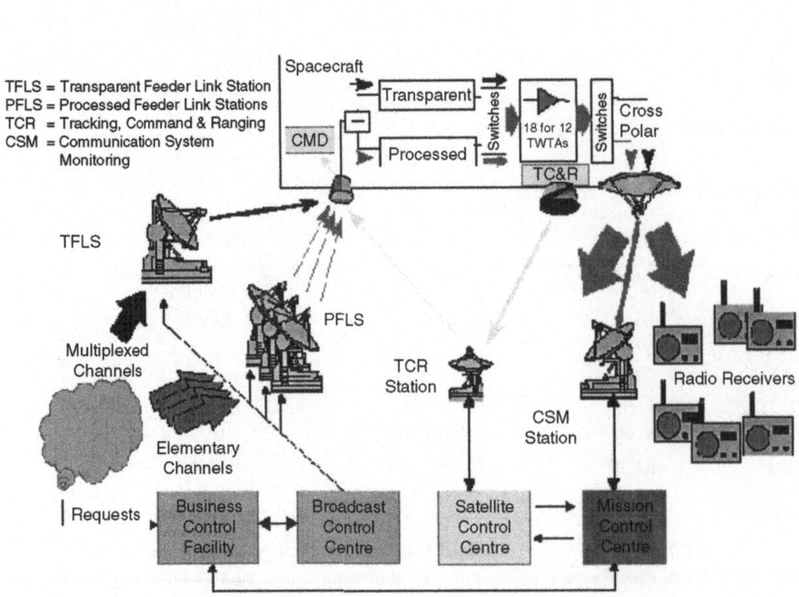

Figure 3.10 A typical architecture of satellite radio. (Source: EBU Technical Review.)

Table 3.8 Comparison of XM and Sirius systems

	XM	Sirius
Constellation	2 Satellites	3 Satellites
Satellite type	Boeing 702	SS/Loral FS-3000
Terrestrial repeaters	1500 in 70 areas	105 in 46 areas
Satellite costs (million)	\$439	\$120
Transmission rate	4 Mbps	4.4 Mbps
Uplink frequencies	7.05–7.075 GHz	7.06–7.0725 GHz
Downlink frequencies	2.3325–2.345 GHz	2.32–2.3325 GHz

XM, which is at the zenith only at the equator. Sirius charges \$12.95 a month for its service. The systems of both satellite radio companies are compared in Table 3.8.

Besides XM Satellite Radio and Sirius Satellite Radio that operate in the United States, WorldSpace is another radio satellite broadcasting company already broadcasting in Africa and Asia. With a constellation of three satellites (AfriStar to cover Africa and Middle East, AmeriStar to serve Latin America and the Caribbean, and AsiaStar to serve nearly all Asia), WorldSpace intends to touch all or parts of the four continents, especially those areas of the world that most conventional radio stations cannot reach.

As a new technology, satellite radio is not without its own peculiar problems. First, people are not yet used to paying for radio programming. If the programming of the satellite broadcasters is not better than what people are getting free from regular, terrestrial radio, they will be reluctant to pay. So the real question is: How many people are going to subscribe? Second, satellite broadcasting requires a near-omnidirectional receive antennas for cars, which in turn requires a powerful signal from the satellite. Third, some believe that the two companies will face a big hurdle in transforming radio from a local medium into a national one. The many-pie-in-the-sky companies are faced with great risks ahead of them.

Satellite radio may will transform radio industry, which has seen little technological change since the discovery of FM, some 40 y ago. Receiving digital-quality music from radio satellite is a major technical milestone. It is as revolutionary to the entertainment industry as was the invention of radio itself. The future of radio by satellite is exciting but uncertain [40–42].

3.4.5. Satellite-Based Internet

The Internet is becoming an indispensable source of information for an evergrowing community of users. The thirst for Internet connectivity and high performance remains unquenched. This has led to several proposals for integrating satellite networks with terrestrial ISDN and the Internet [43–46].

Several factors are responsible for this great interest in IP-over-satellite connectivity. First, satellites cover areas where land lines do not exist or cannot be installed. Satellites can serve as an access link between locations separated by great distances. Second, developments in satellite technology allow home users to receive data directly from a geostationary satellite channel at a rate 20 times faster than of an average telephone modem. With more power transponders utilizing wider frequencies, commercial satellite links can now deliver up to 155 Mbps. Third, the unique positioning of satellites between sender and receivers lends itself to new applications such as IP multicast, streaming data, and distributed web caching. Fourth, satellite connectivity can be rapidly deployed because trenches and cable installation are unnecessary. Moreover, satellite communication is highly efficient for delivering multimedia content to businesses and homes [47].

As an inherently broadcast system, a satellite is attractive to point-to-multipoint and multipoint-to-multipoint communications especially in broadband multimedia applications. The

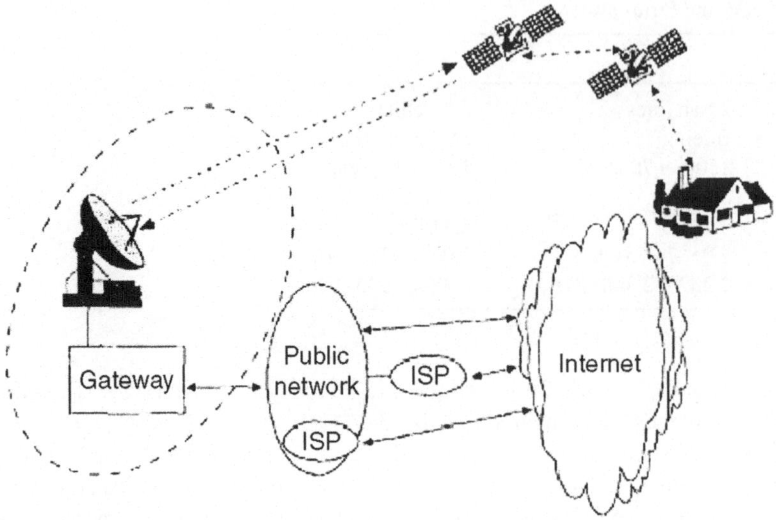

Figure 3.11 A typical configuration for satellite-based Internet.

asymmetrical nature of Web traffic suggests a good match to VSAT systems since the VSAT return link capacity would be much smaller than the forward link capacity.

A typical network architecture for a satellite-based Internet service provider (ISP) is shown in Fig. 3.11, which has been simplified to focus on the basic functionality. It includes its own satellite network and a network of ground gateway stations. The ground gateway stations interface with the public network through which access to the Internet is gained. The number of satellites may vary from dozens to hundreds, and they may be GEO, MEO, or LEO. Thus, the satellite-based Internet has several architectural options due to the diverse designs of satellite systems, orbit types, payload choice, and intersatellite links designs [48].

There is ongoing research into various aspects of implementation and performance of TCP/IP over satellite links. Related issues include the slow start algorithm, the ability to accommodate large bandwidth-delay products, congestion control, acknowledgment, and error recovery mechanisms.

More information about satellite communications systems can be obtained from Refs. 49–53.

REFERENCES

1. Marihart, D.J. Communications technology guidelines for EMS/SCADA systems. IEEE Trans. Power Delivery **April 2001**, *16* (2), 181–188.
2. Calcutt, D.; Tetley, L. *Satellite Communications: Principles and Applications*; Edward Arnold: London, 1994; 3–15, 321–387.
3. Elbert, B.R. *The Satellite Communication Applications Handbook*; Artech House: Norwood, MA, 1997; 3–27, 257–320.
4. Pritchard, W. Geostationary versus nongeostationary orbits. Space Comm. **1993**, *11*, 205–215.
5. Farserotu, J.; Prasad, R. *IP/ATM Mobile Satellite Networks;* Artech House: Boston, MA, 2002; 14.
6. Ha, T.T. *Digital Satellite Communications*; McGraw-Hill: New York, 1990; 1–30, 615–633.
7. Tycz, T.S. Fixed satellite service frequency allocations and orbit assignment procedures for commercial satellite systems. Proc IEEE **July 1990**, *78* (7), 1283–1288.
8. Morgan W.L.; Gordon, G.D. *Communications Satellite Handbook*; Wiley: New York, 1989; 573–589.
9. Martin, J.S. *Communications Satellite Systems*; Prentice Hall: Englewood Cliffs, NJ, 1978; 17–28.
10. Richharia, M. *Satellite Communication Systems*; McGraw-Hill: New York, 1995; 16–49.
11. Fthenakis, E. *Manual of Satellite Communications*; McGraw-Hill: New York, 1984; 31–44.

12. Pratt, T. *Satellite Communications*; Wiley: New York, 1986; 11–51.
13. Wu, W.W. Mobile satellite communications. Proc. IEEE **Sept. 1994**, *82* (9), 1431–1448.
14. Hu, Y.; Li, V.O.K. Satellite-based Internet: a tutorial. IEEE Comm. Mag. **March 2001**, 154–162.
15. Propagation special issue. Int. J. Satellite Comm. **May/June 2001**, *19* (3).
16. Hogg, D.C.; Chu, T.S. The role of rain in satellite communication. Proc IEEE **1975**, *63*, 1308–1331.
17. Bargellini, P.L.; Hyde, G. Satellite and space communications. In *Reference Data for Engineers*; 8th Ed.; SAMS: Carmel, IN, 1993; Chapter 27.
18. Ippolito, L.J. *Radiowave Propagation in Satellite Communications*; Van Nostrand Reinhold: New York, 1986.
19. Gordon, G.D.; Morgan, W.L. *Principles of Communications Satellites*; Wiley: New York, 1993.
20. Elbert, B.R. *Introduction to Satellite Communication*; Artech House: Norwood, MA, 1999; 390–395.
21. Hadjitheodosiou, M.H. Next generation multiservice VSAT networks. Electron. Comm. Eng. J. **June 1997**, 117–126.
22. Abramson, N. VSAT data networks. Proc. IEEE **July 1990**, 78 (7), 1267–1274.
23. Raison, J.C. Television via satellite: convergence of the broadcasting-satellite and fixed-satellite service—the European experience. Space Comm. **1972**, *9*, 129–141.
24. Wood, P. Mobile satellite services for travelers. IEEE Comm. Mag. **Nov. 1991**, 32–35.
25. Abrishamkar, F. PCS global mobile services. IEEE Comm. Mag. **Sept. 1996**, 132–136.
26. Comparetto, G.; Ramirez, R. Trends in mobile satellite technology. Computer **Feb. 1997**, 44–52.
27. Evans, J.V. Satellite systems for personal communications. IEEE Ant. Prop. Mag. **June 1997**, *39* (3), 7–20.
28. Satellite systems for personal communications. Proceedings of the IEEE **June 1997**, *39* (3), 7–20.
29. Satellite communications—a continuing revolution. IEEE Aerospace Electron. Sys. Mag. **Oct. 2000**, 95–107.
30. Pattan, B. *Satellite-Based Cellular Communications*; McGraw-Hill: New York, 1998; 45–88.
31. Lemme, P. Iridium: Aeronautical satellite communications. IEEE AES Sys. Mag. **Nov. 1999**, 11–16.
32. Hubbel, Y.C. A comparison of the iridium and AMPS systems. IEEE Network **March/April 1997**, 52–59.
33. Leopold, R.J.; Miller, A. The iridium communications system. IEE Potentials **April 1993**, 6–9.
34. Hirshfield, E. The Globalstar system: breakthroughs in efficiency in microwave and signal processing technology. Space Comm. **1996**, *14*, 69–82.
35. Dietrich, F.J. The Globalstar cellular satellite system. IEEE Trans. Ant. Prop. **June 1998**, *46* (6), 935–942.
36. Hendrickson, R. Globalstar for the military. Proc. MILCOM **1998**, *3*, 808–813.
37. Poskett, P. The ICO system for personal communications by satellite. Proc. IEE Colloquim (Digest), Part 1 **1998**, 211–216.
38. Ghedia, L. Satellite PCN—the ICO system. Int. J. Satellite Comm. **1999**, *17*, 273–289.
39. Werner, M. Analysis of system parameters for LEO/ICO-satellite communication networks. IEEE J. Selected Areas Comm. **Feb. 1995**, *13* (2), 371–381.
40. Sadiku, M.N.O. XM radio. IEEE Potentials **April/May 2002**.
41. Wood, D. Digital radio by satellite. EBU Tech. Rev. **Summer 1998**, 1–9.
42. Layer, D.H. Digital radio takes to the road. IEEE Spectrum **July 2001**, 40–46.
43. Otsu, T. Satellite communication system integrated into terrestrial ISDN. IEEE Trans. Aerospace Electron. Sys. **Oct. 2000**, *36*(4), 1047–1057.
44. Metz, C. TCP over satellite . . . the final frontier. IEEE Internet Computer **Jan./Feb. 1999**, *3* (1), 76–80.
45. Choi, H.K. Interactive web service via satellite to the home. IEEE Comm. Mag. **March 2001**, 182–190.
46. Hu, Y.; Li, V.O.K. Satellite-based internet: a tutorial. IEEE Comm. Mag. **March 2001**, 154–162.
47. Metz, C. IP over satellite: Internet connectivity blasts off. IEEE Internet Computer **July/August 2000**, 84–89.
48. Cooper, P.W.; Bradley, J.F. A space-borne satellite-dedicated gateway to the Internet. IEEE Comm. Mag. **Oct. 1999**, 122–126.
49. Special issue, Satellite communications. Proc. IEEE **March 1977**, *65* (3).
50. Special issue, Satellite communication networks. Proc. IEEE, **Nov. 1984**, *72* (11).
51. Special issue, Global satellite communications technology and system. Space Comm. **2000**, *16*.
52. Maral, G.; Bousquet, M. *Satellite Communications Systems*, 3rd Ed.; Wiley: New York, 1998.
53. Sadiku, M.N.O. *Optical and Wireless Communications: Next Generation Networks*; CRC Press: Boca Raton, FL, 2002.

4
Optical Communications

Joseph C. Palais
Arizona State University
Tempe, Arizona

4.1. INTRODUCTION

Electromagnetics plays a key role in modern optical telecommunications systems. This chapter emphasizes the electromagnetic phenomena peculiar to fiber optic communications.

Optical communications was first seriously considered just after the invention of the laser in 1960 [1,2]. Atmospheric propagation was proposed and much research was done in the following decade. Problems with weather, line-of-site clearance, beam spreading, and safety (and maybe others) removed free-space optical communication as a major player in the communications area.

In the mid-1960s, guided propagation in a glass fiber was proposed as a strategy for overcoming the many problems of optical atmospheric propagation for telecommunications. In 1970, the first highly transparent glass fiber was produced, making fiber-optic communications practical.

4.1.1. Fiber-Optic System

To help understand the electromagnetic features of fiber-optic communications, we will look briefly at the major components of an optical link. The basic fiber system is pictured in Fig. 4.1. The message is assumed to be available in electronic form, usually as a current.

The transmitter contains a light source that is modulated so that the optical beam carries the message. As an example, for a digital signal, the light beam is electronically turned on (for binary ones) and off (for binary zeros). The optical power in a digitally modulated signal is pictured in Fig. 4.2.

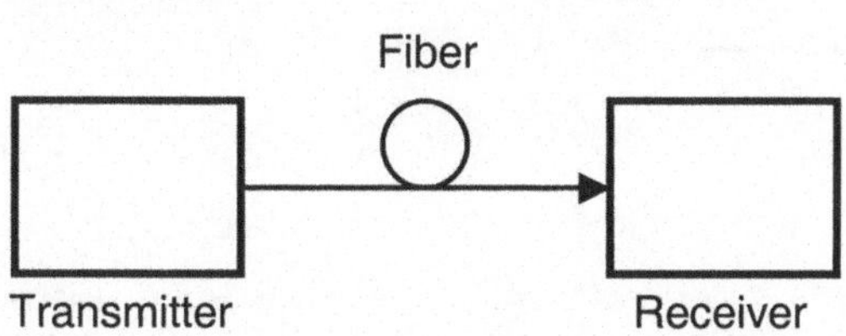

Figure 4.1 Basic fiber communications system.

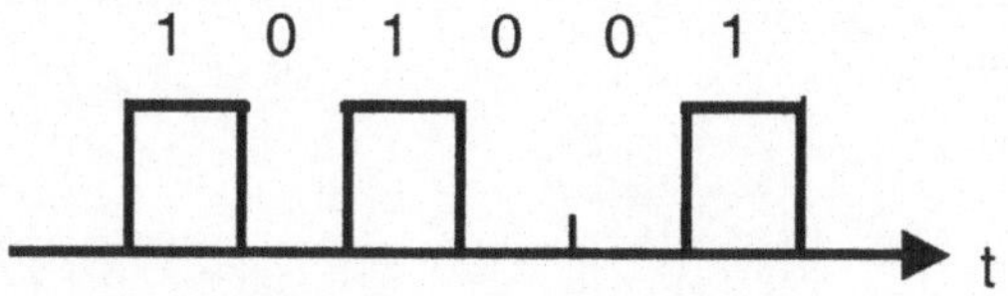

Figure 4.2 Digitally modulated optical signal.

The optical beam is the carrier of the digital message. The most likely choices for fiber-optic light sources are the light-emitting diode and the laser diode. Several characteristics of the light source determine the behavior of the propagating optical wave. Because of that we will briefly describe some of the properties of common sources in a later section of this chapter.

The modulated light beam is coupled into the transmission fiber. At the receiver, the signal is collected by a photodetector, which converts the information back into electrical form. The photodetectors do not affect the propagating properties of the wave but certainly must be compatible with the rest of the system. For completeness then, we will briefly describe properties of the photodetectors in a later section of this chapter.

4.1.2. Optical Spectrum

Fiber communications utilize carrier wavelengths in the optical region of the electromagnetic spectrum. A partial view of the optical spectrum, Fig. 4.3, shows the ultraviolet, visible, and near-infrared portions. Most fiber communications use carriers in the infrared, because that is where glass fiber attenuation losses are lowest. There is some activity in the visible using plastic fibers (which have higher losses than glass) for short paths.

The wavelength regions where fiber systems have been constructed appear in Table 4.1.

The relationship between wavelength (λ) and frequency (f) is

$$\lambda = \frac{c}{f} \tag{4.1}$$

The velocity of light in empty space is $c = 3 \times 10^8$ m/s. As an example, a wavelength of 1.5 μm corresponds to a frequency of 2×10^{14} Hz (and a period of oscillation of 0.5×10^{-14} s). Because the amount of information that can be transmitted is proportional to the carrier frequency, the amount of information that can be transmitted on an optical carrier is enormous. In addition, multiple optical carriers (different wavelengths) can travel the fiber, further enhancing their information carrying capacity.

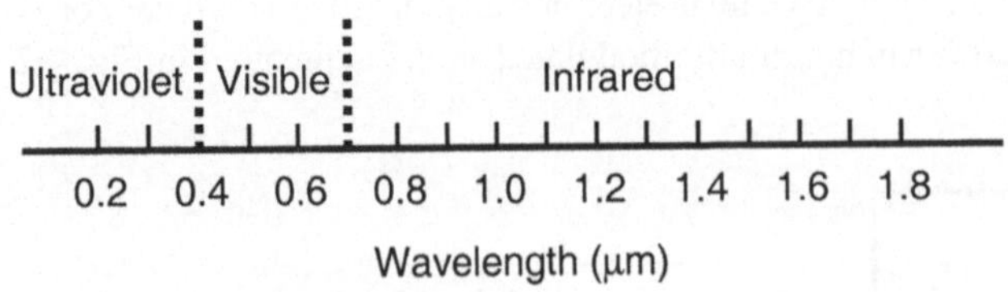

Figure 4.3 Optical spectrum.

Table 4.1 Major Wavelength Regions for Fiber Systems

Wavelength (nm)	Fiber material	Loss (dB/km)
650–670	Plastic	120–160
800–900	Glass	3
1250–1350	Glass	0.5
1500–1600	Glass	0.25

4.1.3. Light Sources

The most commonly used light sources in optical communication are the light-emitting diode (LED) and the laser diode (LD) [3–6].

Light-Emitting Diodes

The LED is a pn junction semiconductor. The LED emits light in the visible or infrared regions when forward biased. Free charges (electrons and holes) injected into the junction region spontaneously recombine with the subsequent emission of radiation.

The electronic driving circuit and output characteristic appear in Fig. 4.4. Ideally the output optical power increases linearly with input current. Thus, the optical power (P) waveform is a replica of the driving current (i). In the forward-biased region, the output power is given by

$$P = a_1 i \tag{4.2}$$

Both analog and digital modulation is possible with the LED. Bandwidth limitations restrict modulation to a few hundred megahertz and a few hundred megabits per second.

Operating voltages are on the order of a volt or 2. Operating currents are on the order of a few tens of milliamps. Output powers are on the order of a few milliwatts.

The output wavelength is determined by the band-gap energy (W_g) of the semiconductor material. In particular, the output wavelength is given by

$$\lambda = \frac{1.24}{W_g} \tag{4.3}$$

In this equation, the wavelength is in micrometers and the band-gap energy is in electron volts. Table 4.2 indicates the materials commonly used for LEDs.

An important property of light sources used in fiber-optic communications is its *spectral width*. This is the range of wavelengths (or frequencies) over which it emits significant amounts of

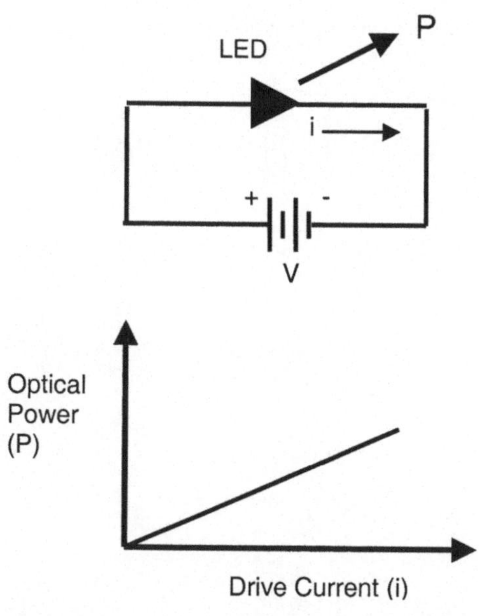

Figure 4.4 LED driving circuit and output characteristic.

power. Ideally a light source would be monochromatic, emitting a single wavelength. In practice, no such light source exists. All radiate over a range. For LEDs, the range is typically on the order of 20 to 100 nm. Coherence refers to how close the source radiation is to the ideal single wavelength. The smaller the spectral width, the more coherent the source.

Laser Diodes

The laser diode shares a number of characteristics with the LED. It is also a pn junction semiconductor that emits light when forward biased. Light amplification occurs when photons stimulate free charges in the junction region to recombine and emit. The light beam is reflected back and forth through the amplifying medium by reflectors at each end of the junction. The amplification together with the feedback produce an oscillator emitting at optical frequencies.

The stimulated recombination leads to radiation that is more coherent than that produced by spontaneous recombination. Spectral widths for LDs are typically in the range of 1 to 5 nm. Specially constructed LDs can be designed to have even smaller spectral widths.

The laser diode is much faster than the LED, allowing for much higher modulation rates. Bandwidths of several gigahertz and several gigabits per second are achievable. For higher rates, external modulation is required.

The driving circuit and the output characteristic appear in Fig. 4.5. Note that the output power does not increase until the input current is beyond a threshold value (I_{TH}). Thresholds are on the

Table 4.2 Materials for Semiconductor Light Sources

Material	Band-gap energy (eV)	Wavelength (μm)
GaInP	1.82–1.94	0.64–0.68
AlGaAs	1.4–1.55	0.8–0.89
InGaAsP	0.73–1.34	0.93–1.7
GaAs	1.4	0.89
InGaAs	0.95–1.24	1.0–1.31

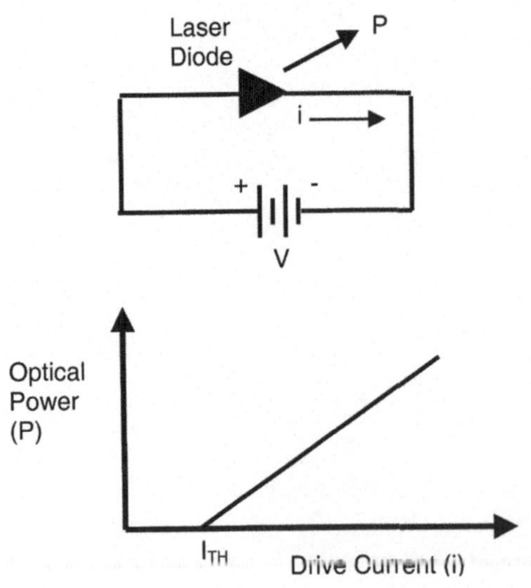

Figure 4.5 Laser diode driving circuit and output characteristic.

order of a few to a few tens of milliamperes. Voltages are on the order of a few volts and output powers of a few milliwatts.

The materials used for LDs are the same as those used in constructing LEDs. Equation (4.3) applies to laser diodes, as does Table 4.2.

4.1.4. Photodetectors

The most common photodetector for fiber communications is the semiconductor junction photodiode. The photodetector converts optical power (P) to an electric current. The received current (i) is [7]

$$i = \rho P \tag{4.4}$$

The term ρ is the photodetector responsivity. A table of photodetector materials, their operating wavelengths, and their peak responsivities appears in Table 4.3. The cutoff wavelength is determined by the band-gap energy and is given by

$$\lambda_c = \frac{1.24}{W_g} \tag{4.5}$$

Just as in Eq. (4.3) for the emission of an LED, the wavelength is in micrometers and the band-gap energy is in electron volts. Only wavelengths equal to or smaller than the cutoff wavelength can be detected.

Because the optical power waveform is a replica of the message current, the receiver current is then a replica of the original signal current. This is ideally the case. Distortions in the waveform caused by transmission and caused by transmitter and receiver irregularities will degrade the signal. We will be describing the degradations due to transmission along the fiber in detail later in this chapter.

The circuit for the simplest type of receiver is shown in Fig. 4.6. According to Eq. (4.4) the photodetector acts like a constant current source. Therefore, the output voltage $v = iR_L$ can be increased by increasing the load resistance R_L. However, the receiver bandwidth is no larger

Table 4.3 Materials for Semiconductor Photodetectors

Material	Wavelength (μm)	Peak response (μm)	Peak responsivity (A/W)
Si	0.3–1.1	0.8	0.5
Ge	0.5–1.8	1.55	0.7
InGaAs	1.0–1.7	1.7	1.1

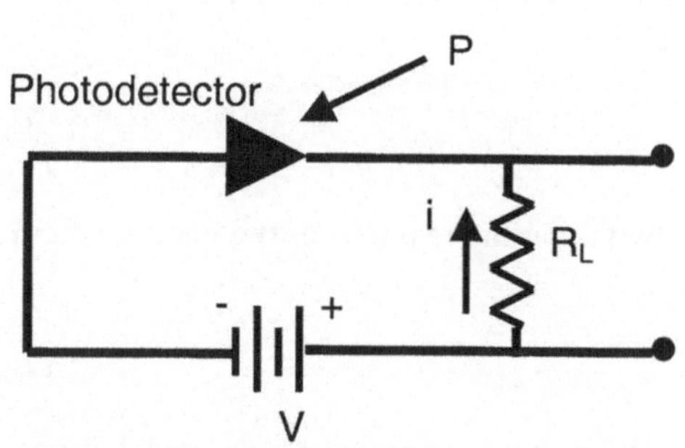

Figure 4.6 Receiver circuit.

than $B = 1/2\pi R_L C_d$, so that doing so decreases the receiver bandwidth. The photodiode's shunt capacitance is C_d.

4.1.5. Electromagnetic Problems

There is a set of electromagnetic problems that bears upon fiber-optic communications. These problems can be described as follows:

- Propagation of a plane wave in unbounded media
- Reflection at a plane boundary
- Waves in an electromagnetic cavity
- Guided propagation in a rectangular dielectric waveguide
- Propagation in an optical fiber

Analyses of these electromagnetic phenomena in the next few sections help explain the design, operation, capabilities, and limitations of fiber transmission lines, fiber-optic components, and fiber-optic systems.

4.2. PROPAGATION

This section describes properties of traveling waves, emphasizing traveling pulses as in an optical digital communication system.

4.2.1. Wave Properties

All electromagnetic fields must satisfy the wave equation [8,9]. It evolves from Maxwell's equations. For dielectric materials it is given by

$$\nabla^2 E = \mu\varepsilon \frac{\partial^2 E}{\partial t^2} \tag{4.6}$$

In rectangular coordinates, the Laplacian is

$$\nabla^2 E = \frac{\partial^2 E}{\partial x^2} + \frac{\partial^2 E}{\partial y^2} + \frac{\partial^2 E}{\partial z^2} \tag{4.7}$$

It is possible to understand many complex electromagnetic wave phenomena by studying the simple case of a plane wave traveling in an unbounded medium. The electric field for such a wave traveling in the z direction can be written (in complex form) as

$$E = E_0 e^{-\alpha z} e^{j(\omega t - kz)} \tag{4.8}$$

The instantaneous form of this field is generated by taking the real part of the complex field. In this case, the result is

$$E = E_0 e^{-\alpha z} \cos(\omega t - kz) \tag{4.9}$$

This is a solution to the electromagnetic wave equation that is appropriate for a wave propagating in an unbounded medium. The amplitude of the field is $E_0 e^{-\alpha z}$, its radian frequency is

$\omega = 2\pi f$, and k is the *propagation factor*. The term $\omega t - kz$ is the *phase* of the wave, while kz is the phase shift over a distance z. In the case being considered, at any instant of time the phase is constant over any plane given by a fixed value of z. Since a fixed value of z defines a plane (parallel to the xy plane), the field described above is a plane wave.

The factor α is the *attenuation coefficient*. It is determined by the losses in the medium. For an ideal (lossless) medium, α would be zero. If α is given in units of km^{-1}, the loss in dB/km is related to it by

$$\text{dB/km} = -8.685\alpha \tag{4.10}$$

The propagation factor is related to frequency and the phase velocity (v) of the wave by

$$k = \frac{\omega}{v} \tag{4.11}$$

The wave velocity in a medium is determined by its *refractive index*, as given by

$$n = \frac{c}{v} \tag{4.12}$$

That is, the *index of refraction* is the ratio of the velocity of light in empty space to that in the medium. Because most media slow light beams, their indices of refraction are greater than unity. Glasses used for fibers have an index close to 1.5. This tells us that a light beam travels in a glass fiber at a speed of approximately

$$v = \frac{c}{n} = \frac{3\times10^8}{1.5} = 2\times10^8\,\text{m/s} \tag{4.13}$$

Fortunately, the travel delay time at this speed is short compared to the response time of human beings for any terrestrial distances. For example, a fiber telephone link including a path under the Pacific ocean may be as long as 10,000 km. The propagation time along this path would be 0.05 s. The two way delay would be 0.1 s, a bit shorter than most humans can detect. We conclude that telephone transmission over fibers appears to be instantaneous to human participants. The wavelength is given in terms of the propagation coefficient by

$$\lambda = \frac{2\pi}{k} \tag{4.14}$$

The electric field of a traveling wave is sketched in Fig. 4.7. The wave is shown at two instants of time, indicating movement to the right. The dotted lines form the envelope of the wave. The peak wave amplitude diminishes with distance traveled because of attenuation. The wavelength, which is the distance between adjacent points of equal phase, is also indicated on the figure.

We know that plane waves travel in unbounded media with the electric field vector lying in a plane perpendicular to the direction of travel. For a plane wave traveling in the z direction, this would be the xy plane. The field given in Eq. (4.8) is the scalar component of the vector field. For simplicity, it can be thought of as the x component of the electric field. It could represent the y component just as well. Any other vector in the xy plane can be resolved into its x and y components. Thus, we say there are two independent ways in which a plane wave can propagate in the unbounded medium. The different ways in which a field can propagate in a given environment are called its *modes*.

Polarization refers to the direction of the electric field. A field that points in just one direction (say the x direction) is *linearly polarized*. A plane wave in an unbounded medium can propagate in two linearly polarized modes.

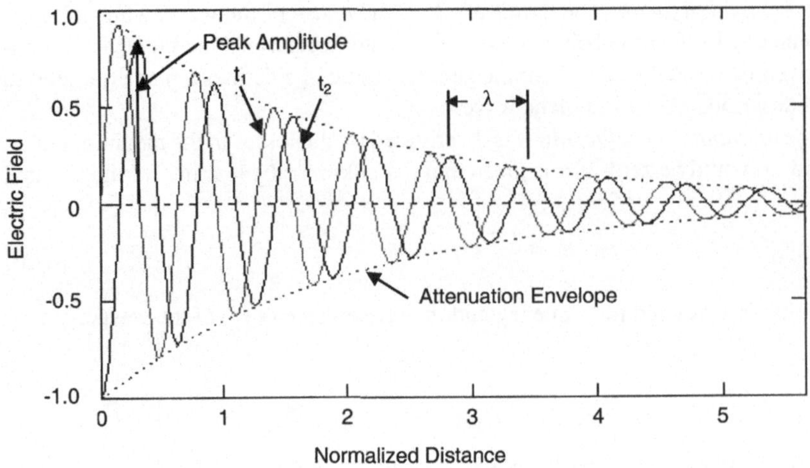

Figure 4.7 Traveling wave at two instances of time, showing peak amplitude, wavelength, and attenuation envelope. Time t_2 is later than time t_1 indicating wave movement to the right.

A *circularly polarized* wave is produced by the simultaneous transmission of two orthogonally polarized waves that are 90 degrees out of phase with each other. For example, the x-directed field

$$E = E_0 e^{-\alpha z} e^{j(\omega t - kz)} \tag{4.15}$$

combined with the y-directed field

$$E = E_0 e^{-\alpha z} e^{j(\omega t - kz + \pi/2)} \tag{4.16}$$

produces a total electric field vector that rotates in a circle at radian frequency ω as time progresses.

If these last two orthogonal fields had a random phase difference (replacing $\pi/2$ in the last equation with a randomly varying function of time), the resulting electric vector would trace out a random pattern with time. This represents a *nonpolarized* (or *unpolarized*) wave. Equations for the fields such as those above imply that the waves are perfectly monochromatic. This is a simplification that produces reliable results in many cases, but is not always applicable. As we mentioned earlier in this chapter, the radiation emitted by practical sources exists over a range of wavelengths. This can be modeled mathematically by including a randomly time varying term in the phase of the wave. For example,

$$E = E_0 e^{-\alpha z} e^{j[\omega t - kz + \phi(t)]} \tag{4.17}$$

where $\phi(t)$ is a random function. In most fiber applications, the light is unpolarized, either because the light source produces unpolarized light or the wave becomes unpolarized during transmission. Bends and twists in the fiber along with other discontinuities in the path (e.g., at connectors and splices) together with random motion of the fiber (e.g., caused by vibrations) are the cause of the depolarization.

4.2.2. Pulse Transmission

A problem in transmission of pulses occurs because of two factors. One is that the source light is not emitted at a single wavelength but exists over a range of wavelengths (the source spectral width). The other factor is that the index of refraction is not the same for all wavelengths. In fact, for glasses used in fiber, the refractive index varies with wavelength. *Dispersion* is the name given to this property of

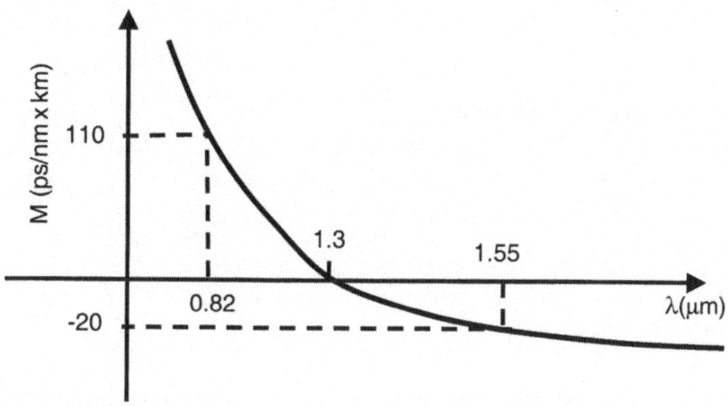

Figure 4.8 Material dispersion for silica glass.

velocity dependence on wavelength. *Material Dispersion* is the appropriate name when the dispersion is due to a property of the material [10].

Dispersion causes a pulse of light to lengthen as it traverses the fiber. This is because each of the component wavelengths travels at a different speed, each arriving with a slight delay with respect to the others. The amount of pulse spreading ($\Delta\tau$) per unit of length of fiber (ΔL) is given by

$$\Delta\left(\frac{\tau}{L}\right) = -M\Delta\lambda \tag{4.18}$$

The factor M is the *material dispersion factor* (or just the *material dispersion*) and is plotted in Fig. 4.8 for pure silica. Note that it is high in the region around 800 nm, goes to zero near 1300 nm, and is small and negative around 1550 nm. Clearly the 1300 nm region is favored to minimize pulse spreading. Spectral widths of common fiber light sources were given earlier in this chapter.

In the range 1200 to 1600 nm, the material dispersion factor can be approximated by

$$M = \frac{M_0}{4}\left(\lambda - \frac{\lambda_0^4}{\lambda^3}\right) \tag{4.19}$$

The constant M_0 is approximately -0.095 ps/(nm^2 × km). The zero dispersion wavelength is λ_0. It is close to 1300 nm for silica fibers. As an example, at 1550 nm the material dispersion factor is close to -20 ps/(nm × km). Using an LED with spectral width of 20 nm, yields a pulse spread per unit length of the transmission path

$$\Delta\left(\frac{\tau}{L}\right) = -M\Delta\lambda = -(-20)20 = 400\ \text{ps/km} \tag{4.20}$$

The problem with pulse spreading is that it limits the information carrying capacity of the fiber. Pulses that spread eventually overlap with neighboring pulses, creating intersymbol interference. This leads to transmission errors and must be avoided. The direct way to avoid this is to place pulses further apart at the transmitter. This means lowering the data rate. The limits on data capacity caused by pulse spreading for non-return-to-zero and return-to-zero pulse codes [11] is

$$R_{\text{NRZ}} \times L = \frac{0.7}{\Delta(\tau/L)} \tag{4.21}$$

$$R_{RZ} \times L = \frac{0.35}{\Delta(\tau/L)} \tag{4.22}$$

Using the numerical values in the preceding example yields

$$R_{NRZ} \times L = 1.75 \text{ Mb/s} \times \text{km} \tag{4.23}$$

and

$$R_{RZ} \times L = 0.875 \text{ Mb/s} \times \text{km} \tag{4.24}$$

Similarly, pulse spreading reduces the bandwidth of an analog system. The 3-dB bandwidth limit is

$$f_{3\text{-dB}} \times L = \frac{0.35}{\Delta(\tau/L)} \tag{4.25}$$

Using the same numerical values as above yields the limit as

$$f_{3\text{-dB}} \times L = 0.875 \text{ MHz} \times \text{km} \tag{4.26}$$

At modulation frequencies much lower than that calculated above, the analog signal propagates without distortion. At the 3-dB frequency the amplitude of the signal diminishes to 50% of what it was at lower frequencies. At modulation frequencies well above the 3-dB value, the signals are attenuated greatly. Pulse spreading causes the fiber to act as a low-pass filter, allowing only the lower modulating frequencies to pass.

4.2.3. Snell's Law and Total Reflection

Reflection of light at a plane boundary is one of the classical problems in optics [12,13]. Results from the analysis of this problem explain the operation of numerous optical components. The physical situation is drawn in Fig. 4.9. A plane boundary exists between two dielectric (insulators) media having refractive indices n_1 and n_2, respectively. Because the materials are insulators, no electric currents will flow. Glass and plastic materials used in fiber-optic waveguides satisfy the assumptions of this model. The incident ray in medium 1 strikes the boundary at an angle θ_i as measured with respect to the boundary normal. There is a reflected wave at angle θ_r and a transmitted wave at angle θ_i. The angle of reflection equals the angle of incidence

$$\theta_r = \theta_i \tag{4.27}$$

and the transmitted angle is determined by Snell's law

$$\frac{\sin\theta_t}{\sin\theta_i} = \frac{n_1}{n_2} \tag{4.28}$$

Figure 4.10 illustrates what happens when a light ray travels from a lower index to a higher index material. The ray bends toward the boundary normal. Figure 4.11 illustrates what happens when a light ray travels from a higher index to a lower index material. The ray bends away from the

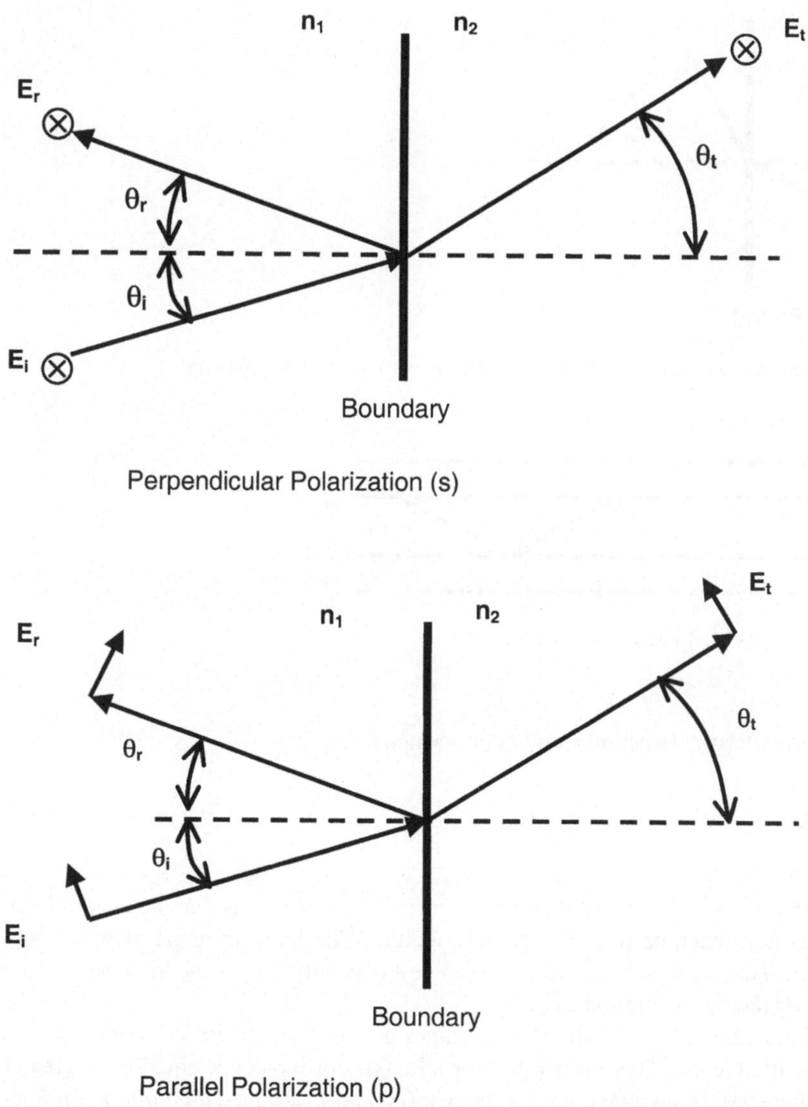

Figure 4.9 Reflection at a plane boundary. Perpendicular and parallel polarizations are illustrated.

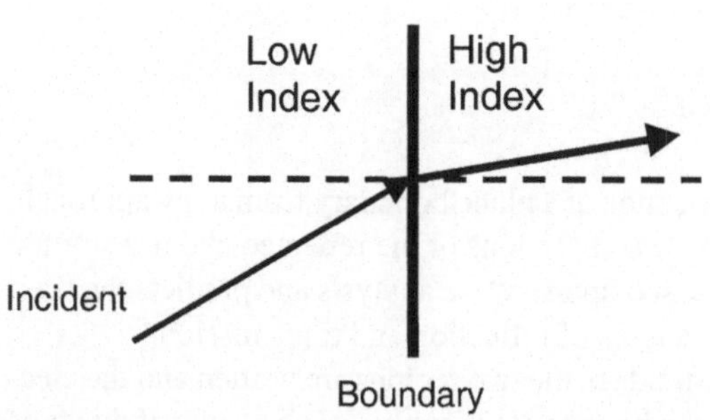

Figure 4.10 Ray bending when a wave travels from a low-index to a high-index material.

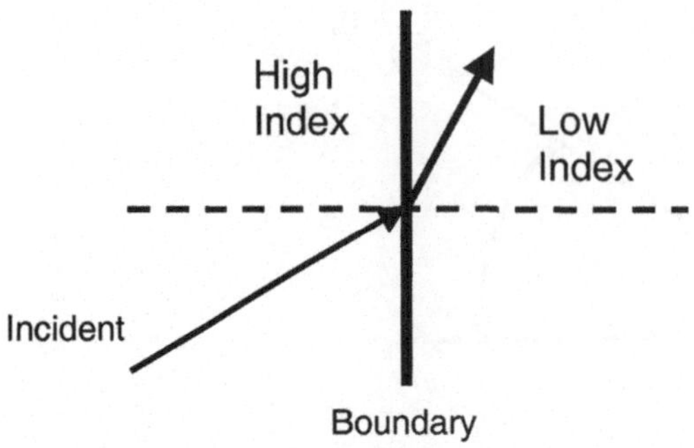

Figure 4.11 Ray bending when a wave travels from a high-index to a low-index material.

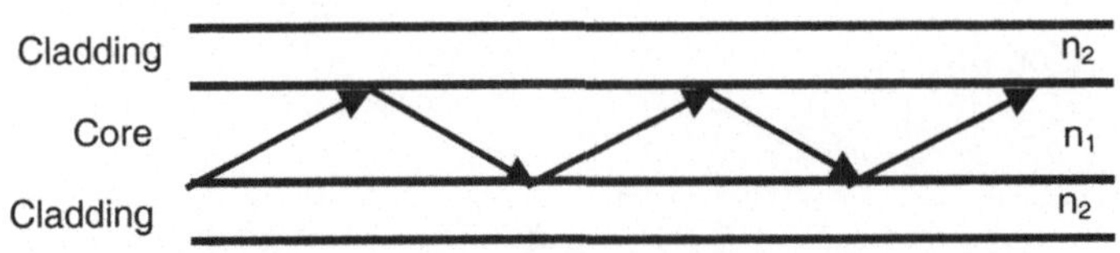

Figure 4.12 Wave guiding by total internal reflection.

boundary normal. A particularly interesting result occurs when

$$\sin \theta_i = \frac{n_2}{n_1} \tag{4.29}$$

At this angle, we find that *sin* $\theta_t = 1$, so that θ_t itself becomes 90°. This means that the energy from the incident beam does not penetrate into the second medium. The incident angle at which this occurs is called the *critical angle*, θ_c. Clearly, a critical angle exists only if $n_1 > n_2$, because the sine function must be unity or less for a solution to exist.

As can be seen from Snell's law, for all incident angles greater than the critical angle, there is no solution for a transmitted angle. This means that for all angles of incidence equal to or greater than the critical angle there will be no transmitted wave. This condition is called *total internal reflection.* It is the basis of wave guiding by an optical fiber, as illustrated in Fig. 4.12. In the figure, the central region is the *core* (having refractive index n_1) and the outer material is the *cladding* (having refractive index n_2). The light stays inside the glass fiber structure by continual reflection at the boundaries. Note that this can only occur (total reflection) if the inner material has a higher refractive index than the outer material so that a critical angle exists.

4.2.4. Reflection at a Boundary

The preceding section considered the problem of reflection at a plane boundary from a ray approach. The only results generated by this type of analysis are the directions of the reflected and transmitted beams. A more complete solution of the problem is based upon a wave analysis and predicts the fraction of light reflected and transmitted as well as the angles of reflection and transmission.

In the wave analysis, the electric and magnetic fields in the two regions are written and the electromagnetic conditions of continuity of the fields at the boundary are applied. The results of this type of analysis are presented in the following paragraphs.

The *plane of incidence* is the plane defined by the normal to the boundary and the direction of travel of the incident wave. That would be the plane of the page in Fig. 4.9. For the problem under consideration, the amount of reflection depends upon the polarization of the incident field. Recall that the two orthogonal linearly polarized modes of a plane wave lie in a plane that is perpendicular to the direction of travel. The corresponding electric fields are either polarized perpendicular to the plane of incidence (this is called s polarization) or parallel to the plane of incidence (this is called p polarization).

The *reflection coefficient* ρ is defined as the ratio of the reflected electric field to the incident electric field when they are written in complex form. The results are known as *Fresnel's laws of reflection.*

For parallel polarization, the reflection coefficient is

$$\rho_p = \frac{-n_2^2 \cos\theta_i + n_1\sqrt{(n_2^2 - n_1^2 \sin^2\theta_i)}}{n_2^2 \cos\theta_i + n_1\sqrt{(n_2^2 - n_1^2 \sin^2\theta_i)}} \tag{4.30}$$

For perpendicular polarization, the reflection coefficient is

$$\rho_s = \frac{n_1 \cos\theta_i - n_1\sqrt{(n_2^2 - n_1^2 \sin^2\theta_i)}}{n_1 \cos\theta_i + n_1\sqrt{(n_2^2 - n_1^2 \sin^2\theta_i)}} \tag{4.31}$$

The reflection coefficients give us the relationship between the incident and reflected electric fields. Because the power is proportional to the square of the field, the fractional reflected power (called the *reflectance*) is determined by taking the magnitude of the square of the reflection coefficient. Thus, the reflectance R is

$$R = |\rho|^2 \tag{4.32}$$

Clearly, the fraction of transmitted power is

$$1 - R = 1 - |\rho|^2 \tag{4.33}$$

For the case of normal incidence ($\theta_i = 0$), both cases of polarization reduce to

$$R = \left(\frac{n_1 - n_2}{n_1 - n_2}\right)^2 \tag{4.34}$$

For an air-to-glass interface ($n_1 = 1$ and $n_2 = 1.5$), the result is $R = 0.04$. In this case, 4% of the light is reflected and 96% of the light is transmitted. Reflectance plots are shown in Figs. 4.13 and 4.14 for air-to-glass and glass-to-air interfaces, respectively. In the first case, the wave goes from a region of lower refractive index to one that is higher. In the second case, the wave goes from a region of higher refractive index to one that is lower. These figures illustrate a few interesting points upon which we will elaborate.

For small angles of incidence, the reflectance does not vary much with a change in the incident angle.

At a certain incident angle, the reflectance is zero for the case of parallel incidence. The angle at which zero reflection occurs is called the *Brewster angle* and is found from Eq. (4.30) to be

$$\tan\theta_B = \frac{n_2}{n_1} \tag{4.35}$$

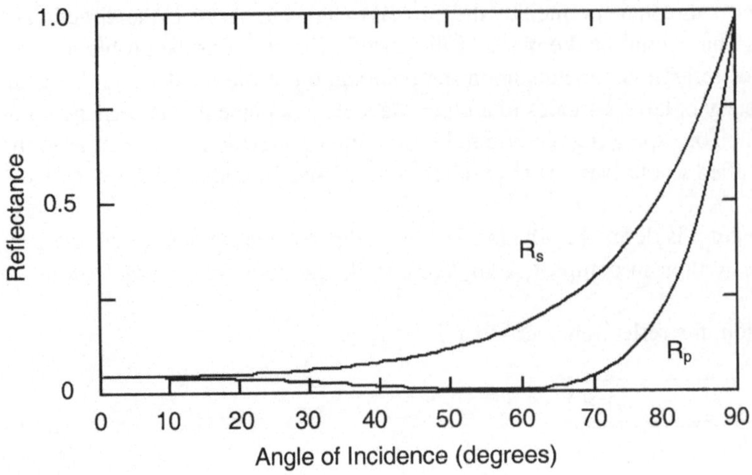

Figure 4.13 Reflectance for an air-to-glass interface.

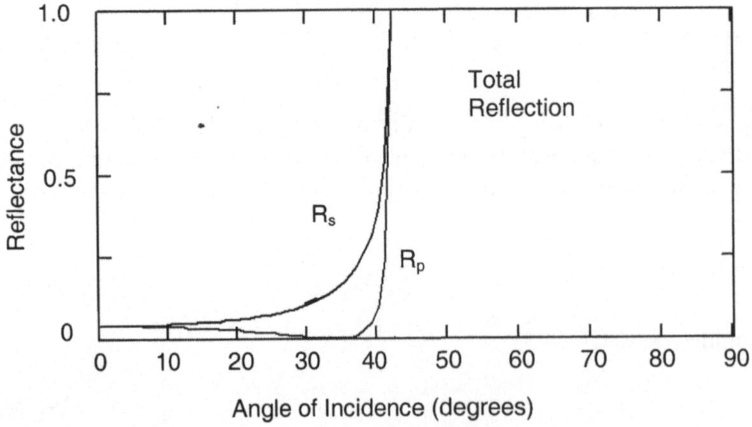

Figure 4.14 Reflectance for a glass-to-air interface.

At the Brewster angle, all the light is transmitted.

As we determined earlier, all the light is reflected when incident at, or beyond, the critical angle. This shows up on Fig. 4.14 where the wave is going from a high-index to a low-index region. The figure shows that for incident angles equal to or greater than the critical angle all the light is reflected. The critical angle is calculated from either of Eqs. (4.30) and (4.31) by setting the term under the square root sign to zero. When that is done, $\rho_p = -1$ and $\rho_s = 1$. The reflectance is unity in either case. The critical angle condition is thus

$$n_2^2 - n_1^2 \sin^2\theta_i = 0 \tag{4.36}$$

The incident angle that satisfies this equation is the critical angle. The solution is

$$\sin\theta_c = \frac{n_2}{n_1} \tag{4.37}$$

just as we found before in Eq. (4.29).

For angles greater than θ_c, we find that $n_1 \sin \theta_i > n_2$, making the term under the square root sign negative and making the second term in the numerator and denominator of both reflection coefficient equations imaginary. Both equations are then of the form

$$|\rho| = \left|\frac{A - jB}{A + jB}\right| \tag{4.38}$$

where A and B are real numbers and j is the imaginary term $j = \sqrt{-1}$. The magnitude of this complex number is unity. We conclude that the reflectance is unity for all angles of incidence equal to or greater than the critical angle. Again, this is the principle of the fiber waveguide. All rays that strike the core-cladding boundary at angles that are equal to or greater than the critical angle are bound to the core.

When we have total internal reflection, you might think there would be no electric field in the second medium. This is not the case, however. The boundary conditions require that the electric field be continuous at the boundary (that is, the field in region 1 and that in region 2, as measured at the boundary, must be equal). The exact solution shows a finite field in medium 2 that decays exponentially away from the boundary and carries no power into the second medium. This is called an *evanescent* field. It is not unlike the field that surrounds an inductor carrying a sinusoidal time-varying current. The stored energy is $Li^2/2$, where L is the inductance and i is the peak current. A magnetic field exists in the region around the inductor (having this same energy), but the energy does not flow away from the inductor. The energy can be captured by the circuit by discharging the inductor.

The field in medium 2 has a decay away from the boundary given by

$$E \propto e^{-\alpha z} \tag{4.39}$$

where the attenuation factor is

$$\alpha = k_0 \sqrt{n_1^2 \sin^2\theta_i - n_2^2} \tag{4.40}$$

and k_0 is the free-space propagation factor. As can be seen, α is zero at the critical angle and increases as the incident angle increases beyond the critical angle. Because α is so small near the critical angle, the evanescent fields penetrate deeply beyond the boundary but do so less and less as that angle increases.

In medium 1, the reflected field interferes with the incident field to produce a standing wave. The envelope of this standing wave and the evanescent wave are pictured in Fig. 4.15.

This chapter is introducing, in a step-wise manner, the necessary electromagnetic fundamentals upon which fiber-optic communications is built. At this point, we can understand how a wave

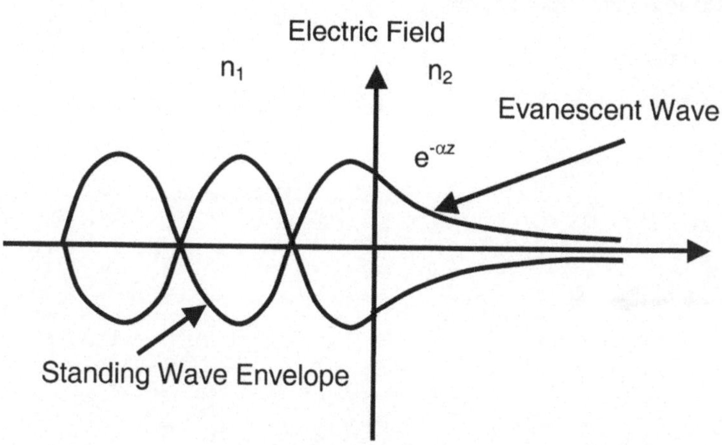

Figure 4.15 Electric field amplitudes near a reflecting boundary.

can be trapped in the fiber core due to total internal reflection at the core-cladding boundary, that a standing wave will exist in the fiber core, and that an evanescent field will exist in the fiber cladding.

4.2.5. Gaussian Beams

The electric field of a plane wave, such as that given by Eq. (4.8), has an amplitude that is the same over any plane given by z = constant. We call this a *uniform* plane wave. While this type of wave does not exactly exist in nature, it is a good approximation in many cases. For example, the radiation from a finite-sized light source (such as a LED or laser diode) spreads out as the observer moves away from the source. At distances far from a source the beam pattern is nearly uniform over a small region near the axis of propagation, approximating a uniform plane wave.

A more common beam distribution is the *gaussian* beam. The gaussian beam is often generated by a laser and is the beam pattern in some fibers. The intensity of a gaussian distribution is given by

$$I = I_0 e^{-2(r/w)^2} \tag{4.41}$$

Intensity is proportional to the power in the wave. Technically, the intensity is the magnitude of the square of the electric field. This beam distribution is plotted in Fig. 4.16. It is cylindrically symmetric. The radial distance from the origin is r, I_0 is the peak intensity (it occurs in the center at $r = 0$), and the term w is called the *spot size*. It is the radial distance at which the intensity decreases to $1/e^2 = 0.135$ of its peak value I_0. That is, when $r = w$, then $I/I_0 = 1/e^2$.

A picture of the gaussian field pattern appears in Fig. 4.17. Since the field is circularly symmetric, the beam appears to be a circular spot of light. The electric field of a gaussian plane wave

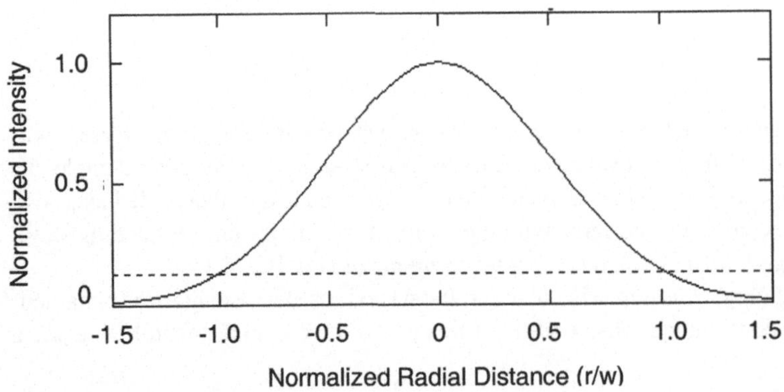

Figure 4.16 Gaussian intensity distribution.

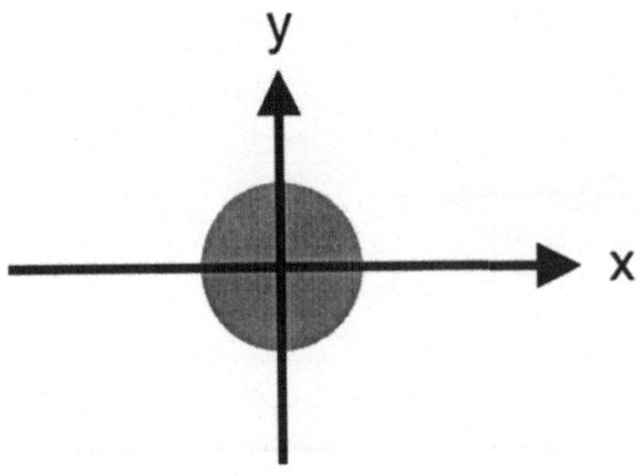

Figure 4.17 Gaussian transverse light pattern.

traveling in the z direction could be written as

$$E = E_0 e^{-r^2/w^2} e^{-\alpha z} e^{j(\omega t - kz)} \tag{4.42}$$

The intensity is

$$I = EE^* = E_0^2 e^{-2\alpha z} e^{-2r^2/w^2} \tag{4.43}$$

where E^* is the complex conjugate of E. The result is as expected, where we recognize $E_0^2 e^{-2\alpha z}$ as the peak intensity at the center of the beam for any position z and E_0^2 as the peak intensity at the origin ($r = 0$, $z = 0$).

4.2.6. Electromagnetic Cavity

A simple electromagnetic cavity is drawn in Fig. 4.18. Its analysis reveals a number of facets of operation of optical devices including the fiber. For simplicity, we consider the propagation of plane waves between infinitely extended perfect mirrors. The waves move along the cavity axis (the z axis) back and forth between the two reflecting mirrors. Assuming the electric fields are linearly polarized in the y direction as indicated on the figure, the forward- and backward-traveling waves are given (respectively) by

$$E_+ = E_1 e^{-jkz} \tag{4.44a}$$

$$E_- = E_2 e^{jkz} \tag{4.44b}$$

where the time variation $e^{j\omega t}$ has been suppressed. The total field at any point in the cavity is then

$$E = E_1 e^{-jkz} + E_2 e^{jkz} \tag{4.45}$$

To be perfectly reflecting, the mirrors must have infinite conductivity. The electromagnetic boundary conditions in such a case require that the total field be zero at the mirrors. That is,

$$E(z = 0) = 0 \tag{4.46a}$$

$$E(z = L) = 0 \tag{4.46b}$$

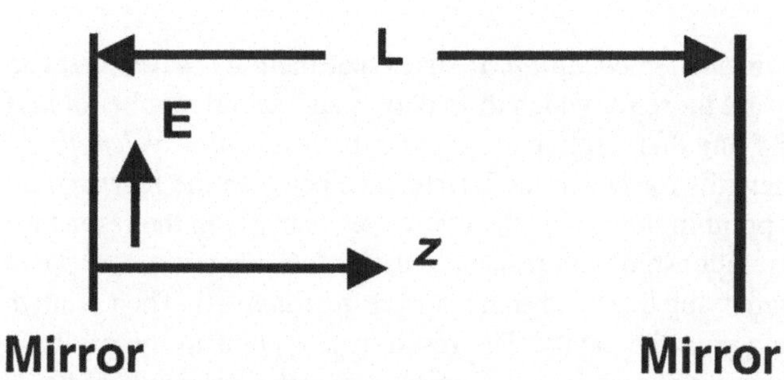

Figure 4.18 Electromagnetic cavity.

Applying the first of these conditions leaves

$$0 = E_1 + E_2 \tag{4.47a}$$

or

$$E_2 = -E_1 \tag{4.47b}$$

We see that the two opposing fields must be equal in magnitude. The minus sign indicates a 180° phase shift at the first mirror needed to make the fields cancel at that mirror.
The total field can now be written as

$$E = E_1 e^{-jkz} - E_1 e^{jkz} \tag{4.48}$$

or

$$E = 2jE_1 \sin kz \tag{4.49}$$

Applying the boundary condition at the second boundary yields

$$0 = 2jE_1 \sin kL \tag{4.50}$$

The conclusion is that

$$0 = \sin kL \tag{4.51}$$

requiring that

$$kL = m\pi \tag{4.52}$$

where m is a positive integer. Since we know that $k = 2\pi/\lambda$, this becomes

$$L = \frac{m\lambda}{2} \tag{4.53}$$

This is a classic result called the *resonance condition*. It states that the only waves that can exist in the steady state within the cavity are those for which the cavity is an integral number of half wavelengths long. The wavelengths satisfying this result are said to be the *resonant wavelengths* of the structure. In fact, they are the wavelengths for which the interference between the forward and backward waves is constructive. At any point in the cavity, the two waves (satisfying the resonance condition) always have the same phase relationship with respect to each other. This is *constructive interference*. Fields at wavelengths not satisfying this condition interfere *destructively*. Their relative phase difference changes for each pass across the cavity. The result at any point in the cavity is the summation of a large number of randomly phased waves. The sum of such a sequence of fields is zero.

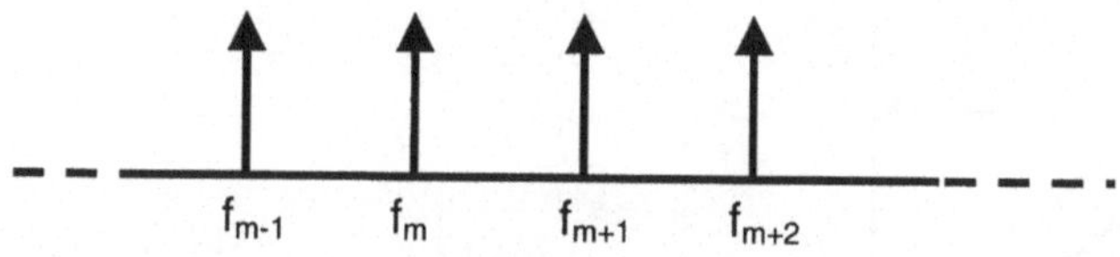

Figure 4.19 Cavity-resonant frequencies.

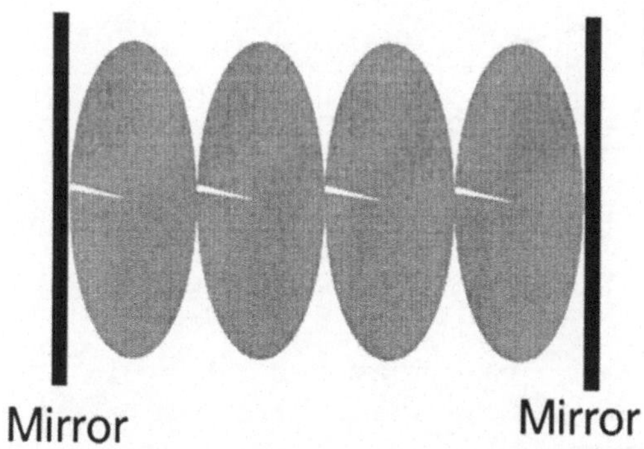

Figure 4.20 Envelope of the cavity standing-wave pattern.

The *resonant frequencies* corresponding to the resonant wavelengths in Eq. (4.53) are given by

$$f = \frac{mc}{2nL} \tag{4.54}$$

where n is the refractive index of the material filling the cavity. A picture of the cavity resonant frequencies appears in Fig. 4.19. The different frequencies that can exist within the cavity are the allowed modes of the cavity.

The total field in the cavity can be written in the simplified form

$$E = E_0 \sin kz \tag{4.55}$$

This represents a standing wave pattern within the cavity. The envelope of this wave is drawn in Fig. 4.20.

The cavity problem is significant for a number of reasons. One is that it is an easily solvable electromagnetic boundary value problem. The solution strategy is to write electric fields that are solutions to the wave equation and that can ultimately be made to satisfy the boundary conditions. In fact, it can be proven that if a field satisfies the wave equation and the boundary conditions of a structure, it is a valid solution. Many electromagnetic boundary value problems are more complex, but they are solved with the same basic strategy. The problems of interest for optical communication are the dielectric slab waveguide and the fiber waveguide. These structures will be considered in later sections of this chapter.

Another reason for studying the resonant cavity is that it is the structure of the laser diode. The amplifying semiconductor fills the cavity. The cavity provides the feedback necessary to produce oscillations. The laser output will be at wavelengths where there is an amplification resonance as in Fig. 4.21. The amplification is indicated on the figure by the dashed curve. The distinct output wavelengths are the *longitudinal modes* of the device. The spectral width of the laser diode is $\Delta\lambda$, as also shown on the figure.

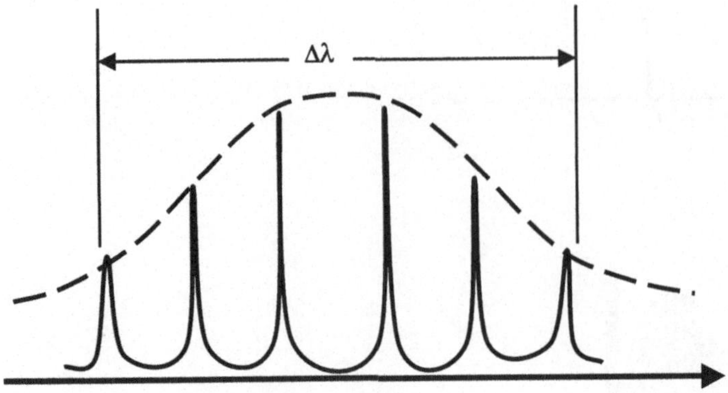

Figure 4.21 Laser diode output spectrum.

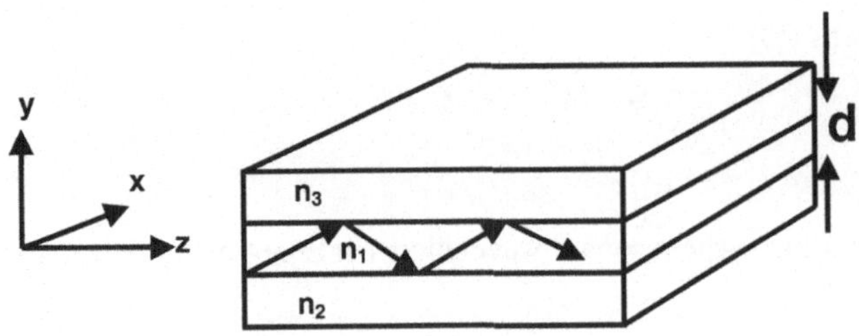

Figure 4.22 The dielectric slab waveguide.

4.3. INTEGRATED OPTICS

In this section we describe the fundamentals of integrated optics. Integrated optics is the technology of constructing optical components on substrates [14–18]. Components that have successfully utilized integrated optics include directional couplers, beam dividers, modulators, phase shifters, and switches.

The study of electromagnetic propagation in the integrated optic structure parallels that of propagation in the fiber. The analysis is simpler to do for integrated optics because of its rectangular geometry as compared to the circular geometry of the fiber structure. Despite the different geometries, the integrated optic analysis to follow tells us a great deal about propagation in the fiber.

4.3.1. Slab Waveguide

The slab waveguide is drawn in Fig. 4.22. It consists of three layers of dielectric materials, having refractive indices n_1, n_2, and n_3. The middle layer is the guiding region and has the largest index of refraction. From our earlier description of total internal reflection, it is apparent how this structure guides optical waves. Rays at, or beyond, the critical angle are reflected at the upper and lower boundaries and cannot escape the structure. The wave zigzags down the waveguide as indicated on the figure. As we expect from our earlier discussion of total internal reflection, evanescent fields in the upper and lower regions exist and will travel along with the wave propagating in the middle guiding layer.

When (as is often the case) the central layer thickness (d) is very small, this layer is referred to as a *film* (or a *thin film*). The critical angles at the lower and upper boundaries, respectively, are given by

$$\sin\theta_{c12} = \frac{n_2}{n_1} \tag{4.56}$$

$$\sin\theta_{c13} = \frac{n_3}{n_1} \tag{4.57}$$

For complete guiding, the ray angles must be equal to, or greater than, the largest of these two critical angles calculated above. Otherwise the wave would not undergo total reflection at one of the two boundaries, and optical energy would escape from the structure.

For integrated optic devices, the upper and lower materials are usually different. In fact, the upper region is commonly air ($n_3 = 1$). This type of structure is *asymmetrical*. If the upper and lower materials are the same ($n_3 = n_2$), the structure is *symmetrical*. The study of the symmetric waveguide most nearly parallels that of the circularly symmetric fiber. The electromagnetic solution of this structure follows the same strategy as that for the electromagnetic cavity: i.e., write equations for fields in the three regions that satisfy the wave equation and apply the boundary conditions. The details are more complicated however. We will sketch the solution, but without going through all the specifics.

As we did in the case of reflection at a plane boundary, we divide the problem into two possible linear polarizations. In Fig. 4.23, the yz plane is the plane of incidence. The perpendicular polarization (s) has the electric field pointing in the x direction. The electric field points in a direction perpendicular to the plane of incidence. We call this *transverse electric* (TE) polarization because the electric field always points transverse to the direction of net travel (the z direction).

The other possible polarization has the electric field parallel to the plane of incidence (p polarization) as indicated in Fig. 4.24. In this case, the magnetic field (which lies perpendicular to both the local direction of travel and the electric field) is polarized in the x direction. It is the magnetic field that now points transverse to the z direction. This is transverse magnetic (TM) polarization.

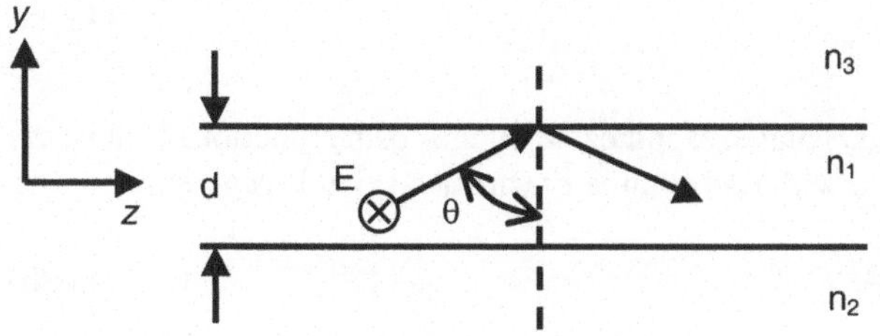

Figure 4.23 TE polarization in the slab waveguide.

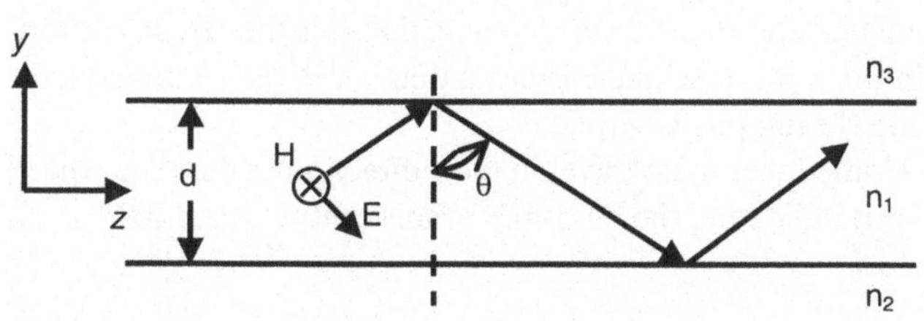

Figure 4.24 TM polarization in the slab waveguide.

4.3.2. TE Mode Chart

Consider a TE wave in a symmetrical waveguide. The electric field points in the x direction. The field in the central region is made up of the superposition of two plane waves, one moving upward at angle θ and one moving downward at that angle. The equation for the upward-traveling plane wave in region 1 is given by

$$E_+ = 0.5E_0 e^{j(\omega t - k_1 y \cos\theta - k_1 z \sin\theta)} \tag{4.58}$$

while the downward wave is given by

$$E_- = 0.5E_0 e^{j(\omega t + k_1 y \cos\theta - k_1 z \sin\theta)} \tag{4.59}$$

where k_1 is the propagation coefficient in the middle layer. The amplitudes of the two waves are the same due to total reflection at the boundaries, just as we found for the fields at the mirrors in the cavity problem.

The total field in the guiding region is the sum of the preceding two waves. Adding the two waves and simplifying yields

$$E_1 = E_0 e^{j\omega t} e^{-jk_1 z \sin\theta} \cos(k_1 y \cos\theta) \tag{4.60}$$

We further simplify by defining

$$h = k_1 \cos\theta \tag{4.61}$$

and

$$\beta = k_1 \sin\theta \tag{4.62}$$

yielding

$$E_1 = E_0 e^{j(\omega t - \beta z)} \cos hy \tag{4.63a}$$

This field is symmetrical in the transverse (xy) plane, as indicated by the cosine function. That is, the field pattern is an even function of y. A field with odd symmetry can also exist. It is given by

$$E_1 = E_0 e^{j(\omega t - \beta z)} \sin hy \tag{4.63b}$$

These equations represents a nonuniform plane wave traveling in the z direction. It is nonuniform because it varies in the transverse plane in the manner indicated by the cosine or sine terms. By comparing this result with the field for the uniform plane wave, we note that β is the effective propagation factor. The pattern of the wave in the transverse plane is a standing wave (as we might have expected) because it is made up of two interfering plane waves.

The field in the region above the middle layer must travel in the z direction at the same speed as the central field. It must also have a term indicating the decaying evanescent wave. A field satisfying this condition is

$$E_2 = A_2 e^{-\alpha(y - d/2)} e^{j(\omega t - \beta z)} \tag{4.64}$$

The last term indicates that the wave travels with respect to the z axis at the same speed (same propagation factor) as does the field in the central region. This field decays away from the central region with attenuation factor α. The field amplitude is A_2 at the boundary ($y = d/2$).

If we substitute this last equation into the wave equation, we find that α must be

$$\alpha = k_0\sqrt{n_1^2 \sin^2\theta - n_2^2} \tag{4.65}$$

just as we found earlier when considering the evanescent wave in the transmitted region for the problem of plane wave reflection at a plane boundary.

The electromagnetic problem now reduces to finding the relative amplitudes of the waves and the allowed values of β, the propagation factor. These are found by applying the boundary electromagnetic conditions, continuity of the tangential electric field. Because of the somewhat complicated structure, the z component of the magnetic fields must also be found and matched at the boundaries. The magnetic fields can be found from the electric fields since they are related by Faraday's law. The results of applying the boundary conditions are

$$A_2 = E_0 \cos\frac{hd}{2} \tag{4.66}$$

so that the evanescent field is now

$$E_2 = E_0 \cos\frac{hd}{2}\, e^{-\alpha(y-d/2)} e^{j(\omega t - \beta z)} \tag{4.67}$$

Note that E_1 and E_2 are now equal at the boundary. In addition, the boundary conditions yield

$$\tan\frac{hd}{2} = \frac{1}{n_1 \cos\theta}\sqrt{(n_1 \sin\theta)^2 - n_2^2} \tag{4.68}$$

This is a transcendental equation that must be solved graphically or numerically. It is called the *characteristic equation* or the *mode equation*. Solving it reveals the allowed propagation angles θ for a given central film thickness d.

An example plot of the mode equation appears in Fig. 4.25 for the case where $n_1 = 3.6$ and $n_2 = 3.55$. Note that there are multiple solutions because the tangent function repeats itself. That

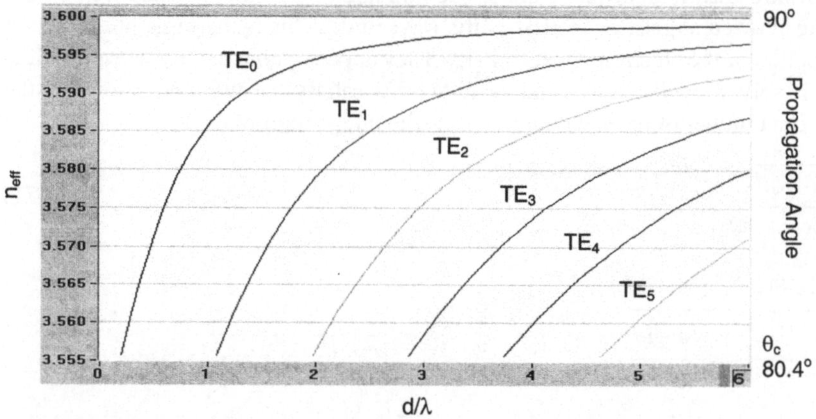

Figure 4.25 Mode chart for a symmetrical slab waveguide with $n_1 = 3.6$ and $n_2 = 3.55$.

is, for a given value of the right-hand side of the mode equation, there is an infinite set of solutions for $hd/2$.

The figure itself is called a *mode chart*. The horizontal axis represents the normalized film thickness d/λ. The vertical axis on the right refers to the allowed propagation angle θ. Because the critical angle for this structure is

$$\theta_c = \sin^{-1}\frac{n_2}{n_1} = \sin^{-1}\frac{3.55}{3.6} = 80.4^\circ$$

the range of angles allowing propagation is between the 80.4° and 90° as indicated on the mode chart. The critical angle is also called the *cutoff angle*, because rays that strike the interface at lesser angles cannot propagate (they are cut off).

The vertical axis on the left is the *effective index of refraction*, defined as

$$n_{\text{eff}} = n_1 \sin\theta \tag{4.69}$$

When the ray angle is 90° (an axial ray), $n_{\text{eff}} = n_1$ and when the angle is equal to the critical angle, $n_{\text{eff}} = n_2$. We see that the range of the effective index of refraction lies between the indices of the two waveguide materials. This is also indicated on the mode chart.

The modes refer to the different ray angles allowed for a fixed film thickness. For example, when $d/\lambda = 2$ the mode chart (Fig. 4.25) shows that three different propagation angles are allowed. The corresponding modes, effective indices of refraction, and ray angles are given in Table 4.4. As indicated, TE_0, TE_1, and TE_2 modes can all travel simultaneously. This represents a multimode waveguide. If the operating wavelength were 1.55 μm in this example, the film thickness would be 3.1 μm. The subscript m in the mode designation (TE_m) is called the *mode order*. The order of the lowest ordered mode is zero in the slab waveguide.

A single-mode waveguide will exist if the film is thin enough. For the structure in this example, if $d/\lambda < 0.836$, the only TE mode that can propagate is the TE_0 mode. The *cutoff condition* is

$$\frac{d}{\lambda} \le \frac{1}{2\sqrt{n_1^2 - n_2^2}} \tag{4.70}$$

For values of thickness that satisfy this inequality, the integrated waveguide is single mode, allowing propagation of only the $m = 0$ mode.

The reason that only specific ray angles are allowed has to do with the interference between the upward- and downward-traveling waves. Only for certain angles will the interference be constructive. These are the allowed angles, or modes, of the waveguide. This is the same phenomenon discussed in the description of the electromagnetic cavity. They appear automatically when the electromagnetic problem is solved explicitly by finding fields that satisfy the wave equation and the boundary conditions. The number of propagating modes is the integer part of

$$N = 1 + \frac{2d\sqrt{n_1^2 - n_2^2}}{\lambda} \tag{4.71}$$

Table 4.4 Modes in the Slab Waveguide

Mode	n_{eff}	Ray angle
TE_0	3.595	87
TE_1	3.58	84
TE_2	3.557	81

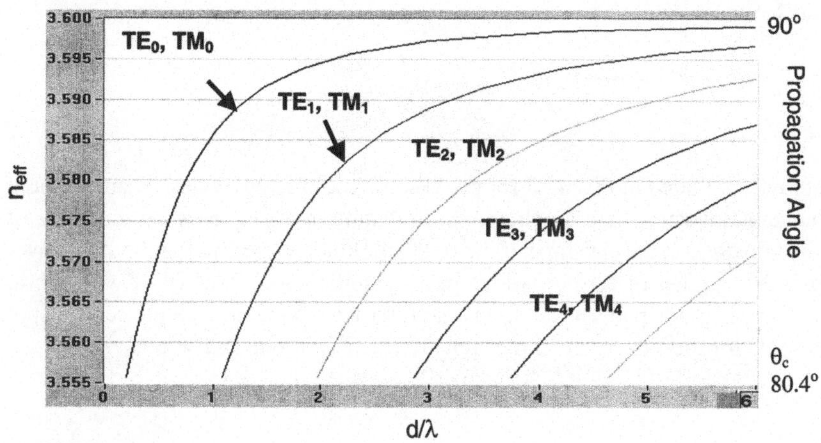

Figure 4.26 Mode chart showing both TE and TM modes.

The number of modes that can propagate decreases as the guiding layer thickness gets smaller and as the two indices of refraction get closer to each other.

As illustrated, a mode chart displays many of the propagation characteristics of a wave-guiding structure. A similar mode chart will be shown for the fiber-optic waveguide to be discussed later in this chapter.

4.3.3. TM Mode Chart

Next consider the mode chart for TM waves. The mode equation for this case is

$$\tan\frac{hd}{2} = \frac{n_1}{n_2^2\cos\theta}\sqrt{(n_1\sin\theta)^2 - n_2^2} \tag{4.72}$$

A mode chart that includes both TE and TM modes appears in Fig. 4.26 using the same values of refractive indices as in the previous example. Because the two indices are so close, both mode equations yield nearly identical results. That is why the TE and TM modes appear to be the same. If two or more modes share the same propagation characteristics, they are *degenerate*. If the two refractive indices were quite different, the TE and TM mode curves would separate from each other and the modes would no longer be degenerate.

In the degenerate case, a single-mode waveguide satisfying the cutoff condition in Eq. (4.70) actually sustains two modes. Both the TE_0 and TM_0 modes propagate, but with the same effective propagation factors. The total number of allowed modes (including both TE and TM) is twice that calculated for the TE case alone.

4.3.4. Mode Field Patterns

The light distribution in the transverse plane is the *transverse mode pattern*. These patterns are particularly important when designing components that connect to, or are built within, the integrated optic structure. The field distributions of all components must match closely to avoid losses.

For the TE mode in the symmetrical waveguide the transverse pattern is given within the film by the term

$$E_1 \propto E_0\cos hy \tag{4.73}$$

and outside the film by

$$E_2 \propto E_0 \cos\frac{hd}{2} e^{-\alpha(y-d/2)} \tag{4.74}$$

The standing wave patterns are plotted in Fig. 4.27 for the four lowest-ordered modes. At any point in the transverse plane the actual electric field amplitude is oscillating at a frequency on the order of 10^{14} Hz. The standing wave pattern is the envelope of the field amplitude variation. For the slab waveguide, the mode order is the number of zero crossings in the standing wave pattern. If we were to view projections of the various mode patterns using visible light we would see a single central spot for the TE_0 mode, two spots for the TE_1 mode, three spots for the TE_2 mode and so on.

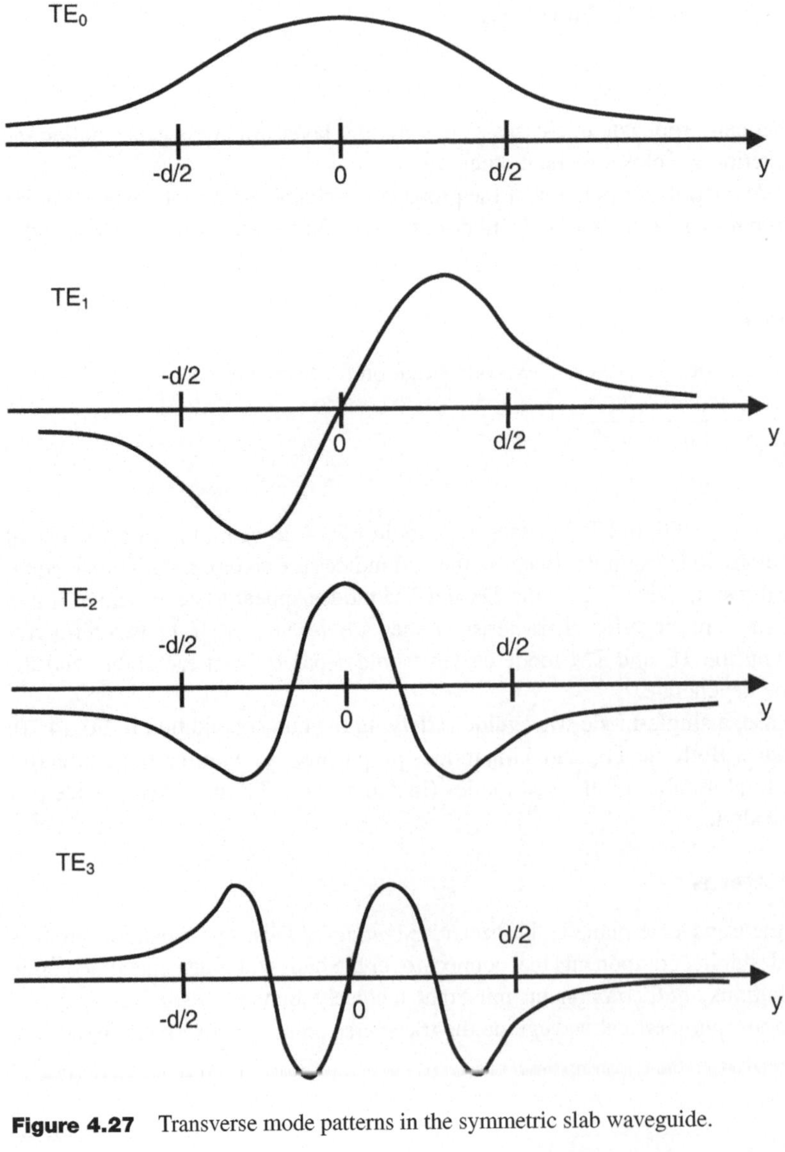

Figure 4.27 Transverse mode patterns in the symmetric slab waveguide.

The evanescent fields outside the central guiding layer are also indicated on the figure. As the mode order m increases, the attenuation factor α decreases and the wave penetrates further into the outer layers. Higher ordered modes travel with ray angles closer to the critical angle than do lower ordered modes. As pointed out in the discussion on reflection from a plane boundary, the attenuation factor decreases as the critical angle is approached accounting for the increased wave penetration.

We can now expand our mode definition. We have been saying that modes refer to the different propagation paths allowed in a waveguide. Now we can also say that modes refer to the different transverse field patterns that are possible in a waveguide.

4.3.5. Asymmetric Waveguide

The equations and the solution for the asymmetric waveguide are more complicated than that for the symmetric waveguide. Here we will simply show a mode chart indicate some of the features of propagation. A mode chart appears in Fig. 4.28 for the case of a zinc sulfide (ZnS) film deposited onto a glass substrate. Air covers the space above the film. Thus, $n_1 = 2.29$, $n_2 = 1.5$, and $n_3 = 0$. The critical angle at the air–ZnS interface is 25.9° and at the glass–ZnS interface it is 41°. Propagation only occurs when the ray angles are greater than the largest of these two values, 41°; otherwise, light will leak from the ZnS film into the glass substrate. Thus, the range of propagating angles is from 41° to 90°. The corresponding range of effective refractive indices (recall that $n_{\text{eff}} = n_1 \sin\theta$) is from n_1 to n_2 (that is, 1.5 to 2.29).

The mode chart shows both TE and TM modes. Because the three indices of refraction are not close, the modes are not degenerate. The TE and TM modes are clearly separate. As found from the mode chart, truly single-mode propagation exists if $d/\lambda < 0.12$ for this structure. This represents the cutoff condition for the TM_0 mode. Only the TE_0 mode can propagate in this case.

The mode patterns are similar to those of the symmetric waveguide except for the lack of symmetry. This is indicated in Fig. 4.29 for a few lower order modes.

4.3.6. Modal Distortion

Earlier in this chapter material dispersion was described as the cause of pulse spreading, ultimately limiting the information capacity of the transmission line. Another cause of pulse spreading can now be described. As illustrated by the mode chart, the different modes travel with different effective indices of refraction and thus with different velocities with respect to the waveguide axis. An input pulse will distribute its energy among all the allowed modes. Because of the different mode velocities, parts of the wave arrive ahead of (or behind) other parts. The result is that the pulse at the receiver is

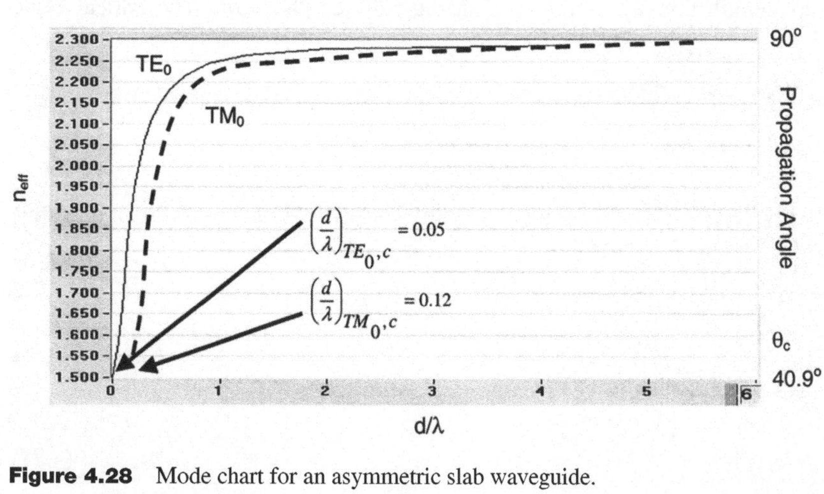

Figure 4.28 Mode chart for an asymmetric slab waveguide.

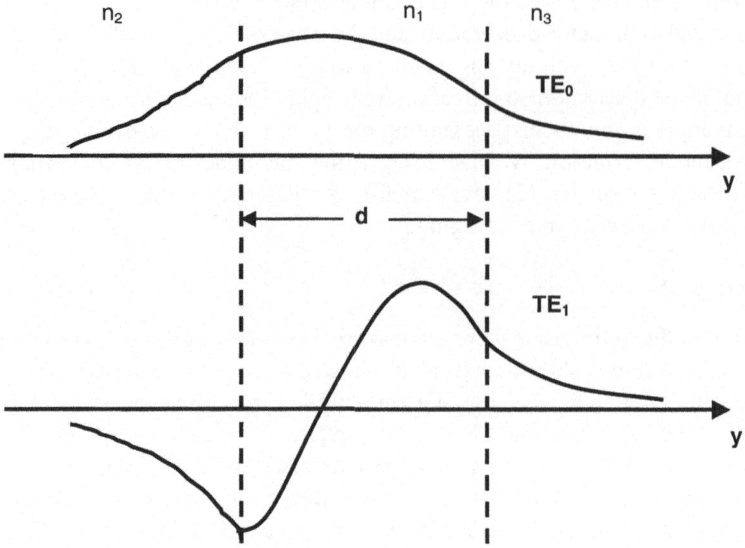

Figure 4.29 Transverse mode patterns in the asymmetric slab waveguide.

wider than that originally transmitted. Once again we have pulse spreading. This is called *modal distortion* or *modal dispersion*. The limitations on bandwidth and data rate previously given apply regardless of the cause of the spreading.

A simple analysis allows calculation of the amount of spreading. The earliest arriving pulse will be that of the lowest order axial ray. Energy in this mode travels straight down the transmission line, a distance L. The last arriving pulse will be that of the highest order mode, traveling at the critical angle. A little geometry shows that this pulse travels a distance Ln_1/n_2. The difference in time of arrival between the fastest and slowest modes will be the pulse spread.

The axial ray travel time will be L/v, where $v = c/n_1$. Thus, for the axial ray

$$t = \frac{Ln_1}{c} \tag{4.75}$$

For the critical angle ray, length L is replaced by the zigzag path length Ln_1/n_2. The critical angle travel time is then

$$t = \frac{Ln_1^2}{cn_2} \tag{4.76}$$

The difference between the two arrival times is the pulse spread. Subtracting and simplifying yields

$$\Delta\left(\frac{\tau}{L}\right) = \frac{n_1(n_1 - n_2)}{cn_2} \tag{4.77}$$

If we define the *fractional refractive index* change as

$$\Delta = \frac{n_1 - n_2}{n_1} \tag{4.78}$$

then the modal pulse spread can be expressed as

$$\Delta\left(\frac{\tau}{L}\right) = \frac{n_1\Delta}{c} \tag{4.79}$$

To generalize, modal distortion can be minimized by designing a waveguide with materials having refractive indices which are close to each other. Notice that modal distortion is independent of the operating wavelength and it is independent of the spectral width of the light source. This is unlike material dispersion, which is highly dependent on wavelength and spectral width.

4.4. FIBER OPTICS

In this section we describe several types of optical fibers and their properties [19–25]. Coverage includes step-index and graded-index fibers and single-mode and multimode fibers. Properties of interest are the modes, attenuation, pulse distortion, and bandwidth limitations.

4.4.1. Step-Index Fiber

The *step-index* (SI) fiber (Fig. 4.30) consists of a central core having radius a and refractive index n_1, surrounded by a cladding having refractive index n_2. In order to have total internal reflection, the core index must be greater than that of the cladding. For analytical purposes, it is convenient to assume that the cladding is infinitely thick. This removes any problems associated with the outer boundary of the cladding. As we already know, there is a decaying evanescent field associated with total internal reflection. This field decays rapidly so that we might expect that effects of a finite cladding thickness are negligible. That is, the field at the outer edges of the cladding are so small there is no chance of interaction with any material placed around the cladding itself. Therefore, the infinite cladding assumption is reasonable.

Propagation in the step-index fiber is very much like propagation within the slab waveguide. Rays zigzag down the core, contained because of total internal reflection. Only discrete modes are allowed because of the requirement for constructive interference between the waves bouncing back and forth off the core-cladding interface. The fiber can transmit many modes if the core is large enough or can restrict transmission to a single mode if the core is small enough. A mode chart describes many of the propagation properties of the fiber. Material dispersion and modal distortion cause pulse spreading, affecting the fiber's ability to transmit unlimited bandwidths and data rates.

These general attributes are known (or at least expected) from the close analogy with the symmetrical slab waveguide. They also become evident from the electro-magnetic solution of this boundary value problem. Unfortunately, the analysis is complicated by the circular symmetry of the fiber requiring the use of cylindrical coordinates. While the solution to the wave equation in rectangular

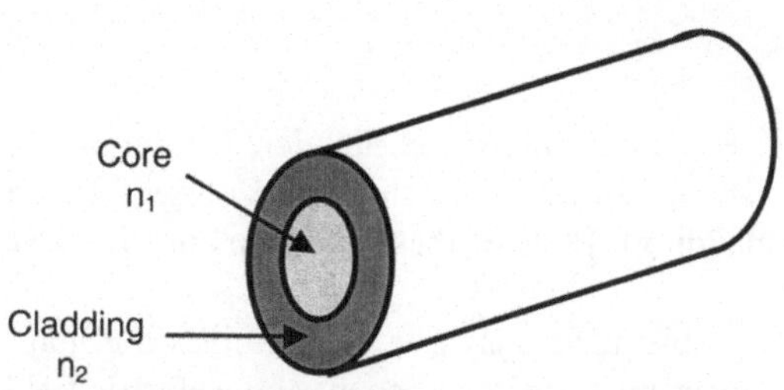

Figure 4.30 Step-index fiber.

coordinates consists of relatively simple trigonometric functions (sines, cosines, and exponentials), in cylindrical coordinates the solutions are Bessel functions.

The solution strategy is the same as described for the electromagnetic cavity and the slab waveguide. Functions for the electric field in the core and in the cladding that satisfy the wave equation in cylindrical coordinates are found. The boundary conditions are then applied. This leads to a characteristic equation from which the mode chart can be constructed.

A linearly polarized (LP) electric field pointing in the y direction can be written as

$$E_y = E_1 J_\ell\left(\frac{ur}{a}\right)(\cos \ell\phi)e^{j(\omega t-\beta z)} \tag{4.80}$$

in the core, and as

$$E_y = E_2 K_\ell\left(\frac{wr}{a}\right)(\cos \ell\phi)e^{j(\omega t-\beta z)} \tag{4.81}$$

in the cladding. In these equations J_ℓ is the Bessel function of the first kind of order ℓ, while K_ℓ is the modified Bessel function of the second kind of order ℓ. For simplicity, we have assumed a lossless transmission fiber. Otherwise, an exponential decay term (of the form $e^{-\alpha z}$) would need to be added to the electric field equations.

The terms u and w are given by

$$u = a\sqrt{n_1^2 k_0^2 - \beta^2} \tag{4.82a}$$

$$w = a\sqrt{\beta^2 - n_2^2 k_0^2} \tag{4.82b}$$

In these equations, k_0 is the propagation factor in free space.

We will be using a term called the *normalized frequency*, derived from u and w, which is given by

$$V = \sqrt{u^2 + w^2} \tag{4.83}$$

This can be rewritten as

$$V = \frac{2\pi a}{\lambda}\sqrt{n_1^2 - n_2^2} \tag{4.84}$$

It is sometimes simply called the *V parameter*.

The Bessel functions are solutions to the wave equation. The Bessel function J_ℓ resembles a sinusoid (appropriate for the field within the core). The modified Bessel function K_ℓ resembles an exponential decay (appropriate for the field in the cladding). Plots of these functions in Fig. 4.31 illustrate this behavior.

The following steps are taken in solving this problem. Faraday's law determines the magnetic fields from the electric fields, and Ampere's law determines the z component of the electric field from the magnetic field. Applying the boundary conditions to these fields yields the characteristic

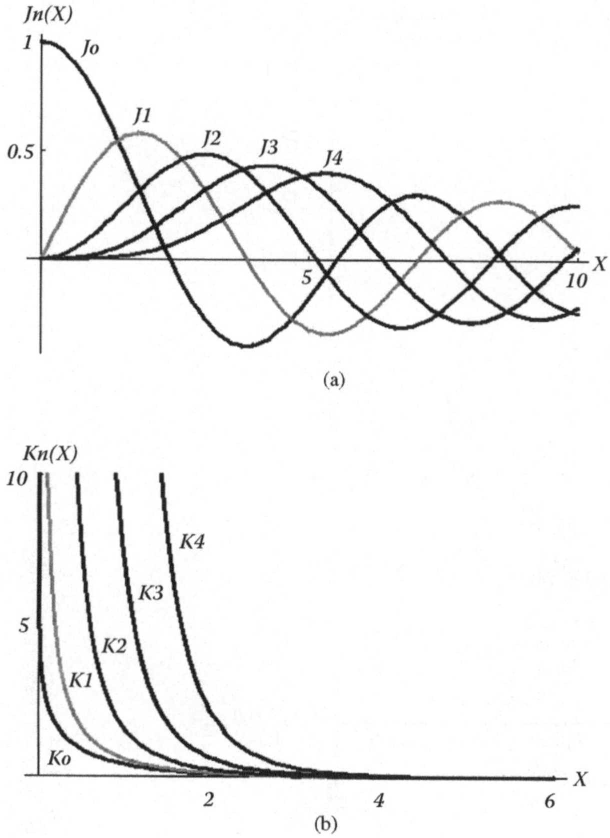

Figure 4.31 Bessel functions.

equation for the linearly polarized (LP) modes

$$u \frac{J_{\ell-1}(u)}{J_{\ell}(u)} = w \frac{K_{\ell-1}(w)}{K_{\ell}(w)} \tag{4.85}$$

Solving this equation yields the mode chart for the step-index waveguide. A few of the lowest-ordered modes are plotted in Fig. 4.32. If the V parameter of the fiber is known, the mode chart reveals the parameter $b = w^2/V^2$. From this, the propagation factor β can be determined from Eq. (4.82), the corresponding propagation angle from Eq. (4.62), and the corresponding effective refractive index from Eq. (4.69).

Several approximations were made in deriving the characteristic equation for the $LP_{\ell m}$ modes. A more exact (and more complicated) approach yields the exact mode chart plotted in Fig. 4.33. Comparison shows that the LP approximation is reasonable and yields good results. The equivalent LP modes are indicated on the exact mode chart. We will continue the discussion with reference to the exact solution. Notice that if $V < 2.405$, only the lowest-ordered mode (HE_{11}) will propagate. This is the condition for design of a single-mode fiber. For larger values of V more than one mode propagates and we have a multimode fiber.

The single-mode fiber has the great advantage of eliminating modal distortion and, consequently, increasing the information capacity (rate length and frequency length). Almost all long-distance optical links use single-mode fibers because of the high capacities and lengths required. Multimode fibers

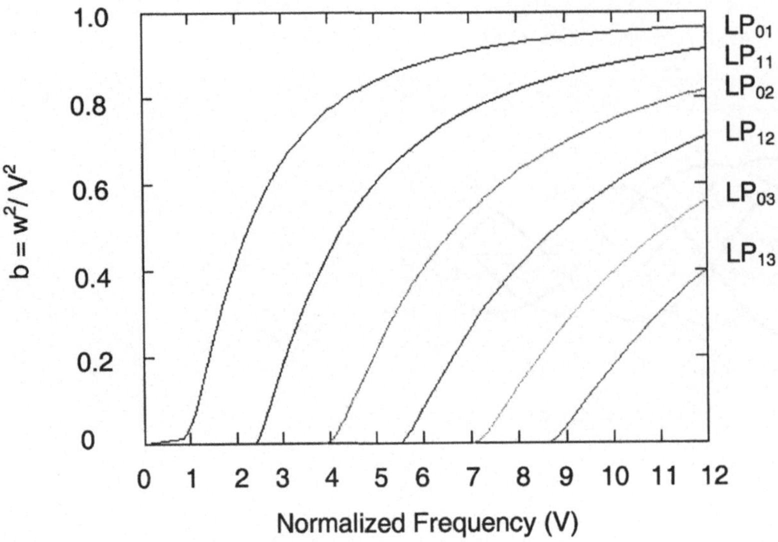

Figure 4.32 $LP_{\ell m}$ mode chart for the step-index fiber.

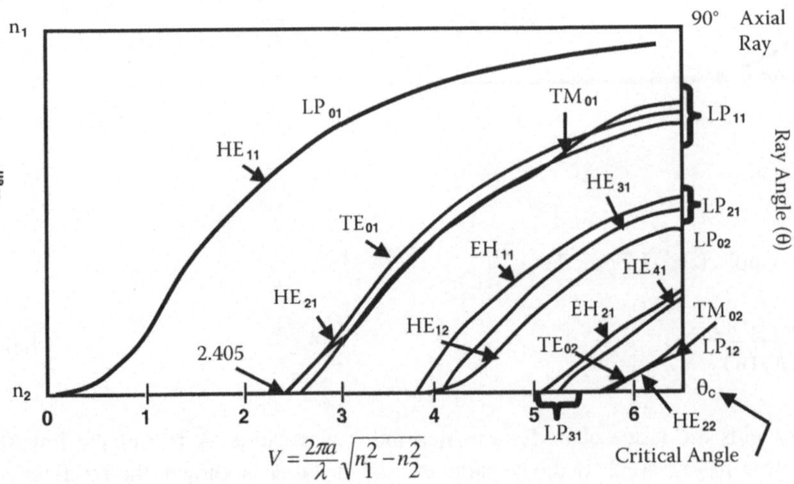

Figure 4.33 Exact mode chart for the step-index fiber.

are sufficient for shorter paths, such as used in LANs. The transverse pattern of the HE_{11} mode is nearly gaussian. It can be written as

$$E_y = E_0 e^{-(r/w)^2} \tag{4.86}$$

where the spot size w is given by

$$\frac{w}{a} = 0.65 + 1.619V^{-3/2} + 2.879V^{-6} \tag{4.87}$$

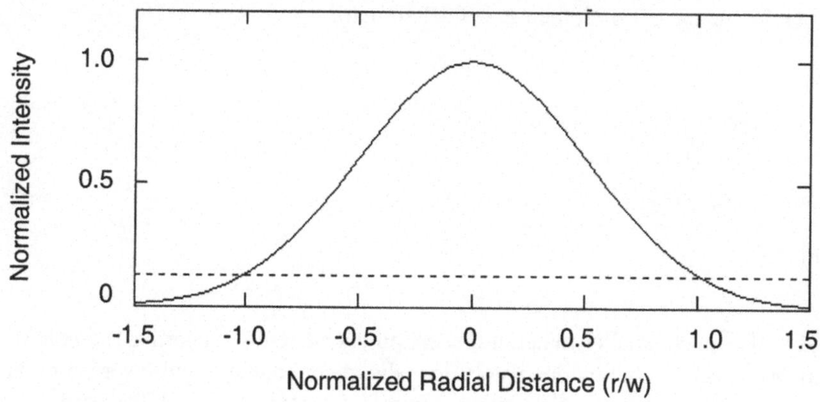

Figure 4.34 HE_{11} mode gaussian intensity distribution.

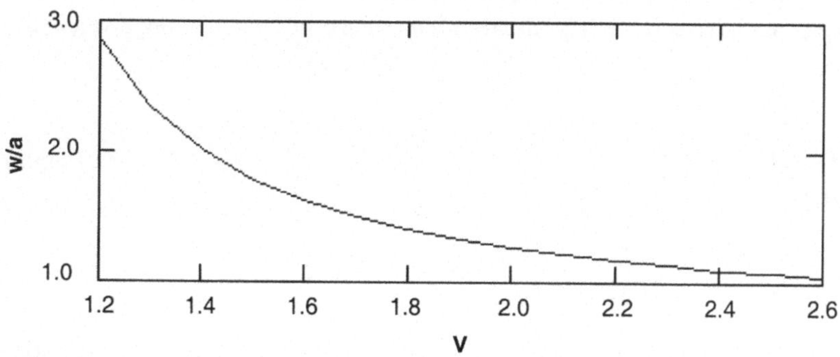

Figure 4.35 Spot size variation with the normalized frequency.

Because the intensity is the square of the electric field, it can be written as

$$I = I_0 e^{-2(r/w)^2} \tag{4.88}$$

A plot of the gaussian beam pattern appears in Fig. 4.34. The spot size variation with normalized frequency appears in Fig. 4.35. When V is close to 2.405, the spot size is only about 10% larger than the core radius. This implies that the energy in the beam is tightly bound to the core of the fiber. For smaller values of V, the spot size increases. This is undesirable as the energy is no longer tightly bound to the core. In this situation, energy can penetrate deeply into the cladding at bends in the fiber, eventually radiating out the sides. Best operation of a single-mode fiber has V in the range from 2.0 to 2.2.

Modal distortion in the multimode step-index fiber can be treated in exactly the same way as was done for the slab waveguide. The total pulse spread in a highly multimode fiber is determined by calculating the difference in arrival times between an axial ray and one traveling at the critical angle. As before, the result is

$$\Delta\left(\frac{\tau}{L}\right) = \frac{n_1 \Delta}{c} \tag{4.89}$$

As an example, if $n_1 = 1.48$ and $n_2 = 1.465$, then $\Delta = 0.01$ so that

$$\Delta \frac{\tau}{L} = 5 \times 10^{-11} \text{ s/m}$$

or

$$\Delta \frac{\tau}{L} = 50 \text{ ns/km}$$

This may be compared to the much smaller material dispersion calculated in an earlier example of 400 ps/km = 0.4 ns/km. We conclude that modal distortion is the major source of pulse spreading in a step-index multimode fiber. That is, modal distortion is much greater than material dispersion.

4.4.2. Graded-Index Fiber

The graded-index (GRIN) fiber (Fig. 4.36) was developed to overcome the large modal distortion in a multimode step-index fiber. It has a refractive index variation across the core and cladding given in the core by

$$n(r) = n_1 \sqrt{1 - 2\left(\frac{r}{a}\right)^{\alpha} \Delta} \tag{4.90a}$$

and in the cladding by

$$n(r) = n_1 \sqrt{1 - 2\Delta} = n_2 \tag{4.90b}$$

If $\alpha = 2$, the refractive index can be reduced to

$$n(r) = n_1 \left[1 - \left(\frac{r}{a}\right)^2 \Delta\right] \tag{4.91a}$$

in the core and

$$n(r) = n_2 = n_1(1 - \Delta) \tag{4.91b}$$

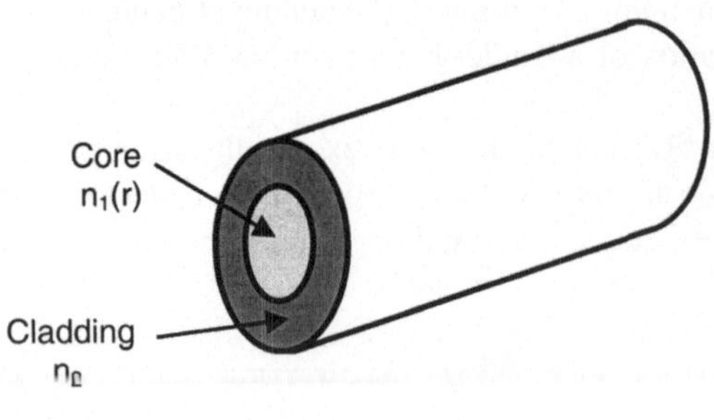

Figure 4.36 Graded-index fiber.

in the cladding. This refractive index distribution is called the *parabolic profile*.

For the parabolic profile, the ray paths in the core are given by

$$r(z) = r_0 \cos(\sqrt{A}z) + \frac{1}{\sqrt{A}} r_0' \sin(\sqrt{A}z) \tag{4.92}$$

where r_0 is the initial ray position (at $z = 0$), r_0' is the initial slope and $A = 2\Delta/a^2$, a property of the GRIN fiber. The ray slopes are given by

$$r'(z) = -\sqrt{A} r_0 \sin(\sqrt{A}z) + r_0' \cos(\sqrt{A}z) \tag{4.93}$$

Several ray paths are illustrated in Fig. 4.37.

Modal distortion still exists in the multimode GRIN fiber because of the different path lengths traversed by the various rays. As with the multimode SI fiber, we can calculate the difference in arrival times between pulses traveling axially (the shortest route) and pulses whose trajectories approach the core-cladding boundary (the longest route). This represents the amount of pulse spreading. Note that in the GRIN fiber case, the index of refraction decreases as the ray moves away from the fiber's axis. Therefore (because $v = c/n$), rays speed up as they move away from the fiber's axis. In doing so they tend to catch up with the axial rays, diminishing the amount of pulse spreading. This is the great advantage of the multimode GRIN fiber. An approximate expression for the pulse spread in a GRIN fiber is

$$\Delta\left(\frac{\tau}{L}\right) = \frac{n_1 \Delta^2}{2c} \tag{4.94}$$

Comparison with the results for the SI fiber shows a reduction in the pulse spread by a factor of $2/\Delta$. As an example, if $n_1 = 1.48$ and $n_2 = 1.465$, then $\Delta = 0.01$ so that

$$\Delta\left(\frac{\tau}{L}\right) = 2.53 \times 10^{-13}\ \text{s/m}$$

or

$$\Delta\left(\frac{\tau}{L}\right) = 0.253\ \text{ns/km}$$

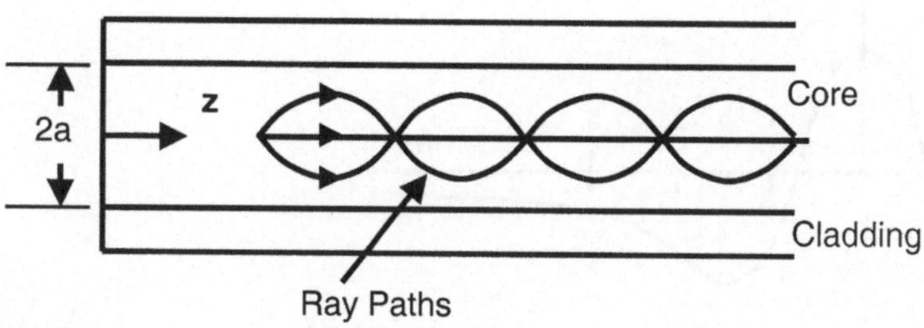

Figure 4.37 Ray paths in a graded-index fiber.

The reduction in pulse spread (and resultant increase in fiber capacity) is close to a factor of 200.

The multimode GRIN fiber is used when path lengths are moderate (such as in LAN applications). Path lengths up to a few kilometers and information rates of a few Gb/s can be accommodated. Longer paths and higher rates require single-mode fibers, where modal distortion is no longer a factor.

For the parabolic GRIN fiber the wave equation can be solved explicitly, without the need for numerical solutions of a characteristic equation. Some of the results follow.

The effective index of refraction for a mode described by positive integers p and q is

$$n_{\text{eff}} = \frac{\beta_{pq}}{k_0} = n_1 - (p+q+1)\frac{\sqrt{2\Delta}}{k_0 a} \tag{4.95}$$

The factors k_0 and β have the same meaning as before.

The lowest ordered mode has $p = q = 0$. Its electric field is

$$E_{00} = E_0 e^{-\alpha^2 r^2/2} e^{j(\omega t - \beta z)} \tag{4.96}$$

where $\alpha = (k_0 n_1/a)^{1/2}(2\Delta)^{1/4}$. The corresponding transverse field plot appears in Fig. 4.38. It is circularly symmetric and gaussian shaped. For simplicity, we have assumed a lossless fiber in writing the filed equations.

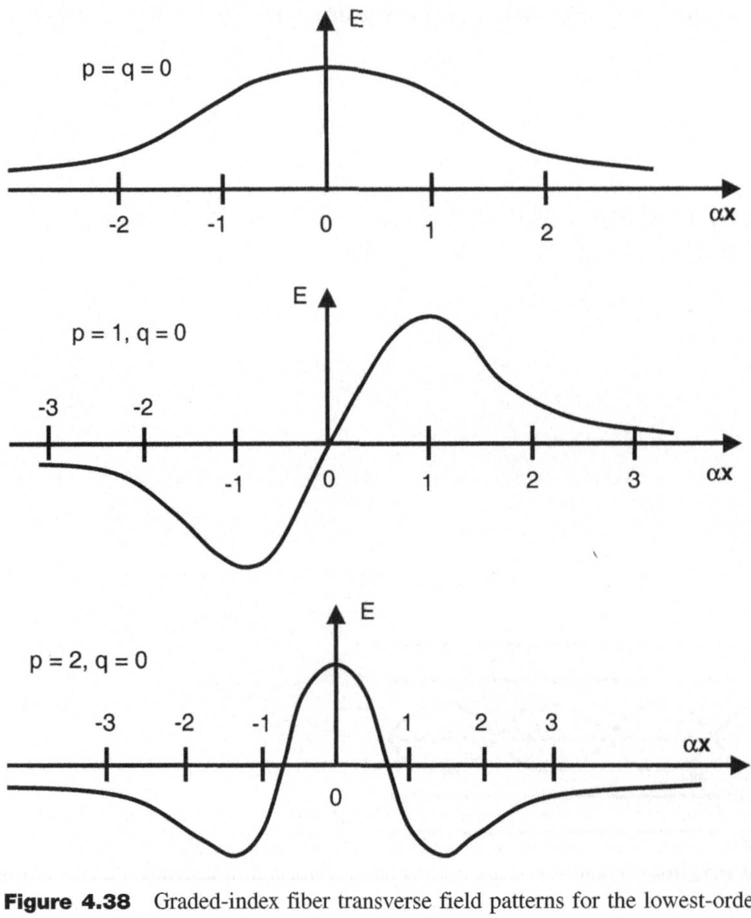

Figure 4.38 Graded-index fiber transverse field patterns for the lowest-ordered ($p = 0$, $q = 0$) mode, the $p = 1$, $q = 0$ mode, and the $p = 2$, $q = 0$ mode.

The $p = 1, q = 0$ and $p = 2, q = 0$ are given, respectively, by

$$E_{10} = E_1 \alpha x e^{-\alpha^2 r^2/2} e^{j(\omega t - \beta z)} \tag{4.97}$$

and

$$E_{20} = E_2[2(\alpha x)^2 - 1] e^{-\alpha^2 r^2/2} e^{j(\omega t - \beta z)} \tag{4.98}$$

These modes are plotted in Fig. 4.38. They are not circularly symmetric nor are they gaussian, although they have a gaussian envelope as indicated by the term $e^{-\alpha^2 r^2/2}$. The mode index p gives the number of zero crossings of the field pattern along the x direction. The mode index q does the same with respect to the y direction.

4.4.3. Attenuation

When we include loss in the equations for the electric fields in fibers, we do so by adding a term of the form $e^{-\alpha z}$, where α is the attenuation coefficient. Typically, fiber attenuation is given in dB/km rather than in terms of the attenuation coefficient. As mentioned earlier, they are related by

$$\text{dB/km} = -8.685\alpha \tag{4.99}$$

where the units of the attenuation coefficient are km^{-1}. For convenience, the minus sign is often omitted when writing the fiber loss in decibels.

A plot of the fiber loss for a silica glass appears in Fig. 4.39. The wavelength regions where fiber systems have been constructed are in the regions around 800–900 nm and from 1250 to 1650 nm. The region around 1400 nm where there is a local increase in attenuation is usually avoided. The nomenclature for the longer wavelength region is presented in Table 4.5.

The local peak in the loss curves near 1380 nm is caused by absorption in the hydroxyl ions (OH) present. This is an impurity whose concentration is minimized during the manufacturing process. The lowest loss region is near 1550 nm. This is where the longest fiber systems are designed

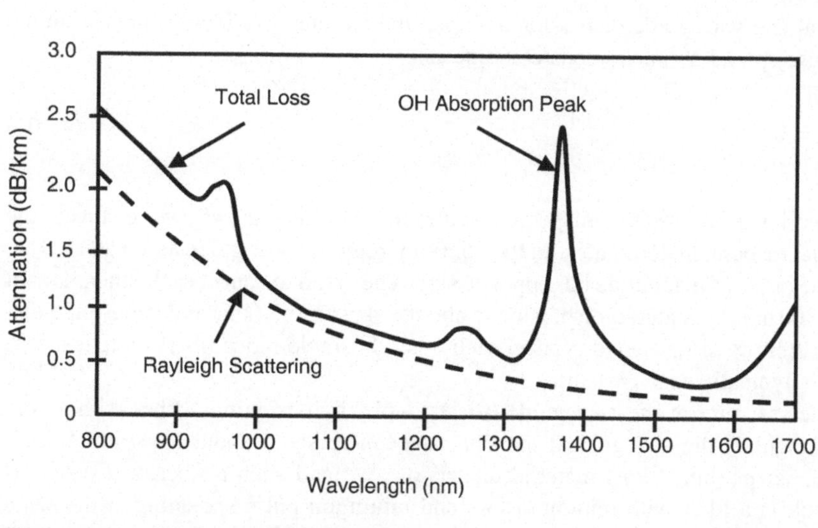

Figure 4.39 Silica glass fiber attenuation.

Table 4.5 Transmission Bands in the Long-wavelength Region

Nomenclature	Descriptor	Range (nm)
O band	Original	1260–1360
E band	Extended	1360–1460
S band	Short wavelength	1460–1530
C band	Conventional	1530–1565
L band	Long wavelength	1565–1625
U band	Ultra-long wavelength	1625–1675

to operate. Shorter paths (on the order of a few hundred meters) can be served in the 800-nm region. Moderately long systems, having path lengths up to a few kilometers, can be served by 1300-nm systems. Recall that this wavelength is advantageous because of the low material dispersion.

4.4.4. Waveguide Dispersion and Polarization-Mode Dispersion

Earlier we presented the concepts behind pulse spreading caused by modal distortion and material dispersion. There are two other major pulse spreading mechanisms, *waveguide dispersion* and *polarization-mode dispersion* (PMD).

Waveguide dispersion arises because different wavelengths travel at different speeds, even if traveling in the same mode. To illustrate this statement refer to the exact mode chart in Fig. 4.33 and consider only the HE_{11} mode. Because the source emits light over a range of wavelengths, the V parameter has a range of values associated with it resulting in a corresponding range of effective refractive indices (and related range of velocities). Just as occurs with material dispersion, component wavelengths travel at a different speeds, each arriving with a slight delay with respect to the others. The amount of pulse spreading is given by an equation very similar to that for material dispersion

$$\Delta\left(\frac{\tau}{L}\right) = -M_g \Delta\lambda \tag{4.100}$$

where $\Delta\lambda$ is the source width and M_g is the *waveguide dispersion*. For the SI fiber the waveguide dispersion looks as in Fig. 4.40.

Because material and waveguide dispersion act upon the various wavelengths present in the same way, the two pulse-spreading phenomena combine as

$$\Delta\left(\frac{\tau}{L}\right) = -(M + M_g)\Delta\lambda \tag{4.101}$$

By comparing the material and waveguide dispersion values in Figs. 4.8 and 4.40, we see that in the 800-nm region material dispersion dominates. In the 1550-nm region, wave-guide and material dispersion are of the same order of magnitude but opposite sign. They tend to cancel each other, but not entirely. Just above 1300 nm the material dispersion is about -4 ps/(nm $\times$ km) and waveguide dispersion is about -4 ps/(nm $\times$ km). They do cancel each other. In single-mode fibers there is a zero dispersion wavelength, typically near 1310 nm.

By changing the structure of the waveguide (for example, by designing a fiber with a core index profile that is triangular), the waveguide dispersion can be increased to about 20 ps/(nm $\times$ km). This just cancels the -20 ps/(nm $\times$ km) material dispersion. We call such a fiber, a *dispersion-shifted fiber*. The result is a fiber with minimum loss and minimum pulse spreading at the same wavelength, a very desirable fiber characteristic for high-rate long-path links.

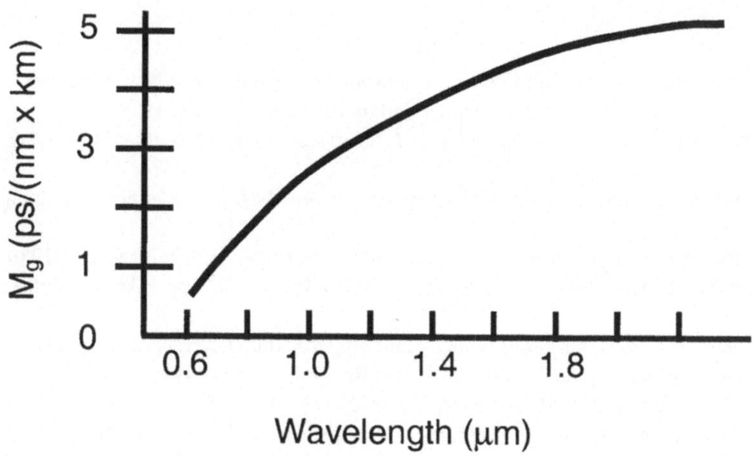

Figure 4.40 Waveguide dispersion in a step-index fiber.

Other index profiles are available that result in other desirable characteristics. A particularly useful one property is a uniform low dispersion over a range of wavelengths. For example, a dispersion of 5 ps/(nm × km) over wavelengths from 1500 to 1600 nm. This is needed when the fiber supports a number of independent carriers in a scheme called *wavelength-division multiplexing*. Tens, and even hundreds, of independent channels can be transmitted simultaneously in this manner. The analytical solution for the fields in waveguides having unusual refractive index profiles can be quite complicated, well beyond the range of what is covered in this chapter.

We have indicated how material and waveguide dispersion add together. If we include modal distortion as well, the total pulse spread is

$$\Delta\left(\frac{\tau}{L}\right) = \sqrt{(\Delta\tau)^2_{\text{modal}} + (\Delta\tau)^2_{\text{dispersive}}} \tag{4.102}$$

The modal spread $(\Delta\tau)_{\text{modal}}$ disappears for a single mode fiber. The dispersive spread $(\Delta\tau)_{\text{dispersive}}$ includes both waveguide and material dispersion.

A final phenomenon causing pulse spreading is polarization-mode dispersion (PMD). PMD occurs in a single-mode fiber because two orthogonally polarized fields can exist simultaneously. For example, we wrote the fields in the step-index fiber for a y-polarized electric field. An x-polarized field can also propagate. Therefore, even in what is called a single-mode fiber ($V < 2.405$), two fields can propagate. In most fibers these two fields will travel at slightly different velocities due to birefringence. *Birefringence* refers to having the index of refraction depend upon the field polarization. Most fibers are birefringent. The birefringence could be caused by an elliptical core (rather than a perfectly circular core). It could also be caused by unequal stresses in the two orthogonal transverse directions occurring during manufacture.

4.5. FURTHER STUDY

This chapter detailed several electromagnetic problems relating to fiber-optic communications. Several books are suggested [26–33] for further study relating to fiber networks and components.

REFERENCES

1. Chaffee, C.D. The rewiring of America. *The Fiber Optics Revolution*; Academic Press: New York, 1987.
2. Hecht, J. *Understanding Fiber Optics*; Prentice-Hall: Upper Saddle River, New Jersey, 1999.
3. Agrawal, G.P.; Dutta, N.K. *Long-wavelength Semiconductor Lasers*; Van Nostrand Reinhold Company: New York, 1986.
4. Kressel, H.; Butler, J.K. (Eds.) *Semiconductor Lasers and Heterojunction LEDs*; Academic Press: New York, 1977.
5. Kressel, H. (Ed.) *Semiconductor Devices for Optical Communications*; Springer-Verlag: New York, 1980.
6. Morthier, G.; Vankwikelberge, P. *Handbook of Distributed Feedback Laser Diodes*; Artech House: Norwood, MA, 1997.
7. Palais, J.C. *Fiber Optic Communications*, 4th Ed.; Prentice-Hall: Upper Saddle River, New Jersey, 1998.
8. Chang, D.K. *Field and Wave Electromagnetics*; Addison Wesley: Reading, Massachusetts, 1983.
9. Born, M.; Wolf, E. *Principles of Optics*, 3rd Ed.; Pergamon Press: New York, 1965.
10. Palais, J.C. *Fiber Optic Communications*, 4th Ed.; Prentice-Hall: Upper Saddle River, New Jersey, 1998.
11. Morris, D.J. *Pulse Code Formats for Fiber Optical Data Communications*; Marcel Dekker: New York, 1983.
12. Chang, D.K. *Field and Wave Electromagnetics*; Addison Wesley: Reading, MA, 1983.
13. Born, M.; Wolf, E. *Principles of Optics*, 3rd Ed.; Pergamon Press: New York, 1965.
14. Coldren, L.A.; Corzine, S.W. *Diode Lasers and Photonic Integrated Circuits*; Wiley: New York, 1995.
15. Ebeling, K.J. *Integrated Optoelectronics*; Springer-Verlag Publishing Co.: New York, 1993.
16. Hunsberger, R.G. *Integrated Optics: Theory and Technology*, 4th Ed.; Springer-Verlag: New York, 1995.
17. März, R. *Integrated Optics: Design and Modeling*; Artech House: Norwood, MA, 1995.
18. Murphy, E.J. *Integrated Optical Circuits and Components, Design and Applications*; Marcel Dekker: New York, 1999.
19. Cherin, A.H. *An Introduction to Optical Fibers*; McGraw-Hill, Inc.: New York, 1983.
20. Ghatak, A.K.; Thyagarajan, K. *Introduction to Fiber Optics*; Cambridge University Press: New York, 1998.
21. Jeunhomme, L.B. *Single-Mode Fiber Optics*, 2nd Ed.; Marcel Dekker: New York, 1990.
22. Marcuse, D. *Theory of Dielectric Optical Waveguides*, 2nd Ed.; Academic Press: New York, 1991.
23. Okamoto, K. *Fundamentals of Optical Waveguides*; Academic Press: New York, 2000.
24. Sodha, M.S.; Ghatak, A.K. *Inhomogeneous Optical Waveguides*; Plenum Press: New York, 1977.
25. Yariv, A. *Introduction to Optical Electronics*; 4th Ed.; Holt, Rinehart and Winston: New York, 1991.
26. Agrawal, G.P. *Fiber-Optic Communication Systems*, 2nd Ed.; John Wiley and Sons: New York, 1997.
27. DeCusatis, C.; Clement, D.; Maass Eric; Lasky, R. *Handbook of Fiber Optic Data Communication*; Academic Press Inc.: New York, 1997.
28. Goff, D.R. *Fiber Optic Reference Guide*; Focal Press: Boston, 1996.
29. Green, P.E. *Fiber Optic Networks*; Prentice-Hall: Englewood Cliffs, New Jersey, 1993.
30. Keiser, G.E. *Optical Fiber Communications*, 3rd Ed.; McGraw-Hill: New York, 2000.
31. Ramaswami, R.; Sivarajan, K.N. *Optical Networks: A Practical Perspective*, 2nd Ed.; Morgan Kaufmann: New York, 2002.
32. Weik, M.H. *Fiber Optics Standard Dictionary*, 3rd Ed.; Chapman & Hall: New York, 1997.
33. Bass, M. (Ed.) *Fiber Optics Handbook*; McGraw Hill: New York, 2002.

5

Numerical Techniques

Randy L. Haupt
Utah State University
Logan, Utah

5.1. INTRODUCTION

Numerical techniques for calculating electromagnetic fields surpassed analytical techniques many years ago. Analytical methods work for only a few basic geometries that do not apply to most practical problems. Most undergraduate electromagnetics texts now contain sections on numerical methods for calculating fields. The IEEE Transactions on Antennas and Propagations have more articles on numerical calculation of fields than on analytical calculation. Professors must skip teaching some of the traditional analytical methods in favor of the newer numerical methods. Classroom and technical presentations make use of electromagnetic movies in which the viewer watches a very colorful display of a gaussian pulse striking an object and scattering. The computer has become a critical part of electromagnetics.

Computational electromagnetics is the simulation of Maxwell's equations and their variations on a computer. Numerical approaches to solving Maxwell's equations find the fields in either the time domain or frequency domain. Time-domain models contain many frequencies and can model transient behavior. On the other hand, frequency-domain methods calculate solutions for one frequency at a time and are appropriate for steady-state behavior. Fourier transforms allow transitioning between the two domains.

Maxwell's equations are a function of space and time. For instance, Faraday's law and Ampere's law in vector form are

$$\nabla \times \boldsymbol{E} = -\mu \frac{\partial \boldsymbol{H}}{\partial t} \tag{5.1}$$

$$\nabla \times \boldsymbol{H} = \sigma \boldsymbol{E} + \varepsilon \frac{\partial \boldsymbol{E}}{\partial t} \tag{5.2}$$

where the time-dependent electric ($\boldsymbol{E}$) and magnetic ($\boldsymbol{H}$) fields are given by

$$\boldsymbol{E}(x, y, z, t) = \mathbf{a}_x E_x(x, y, z, t) + \mathbf{a}_y E_y(x, y, z, t) + \mathbf{a}_z E_z(x, y, z, t) \tag{5.3}$$

$$\boldsymbol{H}(x, y, z, t) = \mathbf{a}_x H_x(x, y, z, t) + \mathbf{a}_y H_y(x, y, z, t) + \mathbf{a}_z H_z(x, y, z, t) \tag{5.4}$$

and μ = permeability, ε = permittivity, and σ = conductivity. Spatial dependence of the material properties and random fluctuations add to the complexity of representing electromagnetic parameters.

The time-dependent forms of Maxwell's equations require boundary values and initial conditions in order to find the fields.

Before finding the behavior of a field at a single frequency, ω, Maxwell's equations must be converted to a form that has a single frequency. Assuming that the time portion of the field is the fundamental harmonic in a Fourier series, the x component of the electric field is

$$E_x(x, y, z, t) = E_x(x, y, z)\cos \omega t = E_x(x, y, z)\mathrm{Re}\{e^{j\omega t}\} \rightarrow E_x(x, y, z)e^{j\omega t} \tag{5.5}$$

Since all the components have the same time factor, it divides out of Maxwell's equations leaving

$$\nabla \times \mathbf{E} = -j\omega\mathbf{B} \tag{5.6}$$

$$\nabla \times \mathbf{H} = \sigma\mathbf{E} + j\omega\varepsilon\mathbf{E} \tag{5.7}$$

The time-harmonic form of Maxwell's equations is a function of space and frequency (frequency terms result from time derivatives in Maxwell's equations) but not time. Consequently, this formulation requires the specification of boundary values. Time behavior of the field comes from calculating the fields at many frequencies and Fourier transforming the results to the time domain.

Unlike analytical methods, computer solutions are not in closed form but are just numbers assigned to grid points. The computer calculates fields and currents at discrete points or grid points specified in the region of interest. Most numerical methods set up a grid of points equally separated in space and time. More grid points increase accuracy but increase computation time as well. The Nyquist rate requires sampling the waveform at twice the highest frequency. Generally, numerical methods limit the maximum spacing between points to be less than $\lambda/10$. For time-domain methods, λ is the wavelength at the center frequency. Grid spacing inside penetrable materials depends on the wavelength inside the material.

Not all numerical methods use uniform grids. A good example is the popular gaussian quadrature formulas for numerical integration. This powerful approach places sample points at the zeros of polynomials and weights and adds the function at those points to find the answer. They have a higher order of accuracy than equally sampled formulas. Some solution domains have small regions where the fields or currents change rapidly. One strategy that maintains accuracy while keeping the number of grid points reasonable is to transition from a coarse grid where fields slowly change to a finer grid where fields rapidly change. Undersampling violates the Nyquist rate resulting in aliasing and the corruption of results.

Realistic radiating objects are very difficult to model with current electromagnetics codes. For instance, accurately modeling the scattered field due to a radar pulse incident on an airplane cannot be done without many reasonable approximations that cut down on the computational load. Some of the more common approximations include

1. Modeling in 1D or 2D instead of 3D. Looking at cuts through a 3D object are often sufficient for many applications.
2. Assuming a surface or wire is infinitely thin. This approach for integral equations and high-frequency methods simplifies the calculations.
3. Simple sources, e.g., plane wave, point source, constant current, constant voltage, and gaussian pulse.
4. Replacing curved lines with straight lines. This assumption can result in stair-step boundaries, straight line instead of a curve, and less complicated math.
5. Ignoring mutual coupling. Only consider mutual coupling between adjacent array elements, using point sources in place of dipoles, discarding small elements in the MOM impedance matrix. Coupling increases storage and calculations.
6. Applying far field approximations. The far-field assumption ignores small amplitude terms.

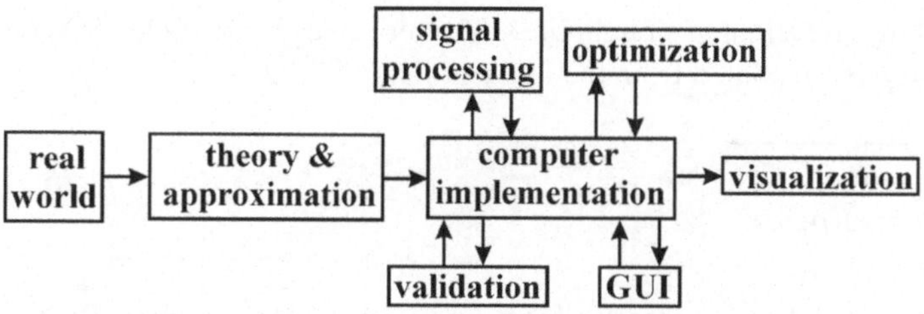

Figure 5.1 Block model of the computational electromagnetics process.

Knowing the approximations used in a given numerical calculation is essential to the proper interpretation of the output.

Figure 5.1 is a flowchart of computer modeling and of this chapter. People develop theory to explain the real world. For electromagnetics, that theory is Maxwell's equations. Approximations save computational effort. Section 5.2 presents the computer implementation of Maxwell's equations in the form of numerical algorithms. The output from the computer models usually goes to an optimization or signal processing algorithm for practical use in system design. These functions are discussed in Sec. 5.3. Results of the computer model must be verified and validated in order to be accepted and used by the electromagnetics community. Graphical user interfaces (GUI) and proper visualization of the output are necessary for a good software package. Programming issues appear in Sec. 5.4.

5.2. SOLVING MAXWELL'S EQUATIONS

The four most commonly used methods for numerically finding electromagnetic fields appear in this chapter. All these methods had their beginnings solving static or frequency-domain problems. Adding time domain capability had to await the computational resources of the 1960s. Finite differencing and integral equation methods proved easiest to convert from frequency to time domain.

5.2.1. High-Frequency Methods

This presentation on high-frequency methods is based on material found in Refs. 1 and 2 with original work coming from Ref. 3. High-frequency methods assume the wavelength approaches zero. This assumption works well when applied to very large objects (at least a few wavelengths across). Geometrical optics, also known as *ray tracing*, forms the basis for these techniques. The electromagnetic rays are orthogonal trajectories to the phase fronts of a wave described by the Eikonal equation:

$$\left(\frac{\partial\zeta}{\partial x}\right)^2 + \left(\frac{\partial\zeta}{\partial y}\right)^2 + \left(\frac{\partial\zeta}{\partial z}\right)^2 = n^2(x, y, z) \tag{5.8}$$

where ζ is known as the *eikonal* or surface of constant phase and n is the index of refraction. Rays are lines perpendicular to the constant phase fronts. As the phase fronts curve due to changes in the index of refraction, the rays correspondingly bend.

When electromagnetic rays impinge on an object, they reflect from and/or transmit through the surface. Geometrical optics (GO) or ray tracing ignores diffraction effects and assumes that an electromagnetic wave is a series of rays traveling in straight lines. The reflected electric field (E^r) at a distance s from the reflection point p is calculated from the incident field (E^i) by

$$\begin{bmatrix} E_{\parallel}^{r}(s) \\ E_{\perp}^{r}(s) \end{bmatrix} = \begin{bmatrix} E_{\parallel}^{i}(p)R_{\parallel} \\ E_{\perp}^{i}(p)R_{\perp} \end{bmatrix} \sqrt{\frac{\rho_1 \rho_2}{(\rho_1 + s)(\rho_2 + s)}} e^{-jks} \tag{5.9}$$

where the subscripts correspond to parallel and perpendicular polarizations. GO treats the reflected wave as a local phenomenon at point p. In other words, the reflected wave is a function of the incident wave, shape of the object (ρ_1 and ρ_2 are the orthogonal radii of curvature of the reflected wavefront), and material makeup of the object ($R_{\parallel}$ and $R_{\perp}$ are the parallel and perpendicular Fresnel reflection coefficients) of the object it strikes. Parallel and perpendicular quantities are referenced to the plane of incidence, which is the plane containing the incident ray and the edge of the object. The total field is the sum of the incident and reflected fields.

$$\mathbf{E} = \mathbf{E}^i + \mathbf{E}^r \tag{5.10}$$

The incident field takes one of the following forms

$$E^i = \begin{cases} e^{-jk\rho_i} & \text{plane waves} \\ \dfrac{e^{-jk\rho_i}}{\sqrt{\rho_i}} & \text{cylindrical waves} \\ \dfrac{e^{-jk\rho_i}}{\rho_i} & \text{spherical waves} \end{cases} \tag{5.11}$$

where ρ_i is the distance from the source to the point of reflection.

The shooting and bouncing ray (SBR) technique [4] represents a plane wave by a bundle of rays that "shoot" into the cavity and bounce around. Each ray is traced using GO as it reflects from conductors and passes through dielectrics. Integrating the exiting rays over the aperture yields the scattered field. This technique works well for very complex shaped cavities that have various materials inside.

Sometimes the surface currents are a more important quantity than the fields. Reflected or scattered fields are then calculated from the induced surface currents. This approach is known as *physical optics* (PO). The PO current only exists where the incident field directly illuminates a surface (lit region) and is found from the tangential component of the incident magnetic field.

$$\mathbf{J}_s = \begin{cases} 2\hat{\boldsymbol{n}} \times \boldsymbol{H}^i & \text{lit region} \\ 0 & \text{shadow region} \end{cases} \tag{5.12}$$

where $\hat{\boldsymbol{n}}$ is the unit normal to the surface. Once the current is found, the reradiated fields are calculated using the appropriate radiation integral.

PO fields are most accurate in the specular direction, since the shadow regions have no surface current. GO and PO work reasonably well for large objects and at angles near the specular direction. PO ignores the edge effects and results in uniform induced currents on the surface. PO and GO are used to derive simple radiation formulas like the radar cross section (RCS) of simple shapes.

Techniques have been developed to supplement GO and PO in order to take into account edge effects. The geometrical theory of diffraction (GTD) adds a diffracted field to the GO approximation. The physical theory of diffraction (PTD) [5], developed independently from GTD, adds a

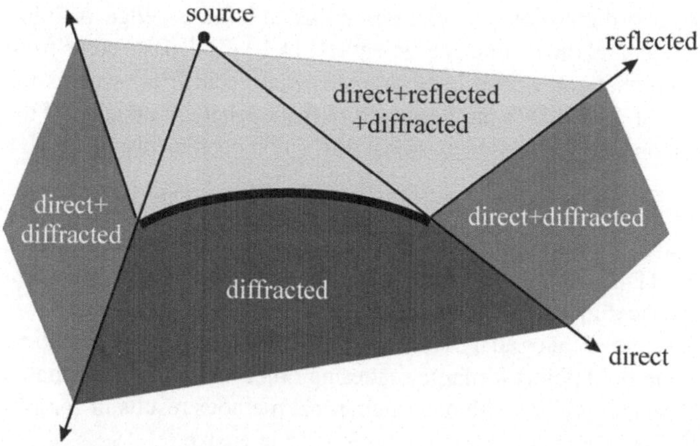

Figure 5.2 A curved object illuminated by a source has three field regions containing the direct, reflected, and/or diffracted fields.

nonuniform or fringe current to the PO approximation. These similar approaches result in the same fields. Only GTD and its extensions are presented here.

Figure 5.2 shows three regions that arise from a wave incident on a finite curved conductor: (1) direct, reflected, and diffracted fields, (2) direct and diffracted fields, and (3) diffracted fields (shadow region). Diffraction results when an incident wave impinges on an edge, corner, or tip and scatters. It also occurs as a creeping ray around a smooth object when the incident field is at grazing incidence. Like the reflected field from a large object, the diffracted field is a localized phenomenon.

The GTD diffracted electric field (superscript d) is given by

$$\begin{bmatrix} E_{\parallel}^{d}(s) \\ E_{\perp}^{d}(s) \end{bmatrix} = \begin{bmatrix} E_{\parallel}^{i}(p)D_{\parallel} \\ E_{\perp}^{i}(p)D_{\perp} \end{bmatrix} A(s)e^{-jks} \tag{5.13}$$

where s = distance from diffraction point to observation point and $D_{\parallel}$ is the parallel and $D_{\perp}$ is the perpendicular polarization diffraction coefficient. The spatial attenuation factor, $A(s)$, is a function of the incident wave (plane, cylindrical, or spherical). Diffraction coefficients exist for various geometries, including

1. Reflection at a plane or curved surface
2. Diffraction at a straight or curved wedge
3. Diffraction at a corner in a plane or doubly curved surface
4. Creeping waves around curved objects like cones, cylinders, and ellipsoids

Adding Eq. (5.13) to the GO field in Eq. (5.9) produces the total electric field. The diffraction coefficient depends on the polarization of the incident wave as well as the geometry and material composition of the object.

Intuition dictates that the discontinuous boundaries inherent in GTD should be smooth transitions. The uniform theory of diffraction (UTD) [6] multiplies the singularities in the diffraction coefficients by transition functions that go to zero at the boundaries resulting in a smooth transition between regions. In addition, UTD takes into account creeping waves that arise from a ray incident tangential to a curved surface by including another term for the scattered field that has a launching coefficient associated with the creeping wave. The creeping wave travels around a geodesic (Fermat's principle) and radiates as it travels around the surface.

UTD and PTD work well for a two-dimensional configuration like an infinite wedge. A finite-length wedge requires the use of incremental diffraction coefficients (ILDCs) [7]. ILDCs come from the closed form diffraction coefficients that correspond to two-dimensional geometries, such as the wedge, strip, polygonal cylinder, and slit. The ILDCs are integrated over the length of an edge. If the edge is infinite, then the result corresponds to the two-dimensional diffraction coefficients. See Ref. 8 for practical applications.

Time-domain UTD models result from Fourier transforming the frequency-domain UTD solutions [9]. The TD-UTD solution is accurate during the time it takes for a ray to propagate from the source to the observation point. The TD-UTD is most useful when the pulse width of the incident field is small compared to the geometric dimensions of the radiating object.

Are high-frequency techniques numerical or analytical? They are presented here as numerical methods because the calculation of the field points for large scattering objects requires a computer. Also, adding time domain to UTD and using UTD with other numerical methods results in complicated numerical algorithms. Finally, the SBR method is computationally intensive.

5.2.2. Integral Equations

Integral equations work well for finding the radiating fields from perfectly conducting objects. Integral equations are derived from the tangential boundary conditions for the electric and magnetic fields

$$\hat{\boldsymbol{n}} \times \boldsymbol{H}^t(r,t) = \hat{\boldsymbol{n}} \times [\boldsymbol{H}^i(r,t) + \boldsymbol{H}^s(r,t)] = \boldsymbol{J}(r,t) \tag{5.14}$$

$$\hat{\boldsymbol{n}} \times \boldsymbol{E}^t(r,t) = \hat{\boldsymbol{n}} \times [\boldsymbol{E}^i(r,t) + \boldsymbol{E}^s(r,t)] = 0 \tag{5.15}$$

where the superscripts t, i, and s stand for total, incident, and scattered, respectively, and $\hat{\boldsymbol{n}}$ is the unit normal. One of the two most commonly used integral equations in electromagnetics is the electric field integral equation (EFIE).

$$\frac{Z}{k}\left[k^2 \iint_S \boldsymbol{J}_s(r')G(\boldsymbol{r}_s,\boldsymbol{r}')ds' + \nabla \iint_S \nabla' \cdot \boldsymbol{J}_s(r')G(\boldsymbol{r}_s,\boldsymbol{r}')ds'\right] = E_t^i(r = r_s) \tag{5.16}$$

where

Z = impedance.
k = wave number.
S = surface of scatterer.
$\boldsymbol{J}_s$ = surface current.
Primed quantities = source points.
r_s = observation point on the surface.
$G(\boldsymbol{r}, \boldsymbol{r}')$ = Green's or transfer function.

It enforces the boundary conditions on the tangential electric field and can be used on open or closed surfaces. The second integral equation is the magnetic field integral equation (MFIE).

$$\boldsymbol{J}_s(r') - \lim_{r \to S}\left\{\hat{\boldsymbol{n}} \times \iint_S \boldsymbol{J}_s(r') \times [\nabla' G(\boldsymbol{r},\boldsymbol{r}')]ds'\right\} = \boldsymbol{H}_t^i(r = r') \tag{5.17}$$

It enforces the boundary conditions on the tangential magnetic field but can only be used on closed surfaces.

The integral operators take into account all surfaces of the computational domain. Calculating the current at one point on an object includes interactions from all other points on all surfaces. Singularities associated with the Green function inside the integrals must be carefully dealt with to

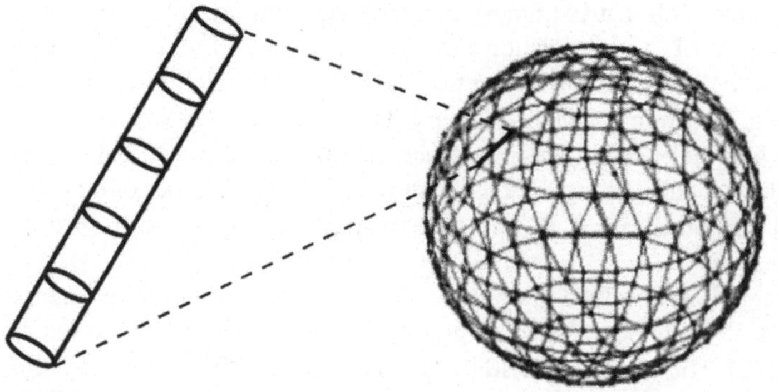

Figure 5.3 A solid perfectly conducting sphere is modeled with a wire grid. Each wire is divided into subsegments. MOM calculates the current induced on each subsegment.

get accurate results. Fields are found from the currents, so only the desired field points need to be calculated. Fields between the current and desired field points are not calculated.

The method of moments (MOM) finds the currents on an object due to an incident wave or an induced voltage or current [10]. It is a way of converting integral equations to matrix equations. This technique works best for wires and flat plates. More complex systems are assembled from wires and/or metal plates. Each wire or metal plate is further subdivided into wire segments or patches that are small compared to the frequency's wavelength. Figure 5.3 shows an example of a solid sphere modeled using wires. Currents on a thin wire are easier to calculate than currents on a thick wire, since the current only flows in one dimension on a thin wire. The wire radius must be large enough that the total surface area of the wires equals the total surface area of the true structure. The wires should be less than a quarter wavelength, with lengths of less than 0.1 wavelength very common. Certain regions, particularly near an induced source, need the wires broken into even shorter segments. The MOM technique determines the current on every wire segment and surface patch due to the sources and all the other currents on the other wire segments and surface patches. Once these currents are known, then the electric field at any point in space is found by integrating the contributions from all the wire segments and surface patches.

The MOM formulation begins by representing the current as the sum of weighted (a_n) simple functions known as basis functions (F_n)

$$J(r') = \sum_{n=1}^{N} a_n F_n(r') \tag{5.18}$$

The basis functions may be full wave functions like sine and cosine or piecewise functions like pulses or triangles. This expansion is substituted into the EFIE or MFIE, then the inner product is taken with N weighting functions. When the weighting functions are the same as the expansion functions the approach is known as *Galerkin's method.* If the weighting functions are delta functions, then the technique is known as *point matching or collocation.*

The MOM results in a $N \times N$ matrix known as the impedance matrix (Z). The unknowns in the vector I are the coefficients, a_n, in Eq. (5.18), and the right-hand side of the matrix equation is the source vector, V.

$$ZI = V \tag{5.19}$$

The induced voltage is either an applied voltage or incident electric field. This equation is solved for the a_n in the I vector. Then, the current on the object is found from Eq. (5.18).

The impedance matrix is usually a full matrix (almost all matrix elements are nonzero). A standard routine for solving this equation is LU decomposition and back substitution. The decomposition part takes about N^2 operations, while the back substitution takes N operations. Multiple right-hand sides use the same decomposition; so finding I takes N steps. Thus, each of multiple incidence angles takes only N operations to find the current. The impedance matrix localization (IML) method uses basis and testing functions that localize strong interactions to only a small number of elements within the impedance matrix [11]. The other matrix elements are small (typically 10^{-4} to 10^{-6} in relative magnitude) and set equal to zero.

The EFIE and MFIE formulations produce spurious currents for cavities near resonance. These result when the eigenvalues of the integral equation go to zero and an ill-conditioned matrix in the MOM formulation results [12]. The combined field integral equation (CFIE) formulation linearly combines the EFIE and MFIE formulations in order to reduce the spurious currents. The EFIE equation times a constant α where $0 \leq \alpha \leq 1$, plus the MFIE times $1 - \alpha$ produce the CFIE. More recently, Shore [13] advocates dual surface MFIE and EFIE to eliminate the resonant problems still associated with some formulations of the CFIE.

The most well-known electromagnetic modeling code is NEC (Numerical Electromagnetics Code) and its commercial variations [14]. This code combines an integral equation for smooth surfaces with one for wires in order to model a variety of antenna structures. The antenna model can have nonradiating networks and transmission lines connecting parts of the structure, perfect or imperfect conductors, and lumped-element loading. The NEC program uses both EFIE and MFIE. The EFIE works best for thin-wire structures of small conductor volume, while the MFIE (does not work for the thin-wire case) works best for voluminous structures with large smooth surfaces. The EFIE models thin surfaces very well. Although the EFIE is specialized to thin wires in NEC, it is frequently used to represent surfaces that may be modeled by wire grids with reasonable success for far-field quantities but with variable accuracy for surface fields.

5.2.3. Finite Difference Methods

Maxwell's equations may be directly solved by replacing the derivatives with finite difference formulas. For instance, replacing a spatial derivative with a central difference approximation of a function, F, on an equally spaced grid along the x axis yields

$$\frac{df(x_i)}{dx} = \frac{F(x_{i+1}) - F(x_{i-1})}{2h} + O(h^2) \tag{5.20}$$

where x_n is between x_{n-1} and x_{n+1} and the grid spacing is h. The central difference approximation is a second order approximation because the error is on the order of h^2. Higher order derivatives use more grid points in the finite difference approximation. Assuming $h \ll 1$, a fourth-order finite difference formula is approximately two orders of magnitude better than a second-order formula.

For example, consider approximating the Poisson equation with second order differencing of the scalar function, v, and source, s

$$s = \nabla^2 v = \frac{\partial^2 v}{\partial x^2} + \frac{\partial^2 v}{\partial y^2} \simeq \frac{v_{i+1,j} - 2v_{i,j} + v_{i-1,j}}{h^2} + \frac{v_{i,j+1} - 2v_{i,j} + v_{i,j-1}}{h^2} \tag{5.21}$$

Solving for v at the grid point (i, j) yields the Jacobi iterative formula

$$v_{i,j}^{n+1} = \frac{1}{4}(v_{i-1,j}^n + v_{i+1,j}^n + v_{i,j-1}^n + v_{i,j+1}^n) - \frac{sh^2}{4} \tag{5.22}$$

New values of v (represented by $n + 1$) are found by averaging the previous (represented by n) four adjacent values of v. Equation (5.22) is a local operator, since it only uses nearby grid points. Local operators allow more detail in the model, such as changing material properties, detailed shapes, and nonconducting objects. Gauss Seidel iteration uses updated values of v on the right-hand side of Eq. (5.22) when available; so it is preferred over the Jacobi formulation.

An example of finite differencing without time dependence is the Laplace equation ($s = 0$) over a square grid with the top of the square at 5 V and the other three sides at 0 V. If the grid of unknown voltages is $N \times N$, then there are N^2 equations and N^2 unknowns. The matrix equation takes the form

$$\begin{bmatrix} -4 & 1 & 0 & \cdots & 0 & 1 & 0 & \cdots & 0 \\ 1 & -4 & 1 & 0 & \cdots & 0 & 1 & \ddots & \vdots \\ 0 & 1 & -4 & 1 & 0 & \cdots & 0 & \ddots & 0 \\ \vdots & 0 & 1 & \ddots & \ddots & \ddots & \vdots & \ddots & 1 \\ 0 & \vdots & \ddots & \ddots & -4 & 1 & 0 & \cdots & 0 \\ 1 & 0 & \cdots & 0 & 1 & -4 & 1 & \ddots & \vdots \\ 0 & 1 & 0 & \cdots & 0 & 1 & \ddots & \ddots & 0 \\ \vdots & \ddots & \ddots & \ddots & \vdots & \ddots & \ddots & -4 & 1 \\ 0 & 0 & 0 & 1 & 0 & \cdots & 0 & 1 & -4 \end{bmatrix} \begin{bmatrix} v_{2,2} \\ \vdots \end{bmatrix} = \begin{bmatrix} c_{1,1} \\ \vdots \end{bmatrix} \tag{5.23}$$

A sparse matrix, like this one, has most of its elements equal to zero. If there is some well-defined pattern to the elements in the matrix, then storage becomes simpler and solution of (5.23) is by an optimized direct method or by iteration using an algorithm like conjugate gradient. See Press et al. [15] for more details.

Multigrid is an important breakthrough in quickly solving boundary value problems like the Poisson equation model. Multigrid iteratively solves the problem on a coarse grid (spacing between grid points is $4h$)

$$\nabla_{4h}^2 V_{4h} = s_{4h} \tag{5.24}$$

then interpolates this solution to a finer grid (spacing between grid points is $2h$). Iteration finds the solution on the finer grid. This process continues until reaching the finest grid (h). Next, the residual (r) of the equation on the fine grid (difference between the right and left hand sides) is restricted or converted back to the coarse grid where the error (e) on the coarse grid is found through iteration.

$$\nabla_{2h}^2 e_{2h} = r_{2h} \tag{5.25}$$

Interpolating the error to the fine grid and added to the fine grid solution. This completes a "V" cycle (Fig. 5.4). Multigrid works by reducing the low-frequency components of the error on the coarse grids while reducing the high-frequency error on the fine grids. Normal iterative techniques work with only a fine grid and take a long time to reduce the low frequencies in the error. Reference 16 provides a tutorial on applying multigrid to some electrostatic problems.

Maxwell's equations have a time dependence in addition to the spatial dependence. Consequently, a temporal grid exists in conjunction with the spatial grid. Time implies initial conditions, and space implies boundary conditions. The most common approach to the finite difference solution of Maxwell's equations is the finite difference time-domain (FDTD) method. Excellent references for FDTD include Refs. 1, 17, 18, and the original work by Yee in Ref. 19.

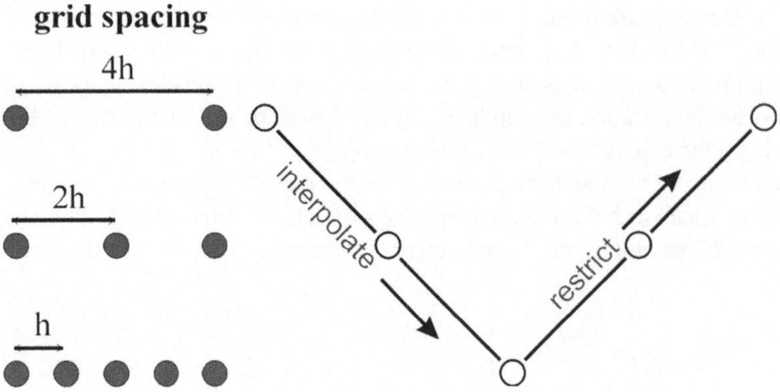

Figure 5.4 A V cycle for multigrid when solving the Poisson equation.

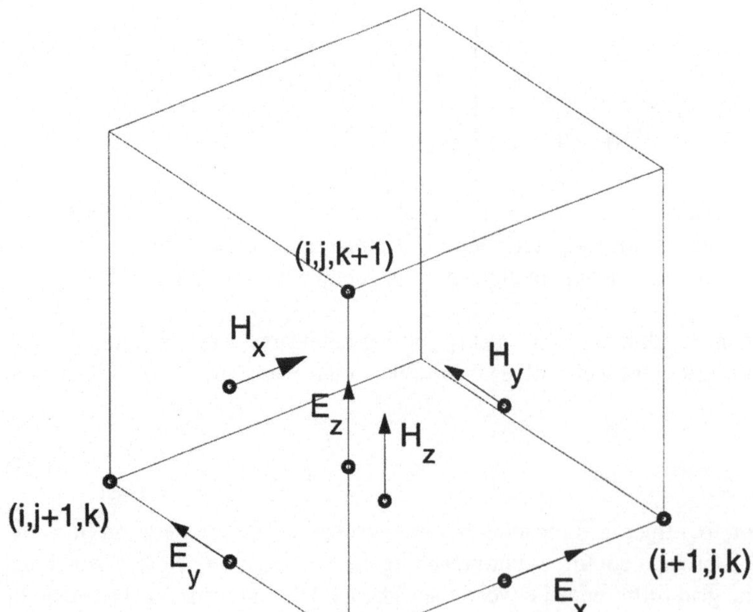

Figure 5.5 The positions of the six field components for the FDTD cube.

FDTD replaces the curl on the left side of Ampere's and Faraday's laws and the partial derivative with respect to time on the right side with second-order finite difference approximation. The resulting grid is complicated and difficult to picture. It is useful to visualize the field components and locations using the smallest complete unit called the *Yee cube* as shown in Fig. 5.5. Note that each of the six field components has a different location in space. Stacking these cubes in the three orthogonal directions covers the computational domain with a three-dimensional grid. The spatial location of grid points is given by $i\Delta x$, $j\Delta y$, and $k\Delta z$ where i, j, and k are integers. Yee's three-dimensional, lossless, source-free ($\mathbf{J} = 0$) equations are written as

$$H^{n+1/2}_{x,i,j+1/2,k+1/2} = H^{n-1/2}_{x,i,j+1/2,k+1/2} + \frac{\Delta t}{\mu_{i,j,k}} \times \left(\frac{E^n_{y,i,j+1/2,k+1/2} - E^n_{y,i,j+1/2,k-1/2}}{\Delta z} - \frac{E^n_{z,i,j,k+1/2} - E^n_{z,i,j+1,k+1/2}}{\Delta y} \right) \tag{5.26}$$

$$H^{n+1/2}_{y,i+1/2,j,k+1/2} = H^{n-1/2}_{y,i+1/2,j,k+1/2} + \frac{\Delta t}{\mu_{i,j,k}} \left(\frac{E^n_{z,i,j,k+1/2} - E^n_{z,i+1,j,k+1/2}}{\Delta x} - \frac{E^n_{x,i+1/2,j,k} - E^n_{x,i+1/2,j,k+1}}{\Delta z} \right) \tag{5.27}$$

$$H^{n+1/2}_{z,i+1/2,j+1/2,k} = H^{n-1/2}_{z,i+1/2,j+1/2,k} + \frac{\Delta t}{\mu_{i,j,k}} \left(\frac{E^n_{x,i+1/2,j,k} - E^n_{x,i+1/2,j+1,k}}{\Delta y} - \frac{E^n_{y,i,j+1/2,k} - E^n_{y,i+1,j+1/2,k}}{\Delta x} \right) \tag{5.28}$$

$$E^{n+1}_{x,i+1/2,j,k} = E^n_{x+1/2,i,j,k} + \frac{\Delta t}{\varepsilon_{i,j,k}} \times \left(\frac{H^{n+1/2}_{z,i+1/2,j+1/2,k} - H^{n+1/2}_{z,i+1/2,j-1/2,k}}{\Delta y} - \frac{H^{n+1/2}_{y,i+1/2,j,k+1/2} - H^{n+1/2}_{y,i+1/2,j,k-1/2}}{\Delta z} \right) \tag{5.29}$$

$$E^{n+1}_{y,i,j+1/2,k} = E^n_{y,i,j+1/2,k} + \frac{\Delta t}{\varepsilon_{i,j,k}} \times \left(\frac{H^{n+1/2}_{x,i,j+1/2,k+1/2} - H^{n+1/2}_{x,i,j+1/2,k-1/2}}{\Delta z} - \frac{H^{n+1/2}_{z,i+1/2,j+1/2,k} - H^{n+1/2}_{z,i-1/2,j+1/2,k}}{\Delta x} \right) \tag{5.30}$$

$$E^{n+1}_{z,i,j,k+1/2} = E^n_{y,i,j,k+1/2} + \frac{\Delta t}{\varepsilon_{i,j,k}} \times \left(\frac{H^{n+1/2}_{y,i+1/2,j,k+1/2} - H^{n+1/2}_{z,i-1/2,j,k+1/2}}{\Delta x} - \frac{H^{n+1/2}_{x,i,j+1/2,k+1/2} - H^{n+1/2}_{x,i,j-1/2,k+1/2}}{\Delta y} \right) \tag{5.31}$$

The spatial samples are often at half increments on the grid. Subscripts indicate the field component (x, y, or z) and the rest of the subscript denotes the location of the field component on the grid. The notation takes some time to learn. Try writing Eqs. (5.24) to (5.29) from Maxwell's equations and drawing your own version of the Yee cube to gain a sufficient understanding of the spatial grid.

Superscripts on the field components indicate the time increment, where n is an integer. All magnetic field components are spaced on the half grid and the electric field components are on the whole integer grid for time. All electric field components are Δt apart. Similarly, all magnetic field components are Δt apart.

An example of one-dimensional space and time grid appears in Fig. 5.6. A wave, such as a gaussian pulse, propagates via Eqs. (5.25) and (5.29). A leapfrog scheme that first calculates the electric field from the magnetic field grid then the magnetic field from the electric field grid, updates the fields. The electric field at the current time comes from the electric field at one previous time step and at the same

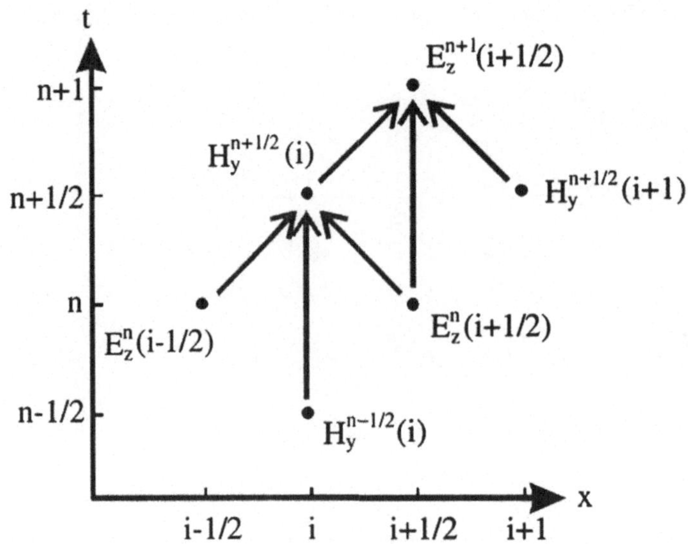

Figure 5.6 A one-dimensional FDTD grid in space and time.

spatial location and the magnetic field at a previous one-half time step and one-half a space step on either side of the electric field. A complementary logic is used to find the magnetic field.

The original Yee FDTD algorithm is second-order accurate in space and time. Numerical dispersion occurs because the phase velocity in the grid is not the same as the phase velocity in the physical problem. Small grid spacing in time and space minimize numerical dispersion. A finer grid implies more unknowns, creating a trade-off between accuracy and computational load. The spatial sampling (in this case Δx) is less than or equal to $\lambda/10$. A full time step for a three-dimensional spatial problem is calculated from the Courant condition given by

$$\Delta t \leq \frac{1}{c\sqrt{1/(\Delta x)^2 + 1/(\Delta y)^2 + 1/(\Delta z)^2}} \tag{5.32}$$

where c is the speed of light. Computationally large problems require many grid points per wavelength to reduce the dispersion error to an acceptable level. One way around this problem is to use higher order derivatives like second-order accuracy in time and fourth-order accuracy in space [20]. Modeling boundary conditions and discontinuities are still a topic of research for the higher order approaches.

Many FDTD applications involve modeling waves traveling in free space, such as with an antenna. In order to model these open region problems accurately, a relatively large free-space area around the objects of interest must be gridded. The end of the gridded space actually forms a numerical boundary that reflects incident waves. Absorbing boundary conditions (ABC) significantly reduce or eliminate these artificial reflections. The perfectly matched layer (PML) technique has proven to be a standard for FDTD [21]. It absorbs the electromagnetic waves from any angle of incidence and of any frequency. Figure 5.7 shows the magnitude of the scattered field from a perfectly conducting metal cube with a sinusoidal incident field. The grid is visible on the interior of the cube because the electric field does not penetrate the cube.

Most of the time, the space between the radiating object and the far field is too large to grid and solve for field values at all the grid points. Instead, near-field data must be transformed into the far field [22]. A transformation boundary surrounds the radiating object while lying within the

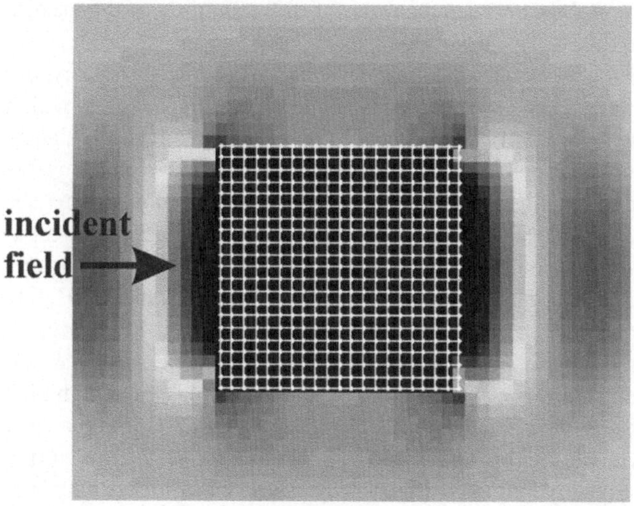

Figure 5.7 Scattered field magnitude due to a sinusoidal incident field on a perfectly conducting cube.

FDTD grid boundaries. Tangential field components calculated using FDTD are converted into equivalent electric and magnetic surface currents on the transformation boundary. Far-field quantities are then calculated from these surface currents.

5.2.4. Finite Element Method (FEM)

R. Courant developed FEM in 1943 [23,24]. He used the Rayleigh-Ritz method for finding approximate solutions to variational problems by replacing the functions with appropriate combinations of basic elements then finding the minimum solution. The first step converts a boundary value problem into an equivalent variational problem. The equation for the variational form is given by

$$F_v(E) = \frac{1}{2}\iiint_V \left[\frac{1}{\mu_r}(\nabla \times \boldsymbol{E})\cdot(\nabla \times \boldsymbol{E}) - k_0^2\varepsilon_r \boldsymbol{E}\cdot\boldsymbol{E}\right] dV + \frac{1}{2}\iint_S [\boldsymbol{E}\cdot(\hat{\boldsymbol{n}}\times\nabla\times\boldsymbol{E})]\,dS + \iiint_V \boldsymbol{E}\cdot\left[jk_0 Z_0 \boldsymbol{J}^i + \nabla\times\left(\frac{1}{\mu_r}\right)\boldsymbol{M}^i\right] dV \tag{5.33}$$

and the equation for the weighted residual method is

$$F_w(E) = \iiint_V \left[\frac{1}{\mu_r}(\nabla \times \boldsymbol{E})\cdot(\nabla \times \boldsymbol{W}) - k_0^2\varepsilon_r \boldsymbol{E}\cdot\boldsymbol{W}\right] dV + \frac{1}{2}\iint_S [\boldsymbol{W}\cdot(\hat{\boldsymbol{n}}\times\nabla\times\boldsymbol{E})]\,dS + \iiint_V \boldsymbol{W}\cdot\left[jk_0 Z_0 \boldsymbol{J}^i + \nabla\times\left(\frac{1}{\mu_r}\right)\boldsymbol{M}^i\right] dV \tag{5.34}$$

where

V = volume containing the unknowns.
S = boundary enclosing the volume.
$\boldsymbol{W}$ = weighting function.
$\boldsymbol{J}^i$ = induced electric current source.
$\boldsymbol{M}^i$ = induced magnetic current source.

Equation (5.32) is also known as the *weak form* because the order of differentiation of the electric field in Eq. (5.32) is less than that of Eq. (5.31), or the strong form.

The FEM then divides an electromagnetic domain into many discrete, easy to analyze, polygon-shaped elements that conform to irregularly shaped subdivisions of the object. FEM begins with a model drawn in 1D, 2D, or 3D space using a preprocessor or a CAD drafting package. Next, automatic mesh generators create triangular/tetrahedral meshes throughout the model. Meshing is the process of breaking up a physical domain into smaller subdomains (elements). Surface domains may be subdivided into triangle or quadrilateral shapes, while volumes may be subdivided primarily into tetrahedra or hexahedra shapes. There are some requirements on the shape of elements. In general, the elements should be as equiangular as possible in equilateral triangles and regular tetrahedra. Highly distorted elements (long, thin triangles, squashed tetrahedra) lead to numerical instability. Manual meshing becomes necessary in regions where an automatic generator fails to create regular meshes. Connecting elements should have the same number of nodes along the common side. Areas with high gradients require a mesh with small elements—the finer the mesh, the better the results. Picking a good mesh density is an art. If the mesh is too coarse, then errors are too large. Alternatively, if the mesh is too fine, then computing time becomes unacceptably long. A fine mesh is necessary in regions having high parameter gradients, whereas a coarse mesh is sufficient elsewhere. The mesh must not have holes, self-intersections, or faces joined at two or more edges, and must conform to the boundary of the domain. Figure 5.8 shows a triangular mesh overlaying a model of a waveguide T junction. Note that the mesh is much finer around the rectangular inset at the top of the T, because the field variations are greater there.

The next step in the FEM is to select the interpolation function that approximates the unknown over an element. Polynomials are the most common because they are simple and have a limited extent. Nodal-based elements come from interpolating function values at the nodes. These elements are generally not used for vector electromagnetic fields, because they produce spurious modes and it is difficult to impose tangential boundary conditions. Edge-based elements overcome these limitations by assigning degrees of freedom to the edges instead of the nodes. The most common two-dimensional elements are rectangles and triangles, while the most common three-dimensional elements are bricks and tetrahedrals.

Assembly is the process of taking all the equations and developing a matrix equation to solve for the weights. All the element equations surrounding a given node are added together to get a single equation. The coefficients of this equation form a row in the matrix. An important step in the assembly process is establishing a global numbering scheme to keep track of all the nodes in the

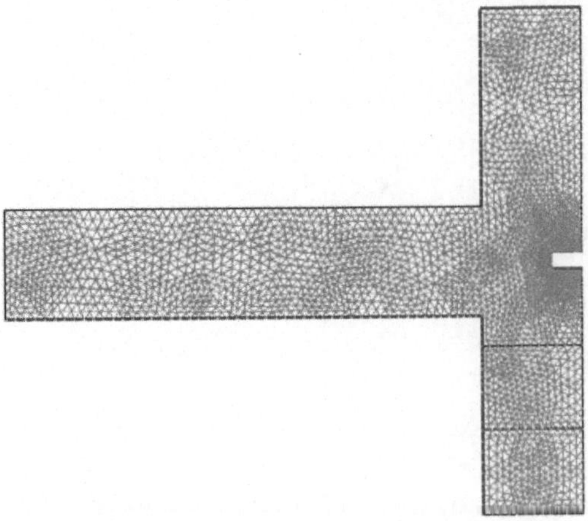

Figure 5.8 A T-junction waveguide is gridded for FEM.

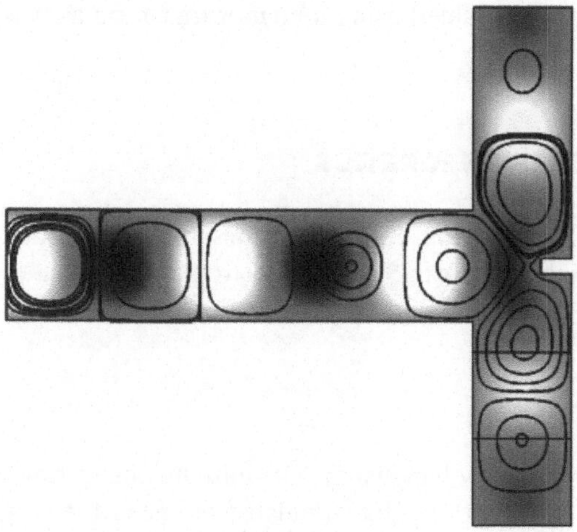

Figure 5.9 The electric field for the T-junction waveguide calculated using FEM.

mesh. The assembled matrix is quite sparse. Finally, the boundary conditions are incorporated and the matrix equation solved using sparse solution methods. Figure 5.9 is a plot of the magnitude of the electric field for the T-waveguide problem as computed by FEM.

5.2.5. Other Techniques for Finding the Fields

In the generalized multipole technique (GMT), the boundaries of the problem are discretized then a number of radiating sources are placed off the boundaries. These sources act as basis functions that are analytical solutions to the field equations in the medium. The sources are weighted such that the boundary conditions are met in a least squares sense. Arranging the sources and boundary points are key to having a well conditioned matrix and good results. The number of boundary points should be much greater than the number of sources.

The transmission line method (TLM) makes use of the fact that Maxwell's equations are analogous to transmission line equations through the following equivalences:

$$E \leftrightarrow V \quad H \leftrightarrow I \quad \varepsilon \leftrightarrow 2C \quad \mu \leftrightarrow L \tag{5.35}$$

where V is voltage, I is current, C is capacitance per unit length, and L is inductance per unit length. TLM models space and objects using a rectangular mesh of transmission lines. Each node has an associated scattering matrix. The reflected voltages are found by multiplying the scattering matrix (S) for the node by the incident voltages at the input ports.

$$\begin{bmatrix} V_1^r \\ V_2^r \\ V_3^r \\ V_4^r \end{bmatrix} = S \begin{bmatrix} V_1^i \\ V_2^i \\ V_3^i \\ V_4^i \end{bmatrix} \tag{5.36}$$

The three-dimensional version of TLM has six ports with two orthogonal polarizations per port. Thus, there are 12 incident and reflected voltages and the scattering matrix is 12 × 12. Although TLM is a time-domain method, frequency-domain information is obtained using a Fourier transform as was done in FDTD.

5.3. RECENT NUMERICAL TOOLS FOR ELECTROMAGNETICS

Section 5.2 presents the most common methods of finding fields. This section presents methods that make use of the field points generated by the numerical models in Sec. 5.2. Electromagnetics makes use of the many numerical methods developed in signal processing.

5.3.1. Model-Based Parameter Estimation (MBPE)

A few years ago, most calculations were done over a narrow bandwidth. The introduction of time-domain methods, wideband antennas, and high Q resonant circuits stimulated the need for very detailed computations to achieve the desired accuracies and not miss important features in the computed output. Many times the output is sampled and connected by straight lines (linear interpolation) to generate the output plots. A closer sampling distinguishes fine features but increases the computation cost. Uniform sampling usually means that some regions with slow varying details are oversampled, while regions with high variations are undersampled. More sophisticated interpolation like splines make use of derivative information to produce a smoother curve through the calculated data points.

MBPE is an interpolation/extrapolation technique for measured or computed data [27]. Unlike splines, polynomials, or Fourier series, MBPE uses interpolating functions derived from physical parameters of the problem. It is smart curve fitting. Complex exponentials are typical solutions to time-domain electromagnetic differential equations, while complex poles are typical solutions to frequency-domain electromagnetic differential equations. Consequently, exponentials and poles seem to be appropriate physically based curve fitting functions for electromagnetic problems:

$$q(t) = \sum_{m=1}^{M} A_m e^{s_m t} + q_{np}(t) \tag{5.37}$$

$$Q(f) = \sum_{m=1}^{M} \frac{A_m}{f - s_m} + Q_{np}(f) \tag{5.38}$$

where

M = number of terms.
q = waveform domain.
Q = transform domain.
A_m = residues.
s_m = poles.
q_{np} = nonpole component of the waveform function.
Q_{np} = nonpole component of the transform function.

The nonpole parts of (5.37) and (5.38) represent the nonresonant response. Equations (5.37) and (5.38) indicate that the transform pairs are the time and frequency domains. In electromagnetics, the frequency-space and space-angle transform pairs are also of great importance.

Prony's method was the original approach to MBPE [27]. This algorithm finds an infinite impulse response (IIR) filter with a prescribed time-domain impulse response. The classical method of Count de Prony models a sequence of $2p$ observations made at equally spaced times by a linear combination of p exponential functions. Prony's ingenious method converts the problem to a system of linear equations. Advances in signal processing have resulted in many other approaches to MBPE.

A closely related technique is the singularity expansion method (SEM) [28]. SEM characterizes an object's response in the time and frequency domains in terms of poles, branch cuts, and entire functions (singularities) in the complex frequency plane. Since most scattering objects have a transient response dominated by a small number of damped sinusoids, the damped sinusoids are poles of the Laplace transformed response. Natural frequencies or resonances are the basic starting ideas for SEM.

5.3.2. Optimization

The numerical techniques discussed so far find a single solution for specific problem parameters. Optimization finds the best set of problem parameters that yield an optimized or desired solution. Bounds or constraints may be placed on the parameters due to physical limitations, prior knowledge, or computational limits. Optimizing implies either finding a minimum or maximum of the output $(y_1,\ldots,y_N)$ from an objective function, F, given the input $(x_1,\ldots,x_N)$.

$$F(x_1,\ldots,x_N) = \{y_1,\ldots,y_N\} \tag{5.39}$$

Objective functions can have multiple inputs and multiple outputs. In computational electromagnetics, the objective function is a numerical model of an antenna, scattering object, microwave circuit, etc. Inputs to an antenna objective function may include parameters such as size, spacing, material properties, etc. Common output variables include gain, null depth, and sidelobe level. More than one solution must be generated in order to find the best set of parameters. Thus, optimization tends to be very time consuming.

Numerical optimization traditionally took two approaches: downhill methods or random methods. The downhill methods primarily rely upon derivative information to find a local minimum and are based on Newton's formula

$$v_{n+1} = v_n - \alpha_n Q_n^{-1} \nabla F(v_n) \tag{5.40}$$

where

v = vector containing the coordinates.
n = iteration number.
α_n = step size.
Q_n = nth approximation to the Hessian matrix = H.

$$H = \begin{bmatrix} \dfrac{\partial^2 F}{\partial x_1 \partial x_1} & \cdots & \dfrac{\partial^2 F}{\partial x_1 \partial x_N} \\ \cdot & \cdots & \cdots \\ \dfrac{\partial^2 F}{\partial x_N \partial x_1} & \cdots & \dfrac{\partial^2 F}{\partial x_N \partial x_N} \end{bmatrix}$$

$\nabla F(v_n)$ = gradient of the objective function.

A myriad of techniques sprouted around solving Eq. (5.37). Some of the more popular include [29]

Steepest descent (in use for over one hundred years): Q_n = identity matrix.
Newton's method: $Q_n = H$ = Hessian matrix.
Conjugate gradient: indirectly constructs Q_n.
Davidon-Fletcher-Powell (DFP): $Q_{n+1} = Q_N + C_{\text{DFP}}$, where C_{DFP} is a correction term.
Broyden-Fletcher-Coldfarb-Shanno (BFGS): $Q_{n+1} = Q_N + C_{\text{DFP}} + C_{\text{BFGS}}$, where C_{BFGS} is a correction term.

The Nelder Mead downhill simplex algorithm [30] is commonly used by software packages like MATLAB, Mathematica, etc. The algorithm iteratively attempts to surround the optimum point with a simplex. A simplex is the most elementary geometrical figure that can be formed in dimension n and has $n + 1$ sides (e.g., a triangle in two-dimensional space). Each iteration creates a new vertex for the simplex. The vertex corresponding to the highest function value is discarded. In this way, the simplex creeps towards the minimum. The simplex shrinks its diameter when it surrounds the minimum. The creeping and shrinking stop when the diameter reaches a specified tolerance. Since the algorithm does not use derivatives, it has a certain robustness that makes it attractive.

Optimization methods are classified as local or global. The downhill minimization algorithms are local because they start at a single point and move downhill to the local minimum. Practical problems often have many local minima. The local minimum found depends upon the initial starting point. Global optimization techniques incorporate random components that allow them to jump out of local minima and explore vast regions of the objective function space. A pure random search is just a guessing game and is rarely used.

Simulated annealing (SA) is random search based on the principles of thermodynamics [31]. The physical process of annealing occurs when a solid melts and its particles try to organize into a low-energy state during the cooling process. The probability that a particle is at a certain energy level is calculated by use of the Boltzmann distribution. As the temperature of the material decreases, the Boltzmann distribution tends toward the lowest energy particle configuration.

SA guesses at the optimum solution and then perturbs that solution. If the new cost (C) is less than the old cost, then it is accepted. If the new cost is greater than the old cost, then it is accepted if $P > p$ and rejected if $p < P$, where P is a uniform random number. The threshold probability, p, is given by

$$p = e^{-C/T} \tag{5.41}$$

The variable, T, corresponds to the temperature in the annealing process. T is slowly reduced so that the probability of accepting a higher cost decreases with time. The formula for reducing T is called the cooling schedule and is critical to the success of SA. Both the step size and T determine the convergence properties of the SA algorithm. Suggested step sizes and T values are approximately 80% of the higher costs accepted.

Another naturally based random search algorithm is the genetic algorithm (GA). The GA is a type of evolutionary algorithm that models the biological processes of genetics and natural selection to optimize highly complex objective functions. A GA helps a population composed of many individuals or potential solutions to evolve under specified selection rules to a state that contains the "most fit" individuals (i.e., minimizes the objective function, assuming it has been written so that the minimum value is the desired solution). The method was developed by John Holland [33] over the course of the 1960s and 1970s and popularized by his student, David Goldberg [34]. Michielssen first applied GAs to the design of radar absorbers in [35], and Haupt first used GAs for antenna design in [36]. An introductory article with the code for a very simple GA helped popularize GAs in electromagnetics [37].

The following explanation follows the flow chart in Fig. 5.10. The first step is defining an objective function with inputs and outputs. A binary GA encodes the value of each input parameter

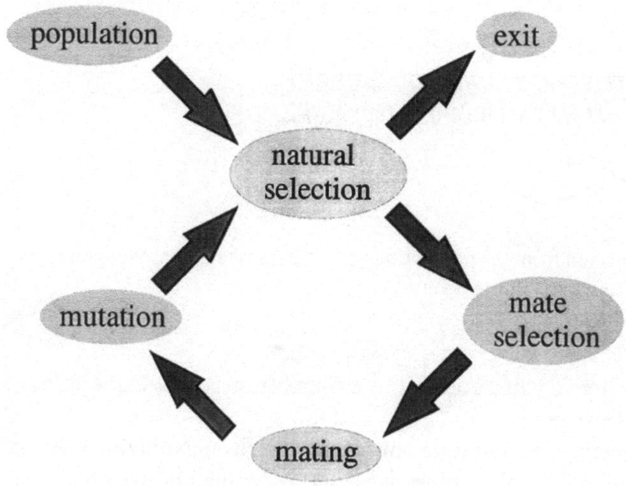

Figure 5.10 Flow chart of a genetic algorithm.

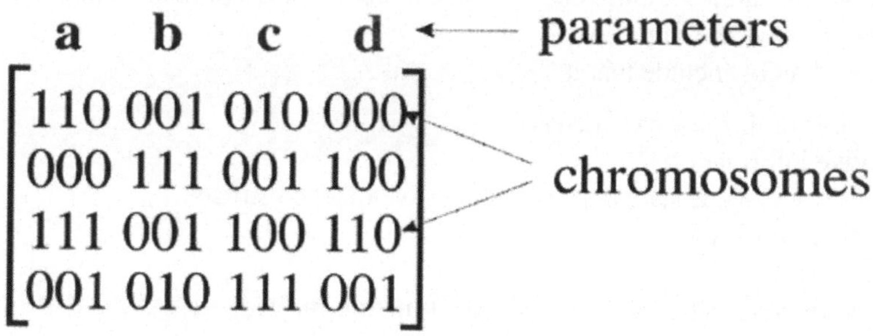

Figure 5.11 The values of the parameters are encoded in a binary representation. All the parameters are placed in a chromosome, and the chromosomes are rows in the population matrix.

(e.g., a, b, c, d) as a binary number. The parameter values are then placed side-by-side in an array known as a *chromosome*. A population is a matrix with each row representing a chromosome. The algorithm begins with a population consisting of random ones and zeros (see Fig. 5.11). These random binary digits translate into guesses to values of the input parameters. Next, the binary chromosomes are converted to continuous values, which are evaluated by the objective function. Mating takes place between selected chromosomes. Mates are randomly selected with a probability of selection greater for those chromosomes yielding desirable output from the objective function (tournament or roulette wheel selection). Offspring (new chromosomes) produced from mating inherit binary codes from both parents. A simple crossover scheme randomly picks a crossover point in the chromosome. Two offspring result by keeping the binary strings to the left of the crossover point for each parent and swapping the binary strings to the right of the crossover point, as shown in Fig. 5.12. Crossover mimics the process of meiosis in biology. Mutations randomly convert some of the bits in the population from "1" to "0" or visa versa. The objective function outputs associated with the new population are calculated and the process repeated. The algorithm stops after finding an acceptable solution or after completing a set number of iterations.

Selecting the best population size, mating scheme, and mutation rate is still an area of controversy. References 38 and 39 address this issue for electromagnetics problems. Since the GA is a random search, a certain population size and mutation rate can give considerably different answers for

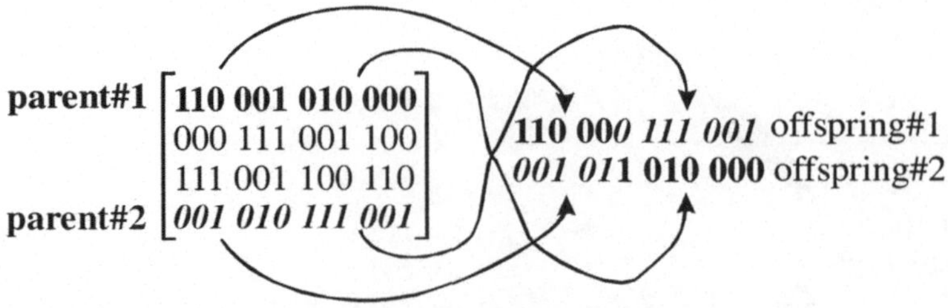

Figure 5.12 Two parents are randomly selected from the population matrix. A random crossover point splits the parents. Two new offspring are formed from parts of the parents.

different independent runs. A GA run will give you a good answer found from a wide exploration of the search space but not necessarily the best answer.

Most real world optimization problems have multiple objectives, such as maximizing gain and maximizing bandwidth for the same antenna. Multiple objectives can be handled by weighting and adding the fitness from each objective. Multiobjective optimization does not have a single optimum solution relative to all objectives. Instead, there is a set of optimal solutions, known as Pareto-optimal or noninferior solutions. A Pareto GA attempts to find as many Pareto-optimal solutions as possible, since all these solutions have the same cost.

Some of the advantages of a GA include that it

Optimizes with continuous or discrete parameters.
Doesn't require derivative information.
Simultaneously searches from a wide sampling of the objective function surface.
Deals with a large number of parameters.
Is well suited for parallel computers.
Optimizes parameters with extremely complex objective function surfaces.
Provides a list of semioptimum parameters, not just a single solution.
May encode the parameters so that the optimization is done with the encoded parameters.
Works with numerically generated data, experimental data, or analytical functions.

5.3.3. Wavelets

Traditionally, we have thought of time domain signals as lasting forever. A function starts at $t = -\infty$ and continues until $t = +\infty$. Consequently, the Fourier transform is an efficient method of finding the spectrum or amplitudes of the frequency components

$$\Im(\omega) = \int_{-\infty}^{\infty} F(t)e^{j\omega t}\,dt \tag{5.42}$$

where $\omega = 2\pi f$. In reality, signals last for a finite duration. Even if they continue for a long period, our patience and computer limits stipulate that only a portion of the signal can be examined at any one time. Thus, most engineering problems do not use Eq. (5.42).

The more practical alternative to the Fourier transform is the short-time Fourier transform (STFT) or the windowed Fourier transform. The STFT windows or works with finite segments of the data.

$$\mathrm{STFT}(t, \omega) = \int F(\tau)w^*(\tau - t)e^{-j\omega t}\,d\tau \tag{5.43}$$

Many different windows (w) have been developed for various purposes. As an example, consider the linear chirp signal plus impulse at $t = 0.1$ s.

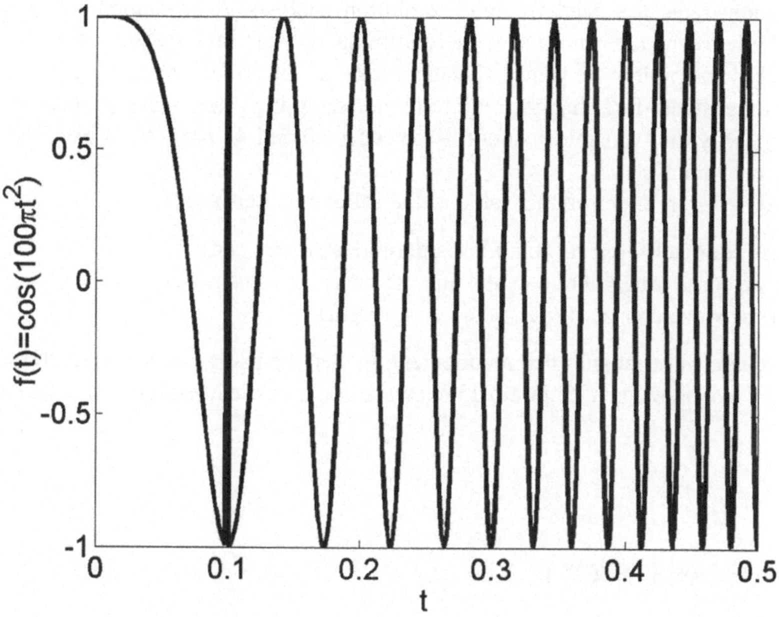

Figure 5.13 Linear chirp signal with an impulse at $t = 0.1$ s.

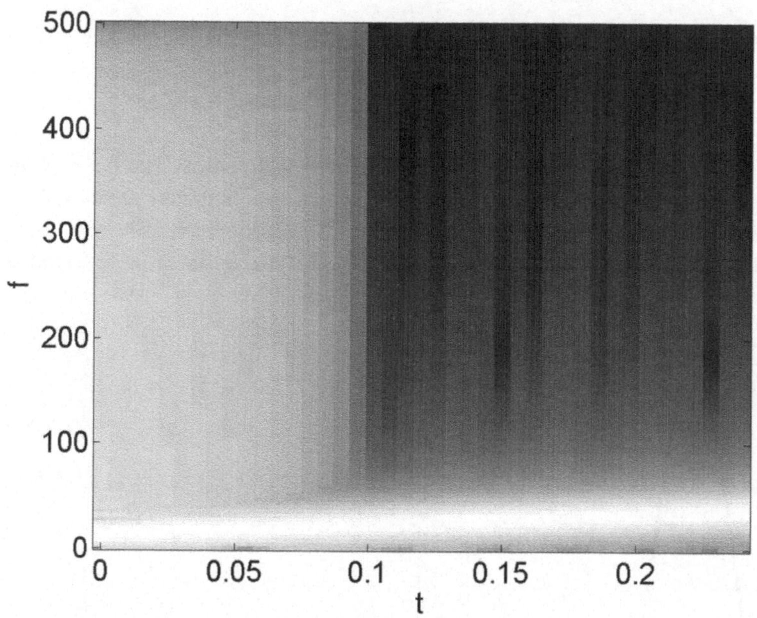

Figure 5.14 STFT of the linear chirp plus impulse signal.

$$F(t) = \cos(100\pi t^2) + 2\delta(t - 0.1) \tag{5.44}$$

as shown in Fig. 5.13. Note that the frequency increases with time and there is an impulse function at $t = 0.1$ s. Figure 5.14 shows the STFT of (5.44). This plot nicely shows a linear increase in frequency with time but cannot accurately show the location of the impulse.

The STFT plot demonstrates the need for multiresolution analysis. A new approach was needed that could specify low-frequency signals accurately in frequency and high-frequency signals accurately in time. Precisely locating low-frequency signals in time is not critical, because they are slowly changing. On the other hand, fast changes require higher sampling rates. Wavelets provide the variable sampling capability that sinusoids cannot. References 40 and 41 provide an excellent introduction to time-frequency analysis.

A *wavelet* is defined to be any function that satisfies the following constraints:

1. The function has compact support (i.e., it has a definite start and end).
2. The area under the curve equals zero (i.e., the functions average value is zero).
3. The area under the wavelet is zero (i.e., no dc component).

A single cycle of a square wave satisfies that requirement. In fact, the first wavelet was the Haar wavelet and is one cycle of a square wave. Figure 5.15 shows a graph of the Mexican Hat wavelet given by the equation

$$\Psi(x) = \frac{2}{\sqrt[4]{\pi}\sqrt{3}}(1 - x^2)e^{-x^2/2} \tag{5.45}$$

The continuous wavelet transform (CWT) is given by

$$\text{CWT}_{a,b} = \frac{1}{\sqrt{|a|}}\int f(t)\Psi^*\left(\frac{t-b}{a}\right)dt \qquad a \neq 0 \tag{5.46}$$

where a is the scale index (inverse of frequency) and b is the time shift (translation). The CWT of the chirp signal in Eq. (5.44) is shown in Fig. 5.16. This plot shows the increase of frequency over time (remember that $a \propto 1/f$) and shows the precise location of the impulse at $t = 0.1$ s.

The wavelet is an appropriate basis function when the scattering object is large and contains features with scales ranging from fractions of a wavelength to many wavelengths. The basis functions are shifted and dilated forms of a mother wavelet. Wavelets with a short expanse are used near edges or small features, while wavelets with a long expanse are used over large smooth features of the object. The resulting MOM matrix has many small elements. By applying a threshold level to the

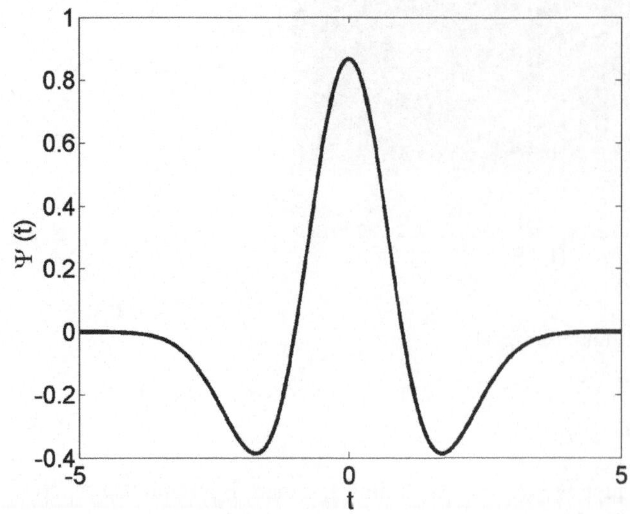

Figure 5.15 Mexican hat wavelet.

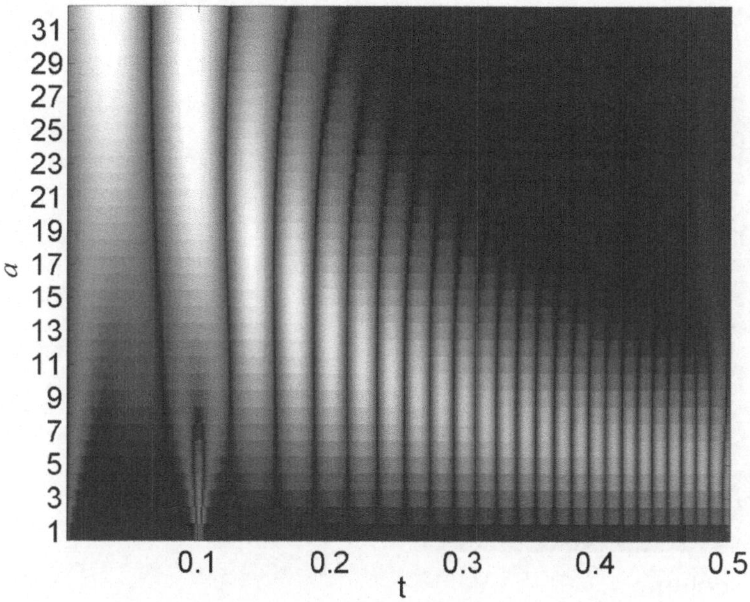

Figure 5.16 Continuous wavelet transform (with Mexican hat wavelet) of the linear chirp plus impulse signal.

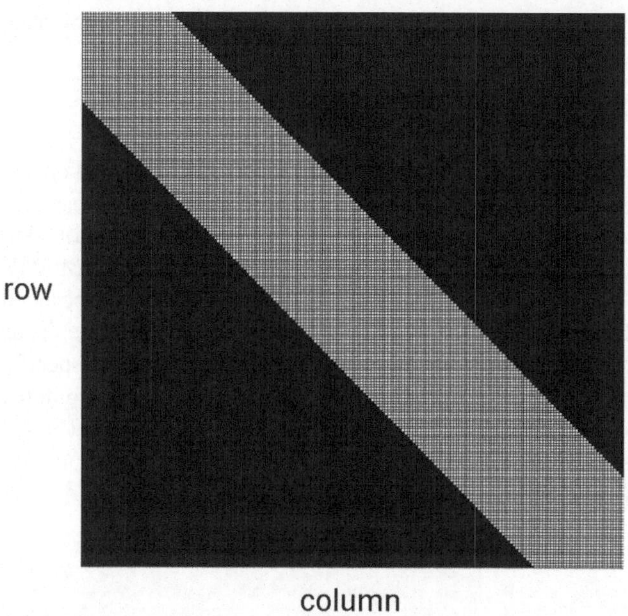

Figure 5.17 Magnitude of a typical MOM impedance matrix. The white elements have a magnitude that is at least 10% of the maximum magnitude element.

matrix elements, many can be set equal to zero. The resulting sparse matrix is of a form that can be quickly solved [42]. Figure 5.17 is a plot of a 200 × 200 impedance matrix with white indicating elements having a magnitude of at least 10% of the maximum magnitude in the matrix. This matrix has 11,944 elements with a magnitude of at least 10% of the maximum. A wavelet transform of that

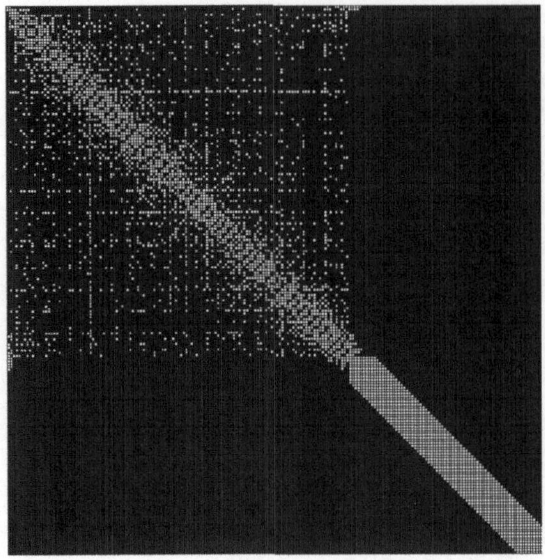

Figure 5.18 Magnitude of a wavelet transform of a typical MOM impedance matrix. The white elements have a magnitude that is at least 10% of the maximum magnitude element.

matrix has 4416 elements with a magnitude of at least 10% of the maximum (see Fig. 5.18). This 63% savings increases as the matrix gets bigger.

5.3.4. Hybrid Methods

Hybrid methods combine two or more of the solution methods described earlier. Building on the strengths of two techniques allows the modeling of more complex structures. Hybrid methods can also include the weaknesses inherent in both techniques. A combination of MOM and GTD is described in Ref. 1. Combining the frequency-domain MOM and the FDTD methods takes advantage of MOM's ability to solve exterior problems using patch models and of the ability of FDTD to model localized regions containing metal structures, dielectrics, permeable media, anisotropic or nonlinear media, as well as wires [43]. Another approach is to combine ray tracing and FDTD methods for site-specific modeling of indoor radio-wave propagation [44]. FDTD is only used to study areas close to complex discontinuities where ray-based solutions are not accurate. Since MOM and PO are current based methods, combining these approaches to solve for currents on large complex objects results in suitable accuracy in a reasonable computation time [45].

5.4. SOFTWARE CONCERNS

Developing the numerical method to solve an electromagnetic problem is the first step toward creating a useful computer program. Next, the programming language must be selected from the myriad available. Proper visualization of the results is essential to proper interpretation of the results. Finally, the code must be verified and validated in order to be accepted by users.

5.4.1. Programming Languages

The numerical solution of an electromagnetics problem may involve a programming language, general-purpose software, or specialized software. A programming language has the advantages of portability, fast execution, wide usage, and cheap or free software. The more popular languages include Java, C, C++, Fortran, and BASIC. Fortran and Basic are the primary languages used for various versions of the Numerical Electromagnetics Code (NEC). For the most part, programming languages must be compiled and take a long time to write and debug. Java is the newest of these languages and has gained popularity. Java is a good object-oriented language for quickly writing programs that run on multiple platforms and has found extensive use on the internet. Unfortunately, it is slower at mathematical operations than the other languages and lacks the extensive library functions for various numerical analysis routines.

General-purpose software has the advantages of fast program design, extensive prewritten routines, fast debugging, and excellent graphics. This type of software is designed to do basic mathematical operations and graphics. The most popular versions include MATLAB, Mathematica, MathCad, and Maple. These programs are interpreted, so they do not have to be compiled. Their advantages include very fast development time; run times as fast as programming languages; extensive mathematical, science, and engineering functions; excellent graphics; and symbolic mathematical manipulations. They tend to be very expensive except for steep academic discounts for student use. Portability can be an issue between different general-purpose software or even between old and new versions of the same software package.

Specialized software has excellent graphics, limited applications, few commands to learn, and usually operates with a GUI (graphical user interface). Specialized software is difficult to link with other software. For instance, combining a UTD code with a MOM code from two different vendors is at best a difficult endeavor. In addition, using a programming language or general-purpose software package to optimize the output of a specialized software package is difficult.

Developing your own software package today requires a GUI for easy interaction with users. An outstanding GUI encourages use by people that did not develop the code. Some pitfalls with GUI design include

Assuming the user knows too much
Limiting user access to the application
Placing too many features at the top level
Terms that are unclear and inconsistent
Being too verbose

Good GUIs are intuitive, consistent, and fast. Users also appreciate knowing how much longer a given operation will take before they can enjoy the fruits of their patience. Easy to use online help is necessary.

5.4.2. Visualization

Not long ago, visualization of electromagnetic fields required good abstract thinking. Today visualization means displaying the physical characteristics of the model as well as the electromagnetic fields associated with the model. Computer graphics quickly convey verbal and numerical information through imagery. A well-designed graphic should [46]

Show the data.
Induce thinking about the substance rather than about methodology, graphic design, or technology of graphic production.
Avoid distorting what the data have to say.
Make large data sets coherent.
Encourage the eye to compare different pieces of data.
Examine the data at several levels of detail.
Serve a reasonably clear purpose.
Be closely integrated with the statistical and verbal descriptions of a data set.

Common pitfalls in visualization are data distortion and putting too much data on a single plot. Distortion is easy to fall prey to with computer graphics that autoscale data. A sphere can look like an ellipsoid if not all axes are to the same scale. Graphics software does not warn you that there are too many lines on the plot or that fine detail of interest is obscured by the rest of the data.

Electromagnetics models are particularly difficult to visualize. A single graph cannot show

Three spatial dimensions
Time
All polarization components
Material properties of the objects
Currents, fields, and charge

Consequently, the software designer and user must decide how to represent the data. Some tradeoffs include still shots vs. movies, 2D vs. 3D, dB vs. magnitude, and color vs. symbol.

5.4.3. Verification and Validation

Any computer model must be validated and verified. Validation compares the computer output with known physical results. Equations and approximations along with other aspects of modeling the physical problem with a computer algorithm must be checked. Validation is the engineering and science part of the computer model. Accepted validation standards include mathematical expressions, experimental results, and other computer models. Validation is a continuous process that compares the computer output to new information as it becomes available.

Verification is the process of correctly solving the equations developed for the model. Unlike validation, it neglects errors caused by the choice of equation and parameters of the equation. It is the numerical analysis part of the computer modeling. Changes to the coefficients of an ill-conditioned numerical model result in large changes in the solution. A large condition number for a matrix implies that solving for the unknown vector in an equation that contains that matrix may result in significant errors. Iterative solutions are prone to chaotic behavior due to the highly nonlinear formulation of the equation.

Experimentalists are used to plots that use error bars and solutions that have a degree of accuracy assigned. Numerical models rarely contain similar information. Many a novice falls prey to giving a numerical model some input and believing the output. Can you justify your results? Experience plays a major role. Verifying and validating numerical results should be a standard requirement for numerical models.

5.5. CONCLUSIONS

This chapter presented the four major numerical approaches to electromagnetics. The advantages and disadvantages of these methods are summarized in Table 5.1. No one technique is perfect. Hybrid methods are becoming more popular for very complicated problems due because they have the advantages of two or more of the numerical approaches.

Many challenges remain in computational electromagnetics. Computation speed is still too slow for most practical problems. Making use of specialized codes, parallel computers, and increasing clock speeds gradually move us toward modeling complex objects. Three-dimensional problems are still tricky and slow for most codes. Some progress has been made in using signal processing techniques to compress data and speed calculations. Cross-fertilization between these two fields needs to continue. Genetic algorithms have opened the possibility of not only modeling electromagnetic behavior but optimizing the design of electromagnetic systems as well. Visualization of results will continue to improve with even the incorporation of artificial intelligence and virtual reality to help.

Table 5.1 Characteristics of the Four Main Numerical Methods Used in Electromagnetics. E = excellent, G = good, and F = fair

	High frequency	MOM	FD	FEM
Single frequency	E	E	G	G
Transient	F	G	E	E
Materials	F	F	E	E
Thin objects	E	E	F	F
Far field	E	E	G	G
Large objects	E	F	F	F
Small objects	F	E	E	E
Ease of formulation	G	G	E	F
3D objects	F	F	E	E
Bandwidth	G	G	E	E

As codes become more complex, verifying and validating their results will become more challenging. Improved experimental measurements and the general availability of the measurement data will be extremely important.

As a closing note, there are many books available for the budding computational electromagneticist besides the ones already referenced.

REFERENCES

1. Stutzman, W.L.; Thiele, G.A. *Antenna Theory and Design*; Wiley: New York, 1998.
2. Balanis, C.A. *Advanced Engineering Electromagnetics*; Wiley: New York, 1989.
3. Keller, J.B. Geometrical theory of diffraction. J. Opt. Soc. Am. **1962**, *52*, 116–130.
4. Ling, H.; Chou, R.; Lee, S.-W. Shooting and bouncing rays: calculating the RCS of an arbitrarily shaped cavity. IEEE Trans. Antennas Propagat. **1989**, *37*, 194–205.
5. Ufimtsev, P.Ia. Approximate computation of the diffraction of plane electromagnetic waves at certain metal bodies: PT. I. Diffraction patterns at a wedge and a ribbon. Zh. Tekhn. Fiz. (USSR) **1957**, *27*, 1708–1718.
6. Kouyoumjian, R.G.; Pathak, P.H. A uniform theory of diffraction for an edge in a perfectly conducting surface. Proc. IEEE **1974**, *62*, 1448–1461.
7. Mitzner, K.M. Incremental length diffraction coefficients. Technical Rep. No. AFAL-TR-73-296, Northrop Corp., Aircraft Division, Apr 1974.
8. Shore, R.A.; Yaghjian, A.D. Incremental diffraction coefficients for planar surfaces. IEEE AP-S Trans. **1988**, *36*, 55–70.
9. Veruttipong, T.W. Time domain version of the uniform GTD. IEEE AP-S Trans. **1990**, *38*, 1757–1764.
10. Harrington, R.F. *Field Computation by Moment Methods*. Robert E. Krieger Publishing Co.: Malabar, FL, 1968.
11. Canning, F.X. The Impedance Matrix Localization (IML) method for moment-method calculations. IEEE Antennas Propagat. Mag. **1990**, *32*, 17–30.
12. Peterson, A.F.; Ray, S.L.; Mittra, R. *Computational Methods for Electromagnetics*; IEEE Press: New York, 1998.
13. Shore, R.A.; Yaghjian, A.D. Dual surface integral equations in electromagnetics. International Union of Radio Science XXVIIth General Assembly, Maastricht, Netherlands, 2002.
14. Burke, J.G.; Poggio, A.J. Numerical Electromagnetic Code (NEC)—Method of Moments Parts I, II, and III. Technical Document No. 116, Lawrence Livermore National Laboratory, U.S.A., 1981.
15. Press, W.H.; Teukolsky, S.A.; Vetterling, W.T.; Flannery, B.P. *Numerical Recipes in C.*; Cambridge University Press: Cambridge, U.K., 1997, 71–89.
16. Haupt, R.L.; Haupt, S.E. An introduction to multigrid using matlab. Computer Appl. Engg. J. **1994**, *2*, 421–431.

17. Kunz, K.S.; Luebbers, R.J. *The Finite Difference Time-Domain Method for Electromagnetics*; CRC Press: Boca Raton, FL, 1993.
18. Taflove, A. *Computational Electrodynamics: The Finite Difference Time-Domain Method*; Artech House: Boston, 1995.
19. Yee, K.S. Numerical solution of initial boundary value problems involving Maxwell's equations in isotropic media. IEEE AP-S Trans. **1966**, *14*, 302–307.
20. Georgakopoulos, S.V.; Birtcher, C.R.; Balanis, C.A.; Renaut, R.A. Higher order finite difference schemes for electromagnetic radiation, scattering, and penetration, part 1: theory. IEEE AP-S Magazine **2002**, *44*, 134–142.
21. Berenger, J.P. A perfectly matched layer for the absorption of electromagnetic waves. Journal of Computational Physics **1994**, *114*, 185–200.
22. Furse, C.M. Faster than Fourier: ultra-efficient time-to-frequency-domain conversions for FDTD simulations. IEEE AP-S Magazine **2000**, *42*, 24–33.
23. Jin, J. *The Finite Element Method in Electromagnetics*; Wiley: New York, 1993.
24. Volakis, J.L.; Chatterjee, A.; Kempel, L.C. *Finite Element Method for Electromagnetics*; IEEE Press: New York, 1998.
25. Ludwig, A.C. A comparison of spherical wave boundary value matching versus integral equation scattering solutions for a perfectly conducting body. IEEE Trans. On Antennas Propagat. **1986**, *34*, 857–865.
26. Christopoulos, C. *The Transmission-Line Modeling Method*; IEEE Press: Piscataway, NJ, 1995.
27. Miller Edmund, K.; Sarkar Tapan, K. Model-order reduction in electromagnetics using modelbased parameter estimation. In *Frontiers in Electromagnetics*; Werner, D.H., Mittra, R. Eds.; IEEE Press: NY, 1999, 371–436.
28. Baum, C.E. The singularity expansion method: background and developments. IEEE AP-S Mag. **1986**, *28*, 15–23.
29. Luenberger, D.G. *Linear and Nonlinear Programming*; Addison-Wesley: Reading, MA, 1984.
30. Nelder, J.A.; Mead, R. Computer J. **1965**, *7*, 308–313.
31. Metropolis, N.; Rosenbluth, A.; Rosenbluth, M.; Teller, A.; Teller, E. Equation of state calculations by fast computing machines. J. Chem. Phys. **1953**, *21*, 1087–1092.
32. Kirkpatrick, S.; Gelatt Jr. C.D.; Vecchi, M.P. Optimization by simulated annealing. Science **1983**, *220*, 671–680.
33. Holland, J.H. *Adaptation in Natural and Artificial Systems*; University of Michigan Press: Ann Arbor, 1975.
34. Goldberg, D.E. *Genetic Algorithms in Search, Optimization, and Machine Learning*; Addison-Wesley: Reading, MA, 1989.
35. Michielssen, E.; Sajer, J.M.; Ranjithan, S.; Mittra, R. Design of lightweight broad-band microwave absorbers using genetic algorithms. IEEE Trans. Microwave Theory Tech. **1993**, *41*, 1024–1031.
36. Haupt, R.L. Thinned arrays using genetic algorithms. IEEE AP-S Transactions, **1994**, *42*, 993–999.
37. Haupt, R.L. An introduction to genetic algorithms for electromagnetics. IEEE Antennas Propagat. Mag. **1995**, *37*, 7–15.
38. Haupt, R.L. *Haupt Sue Ellen. Practical Genetic Algorithms*; Wiley: New York, 1998.
39. Haupt, R.L. Haupt, S.E.; Optimum population size and mutation rate for a simple real genetic algorithm that optimizes array factors. Appl. Computat. Electromagnet. Soc. J. **2000**, *15*, 94–102.
40. Rioul, O.; Vetterli, M. Wavelets and signal processing. IEEE Signal Proc. Mag. **Oct. 1991**, *11*, 14–38.
41. Strang, G. Wavelets. American Scientist **1994**, *82*, 250–255.
42. Steinberd, B.Z.; Leviatan, Y.; On the use of wavelet expansions in the method of moments. IEEE AP-S Trans. **1993**, *41*, 610–619.
43. Taflove, A.; Umashankar, K. A hybrid moment method/finite difference time domain approach to electromagnetic coupling and aperture penetration into complex geometries. IEEE Trans. Antennas Propagat. **1982**, *30*, 617–627.
44. Wang, Y.; Safavi-Naeini, S.; Chaudhuri, S.K. A hybrid technique based on combining ray tracing and FDTD methods for site-specific modeling of indoor radio wave propagation. IEEE Trans. Antennas Propagat. **2000**, *48*, 743–754.
45. Jakobus, U.; Landstorfer, F.M. Improved PO-MM hybrid formulation for scattering from three-dimensional perfectly conducting bodies of arbitrary shape. IEEE Trans. Antennas Propagat. **1995**, *43*, 162–169.
46. Miller, E.K.; Shaeffer, J. Theory, techniques and applications of electromagnetic visualization. Short Course Notes from the IEEE AP-S Symposium, Salt Lake City, UT, July 2000.

6
Biological Effects of Electromagnetic Fields

Riadh Habash
University of Ottawa
Ottawa, Ontario, Canada

6.1. INTRODUCTION

Electromagnetic (EM) fields have become a driving force of our civilization through their numerous applications. However, there are concerns about the hazards that might exist due to exposure to such fields. Actually, such concerns began as early as the eighteenth century, which saw rapid developments in medical applications and physiological effects of electricity and magnetism.

EM field is classified as either nonionizing or ionizing. There is a fundamental distinction made between ionizing field, which has enough energy to physically break chemical bonds at the molecular level, and nonionizing field, which does not. Nonionizing fields (frequencies below the ultraviolet range), which are the subject of this chapter, have photon energy less than 10 eV, a level not enough to produce ions by ejection of orbital electrons from atoms, but still have a strong effect, which is heating.

Investigations started after World War II, and much of the concern was directed toward possible health hazards of radio-frequency radiation (RFR). In the following years, with the help of the media, public concern diverted from RFR to electric and magnetic fields (EMFs). Also, attention shifted from the strong electric fields near highvoltage power lines to those relatively weak magnetic fields produced by distribution lines and electrical appliances. In recent years, concerns regarding RF exposure from mobile phones have grown considerably. These concerns are generated because of the wide use of such equipment and they are largely inflamed by the fact that the mobile phone is placed very close to the user's head.

This chapter traces the various components of the entire subject including interaction mechanisms, safety standards and protection guidelines, sources and exposure scenarios, description of large-scale epidemiological studies involving humans as well as research on exposure of cells and animals relevant to adverse health effects.

6.2. ELECTRIC AND MAGNETIC FIELDS

There are two types of EMFs classified according to the frequency range: extremely low frequency (ELF) fields and very low frequency (VLF) fields. ELF fields are defined as those having frequencies up to 3 kHz. VLF fields cover the frequency range 3–30 kHz. Because of the quasistatic nature of EM fields at these frequencies, electric and magnetic fields act independently of one another and are measured separately. Electric fields created by voltage and measured in volts per meter (V/m), are present whenever an electric appliance is plugged in. The appliance need not be turned on for electric fields to be detected. Magnetic fields, induced by alternating current (AC) and measured using the derived quantity magnetic flux density (B) in tesla (T) or gauss (G), are present when the appliance is turned on. The strength of EMFs decreases as we move away from their sources.

Any residential or occupational site is subject to coincident exposure from many EMF sources external and internal to the site itself. External sources include high-voltage power lines, distribution lines, underground cables, substations, transformers, and transportation systems. In the workplace, sources of EMFs include computers, fax machines, copy machines, fluorescent lights, printers, scanners, telephone switching systems (PBX), motors, induction heaters, electronic article surveillance (EAS), demagnetizers, security systems, and metal detectors. In homes, there are two immediate sources of EMFs. The first type includes internal wiring, meters, service panels, subpanels, and grounding systems. The second type includes electrical appliances such as electric blankets, electric waterbed heaters, hairdryers, electric shavers, television (TV) sets, video display terminals (VDTs), stereo systems, air conditioners, fluorescent lights, refrigerators, blenders, portable heaters, clothes washers and dryers, coffee makers, vacuum cleaners, toasters, and other household appliances.

6.2.1. Interaction Mechanisms

There are several proposed mechanisms for the interaction of EM fields with living systems. These include induced electric currents, direct effect on magnetic biological materials, effects on free radicals, and excitation of cell membranes.

Before discussing these mechanisms, one must understand the relationship between electric and magnetic fields outside and inside biological systems (coupling), which varies greatly with frequency. Electric fields are greatly diminished by many orders of magnitude inside biological tissues from their values in air external to the tissues. Biological tissues are nonmagnetic materials, which mean the magnetic field inside the human body is same outside it.

The first mechanism involves the ability, through magnetic induction, to stimulate eddy currents at cell membranes and in tissue fluids, which circulate in a closed loop that lies in a plane normal to the direction of the magnetic field. The above current can be calculated using only Faraday's law and Laplace's equations, without simultaneously solving Maxwell's equations. Both current and electric fields are induced inside living systems by external time-varying magnetic fields [2].

All living organisms are basically made of diamagnetic organic compounds, but some paramagnetic molecules (e.g., O_2), and ferromagnetic microstructures (hemoglobin core, magnetite) are also present. Biological magnetites are usually found in single domain units, covered with thin membranes called *magnetosomes* (Fe_3O_4). These microstructures behave like small magnets and are influenced by external fields changing their energy content. They are found in bacteria and other small biological elements. Such bacteria and biological elements orient along the applied magnetic fields.

According to Foster [3],

> Low-frequency electric fields can excite membranes, causing shock or other effects. At power line frequencies, the threshold current density required to produce shock is around $10\ A/m^2$, which corresponds to electric field of $100\ V/m$ in the tissue. Electric fields can create pores in cell membranes by inducing electric breakdown. This requires potential differences across the membranes at levels between 0.1 and 1 V, which, in turn, requires electric field in the medium surrounding the cell of at least $10^5\ V/m$.

Many life scientists through series of studies [4–6] believe that the cell membrane plays a principal role in the interaction of EM fields with biological systems. Indications point to cell membrane receptors as the probable site of initial tissue interactions with EM fields for many neurotransmitters, growth-regulating enzyme expressions, and cancerpromoting chemicals.

Scientists theorizing this mechanism conclude that biological cells are bioelectrochemical structures, which interact with their environment in various ways, including physically, chemically, biochemically, and electrically. According to Dr. William Ross Adey at the University of California, Riverside [7], "The ions, especially calcium ions could play the role of a chemical link between EM fields and life processes. The electrical properties and ion distribution around cells are perfect for establishing effects with external steady oscillating EM fields."

The impact of EM fields may also be understood in terms of amplification and/or the cooperative sensing associated with simultaneous stimulation of all membrane receptors. Litovitz et al. [8] hypothesized that oscillating EM fields need to be steady for certain period of time (approximately 1 s) for a biological response to occur. This allows cells to discriminate external fields from thermal noise fields, even though they might be smaller than the noise fields.

6.2.2. Laboratory Studies

Scientists look to laboratory studies as a source of information that will address concerns regarding likely health effects. Laboratory studies on cells or whole organisms play a key role in evaluating the response of different systems of the body. Laboratory studies are easier to control and provide the opportunity to check whether EMFs cause cancer or other illnesses, something that is not possible with human volunteers. However, laboratory studies entail complications especially those related to extrapolation to humans. Numerous health effects from EMFs have been discussed in the literature, but most of the attention has focused on possible relationship with DNA and cancer.

Numerous cellular studies referred to as in vitro have been carried out to find out if EMFs can damage DNA or induce mutations. In general, it is believed that the energy associated with EMFs is not enough to cause direct damage to DNA; however, it is understood that indirect effects might be possible by EMF changing processes within cells that could lead to DNA breakage. Meanwhile, EMFs well above environmental field intensities might enhance DNA synthesis, change the molecular weight distribution during protein synthesis, delay the mitotic cell cycle, and induce chromosome aberrations [9–11]. In contrast, EMFs, according to a number of studies [12–15], are unable to induce chromosomal aberrations even under relatively strong magnetic field exposure.

Studies of animals referred to as in vivo aim to determine the biological effects of EMFs on whole animals. Animal studies are very important because they supplement epidemiological studies and can provide a reliable model in which to look at exactly how EMFs characteristics cause the risk. There has been no absolute evidence in any study that low-level EMFs alone can cause cancer in animals. This is supported by the findings of many studies conducted during the last few years [16–18]. A study of animals treated with a known chemical initiator have shown greater numbers of tumors in those animals subsequently or concurrently exposed to magnetic fields at moderate to high exposure levels [19].

It is clear from the literature that the energy associated with EMF environmental exposures is not enough to cause direct damage to DNA or cause cancer in animals.

6.2.3. Melatonin Hypothesis

One possible interaction hypothesis under investigation is that exposure to EMFs suppresses the production of melatonin, which is a hormone produced by the pineal gland, a small pinecone-shaped gland located deep near the center of the brain. Melatonin is produced mainly at night and released into the blood stream to be dispersed throughout the body. It surges into almost every cell in the human body, destroying the free radicals and helping cell division to take place with undamaged DNA. Melatonin reduces secretion of tumor-promoting hormones. It has the ability to increase cytotoxicity of the immune system's killer lymphocytes; therefore, its production is essential for the immune system, which protects the body from infection and cancer cells. Various cancers might proliferate if melatonin is lowered such as breast cancer, prostate cancer, and ovarian malignancies. Figure 6.1 displays consequences of melatonin reduction.

Several studies [20–22] have found melatonin reduction in cells, animals, and humans exposed to EMFs. The effect varies according to the period of exposure and strength of EMFs.

In contrast, Rogers et al. [23,24] exposed baboons to 60-Hz fields at 6 kV/m plus 50 μT or at 30 kV/m and 100 μT (12 h/d for 6 weeks). They noticed no evidence of any effect on melatonin levels. Graham et al. [25] found also no effects on melatonin levels among young men volunteers exposed on four continuous nights to 60-Hz fields at 28.3 μT.

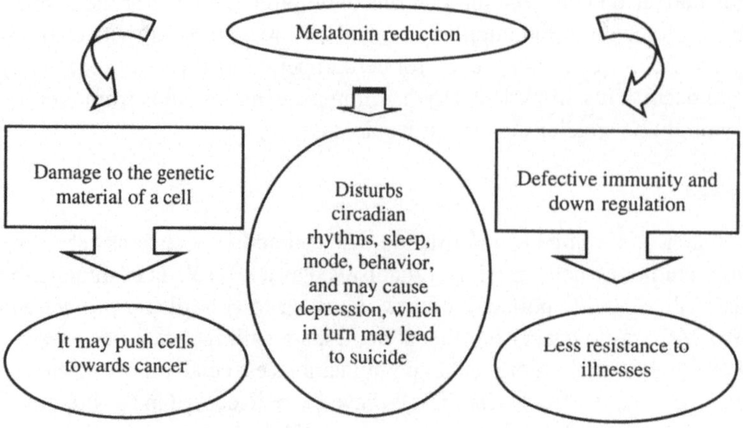

Figure 6.1 Biological consequences of melatonin reduction.

6.2.4. Human Studies

Effects of EMFs might be studied safely and effectively in the laboratory with human volunteers in spite of limitations to the duration of exposure and types of tests that are performed. Laboratory studies on humans have certain advantages. They focus directly on the "right" species, therefore avoiding the problem of extrapolation from data obtained in other species. Even negative results can be of immediate use in addressing public concerns. Such studies may also be used to directly evaluate the effects of exposure on "real-life" functions. The main sources of information in this field are surveys of people and workers living close to potential sources of EMFs, laboratory tests, and epidemiological data.

EMFs may affect the brain and nervous system and may cause effects on normal behavior or cognitive abilities of humans have been a persistent concern. In the early studies of occupational exposure to EMFs [26], switch yard workers in the former Soviet Union who differed in the duration and intensity of their exposure to 50-Hz fields suffered from an abnormally high incidence of neurophysiological complaints. A recent review on behavioral effects of EMFs was conducted by Zenon [27].

Heart rate and blood pressure may assess cardiovascular functions. Current densities of about 0.1 A/m^2 can stimulate excitable tissues, while current densities above about 1 A/m^2 interfere with the action of the heart by causing ventricular fibrillation, as well as producing heat. For example, Sazonova [26] observed that the pulse rates of people among workers with an average exposure of 12–16 kV/m for more than 5 h/d were lower by 2–5 beats/min at the end of the day, although they had been equivalent at the start of the day. According to a review by Stuchly [28], exposure of healthy male volunteers to 20-μT electric and magnetic fields at 60 Hz has been linked to a statistically significant slowing of the heart rate and to changes in a small fraction of the tested behavioral indicators. Korpinen et al. [29] used ambulatory recording techniques to carry out an extensive study on the effects of EM occupational exposure on heart rate. No field-related changes in mean heart rate were found as a result of exposure to 50-Hz fields directly under power lines ranging in intensity from 110 to 400 kV.

6.2.5. Epidemiological Studies

Epidemiological studies address the observed effects of possibly harmful EMF exposure on human health and whether the level of exposure is related quantitatively to the severity of health effects. These studies are limited in the sense that they are indirect experiments where the exposure can only be assessed through different substitute measures. An epidemiological association, if found, might not be related directly to exposure, it may be due to chance, confounding factors, or some unrecognized factors related to the way the data have been collected.

Childhood Leukemia

Childhood is a critical period of rapid cell growth and the cancer development cycle is correspondingly much quicker than adults. In addition, a child's immune system is underdeveloped and melatonin production is lower. Childhood exposure to EMFs has been studied intensively for many decades. However, research into this area gained momentum in 1979, when one of the first epidemiological studies [30] showed an association between exposure to EMFs and cancer among children living near power lines. Many studies have since been conducted but they do not support the notion that EMF exposure increases the risk of childhood cancer [31–36].

The association between EMF exposure and childhood cancer is inadequate and inconclusive. Some studies showed a link but their findings have small risk magnitudes by epidemiological standards with odds ratio (OR) <5 and were unable to exclude other environmental influences.

Adult Cancers

Occupational exposure was studied considering various health problems as well as adult cancers, including brain tumors and leukemia [37–42]. Sahl et al. [37] studied utility workers at Southern California Edison. Comparisons in the cohort study focused on electrical versus nonelectrical workers, and exposure was characterized on the basis of job history. The authors noticed no difference in risk for brain cancer among electrical workers compared to the reference group. However, small but significant increases in brain cancer risk were observed for electricians with risk ratio (RR) = 1.6 and plant operators (RR = 1.6).

Researchers from Canada and France [38] conducted a study of 223,292 workers at three large utilities, two in Canada (Hydro Quebec and Ontario Hydro) and a national utility in France (Electricite de France). The result shows that workers with acute myeloid leukemia (AML) were about three times more likely to be in the half of the workforce with higher cumulative exposure to magnetic fields. In the analysis of median cumulative magnetic field exposure, no significant elevated risks were found for most types of cancer studied.

The elevated risks of leukemia were also seen among senior workers who spent the most time in electric fields above certain thresholds, in the range of 10 to 40 V/m [41]. In a recent Canadian population-based control study, Villeneuve et al. [42] conducted a study among men in eight Canadian provinces, for 543 cases of brain cancer confirmed histologically (no benign tumors included). Astrocytoma and glioblastomas accounted for over 400 of these. Population based controls (543) were selected to be of similar age. They reported a nonsignificant increased risk of brain cancer among men who had ever held a job with an average magnetic field exposure $>0.6\ \mu T$ relative to those with exposures $<0.3\ \mu T$. A more pronounced risk was observed among men diagnosed with glioblastoma multiforme (the most malignant of neuroepithelial neoplasms) (OR = 5.36). There are rather notable differences in adult cancer studies with two kinds of results: (1) null association [37,40] and (2) mixed but in general strongly positive results from Canada-France study [38] and Canadian senior workers Villeneuve et al. [41,42]. RRs in the upper exposure categories were above 2.0 and for the more highly exposed groups between 1.1 and 1.3. RRs of this magnitude are below the level at which a casual association between EMFs and cancer can be assessed.

6.2.6. Safety Standards and Protection Guidelines

Safety standard is a standard specifying measurable field values that limit human exposure to levels below those deemed hazardous to human health. The standard consists of regulations, recommendations, and guidelines that would not endanger human health.

There are many worldwide institutions and organizations that have recommended safety limits for EM exposure. These include the Institute of Electrical and Electronic Engineers (IEEE) [43], the National Radiological Protection Board (NRPB) of the United Kingdom [44], the International

Commission on Non-Ionizing Radiation Protection (ICNIRP) [45], the Swedish Radiation Protection Institute [46], Safety Code 6 of Canada [47], and Australian Radiation Protection and Nuclear Safety Agency (ARPANSA) [48].

Most of the protection guidelines use a two-tier standard, indicating a basic restriction (current density) and corresponding investigation levels or reference levels (external field strengths). The exposure limits range from few microtesla (μT) up to 1300 μT. The levels for those occupationally involved in various electrical industries are set higher than those for the general public.

The IEEE has a standard covering exposures above 3 kHz but, at present, does not have a standard covering the lower frequencies relevant to the electricity power system. However, a new standard is being prepared by Subcommittee 28 that will be based on known interactions of internal electric fields with the different parts of the nervous system.

The recommended NRPB guidelines are same for occupational and public environments. The basic restriction specified by the NRPB is an induced current density of 10 mA/m^2 in the head and trunk, while the investigation levels for electric and magnetic fields at 50 Hz are 12 kV/m and 1600 μT, respectively.

Sweden has been a leader in developing recommended visual ergonomic and EM emission standards for computer displays. The Swedish Confederation of Professional Employees, or TCO, which represents over a million workers, published its own series of guidelines [46], which include guidelines for energy consumption, screen flicker, luminance, and keyboard use.

6.3. RADIO-FREQUENCY RADIATION

As defined by the Institute of Electrical and Electronics Engineers (IEEE), RFR is a band in the electromagnetic spectrum that lies in the frequency range of 3 kHz to 300 GHz. Microwave (MW) radiation is usually considered a subset of RFR, although an alternative convention treats RF and MW as two separate spectral regions. Microwaves occupy the spectral region between 300 MHz to 300 GHz, while RF includes 3 kHz to 300 MHz. Since they have similar characteristics, RF and MW are recognized together, and referred to as RFR throughout this chapter.

Many frequencies of RFR are used in various applications. For example, the frequency range of 5 to 16 kHz is used in AM radio transmission, while 76 to 108 MHz is used for FM radio. Cellular and personal communication uses frequencies between 800 MHz and 3 GHz. The 2.45 GHz is reserved for industrial, scientific, and medical (ISM) applications, mainly microwave cooking.

The interaction of RF fields with living systems, and consequently their related bioeffects, can be considered at various levels including the molecular, subcellular, organ, system level, or the entire body. Biological effects due to RF exposure are classified as highlevel (thermal) effects, intermediate-level (athermal) effects, and low-level (nonthermal) effects.

6.3.1. Thermal Effects

An obvious outcome of RFR absorption by the human body is heating (thermal effect), where the core temperature of the body rises despite the process of thermoregulation by the body. Many of the biological effects of RFR that have significant implications for human health are related to induced heating or induced current. Heating is the primary interaction of RF fields at high frequencies especially above about 1 MHz. Below about 1 MHz, the induction of currents in the body is the dominant action of RFR. Heating from RFR best relates to specific absorption rate (SAR) rather than to incident power density to account for differences in coupling.

Biological systems alter their functions as a result of a change in temperature. It is worth mentioning that most adverse health effects due to RF exposure between 1 MHz and 10 GHz are consistent with responses to induced heating, resulting in raising tissue temperatures higher than 1°C.

Elevated temperatures have obvious effects on humans such as increased blood pressure, dizziness, weakness, disorientation, and nausea.

6.3.2. Athermal and Nonthermal Effects

Controversy surrounds two issues regarding biological effects of intermediate- and low-level RFR. First, whether RFR at such levels can even cause harmful biological changes in the absence of demonstrable thermal effects. Second, whether effects can occur from RFR when thermoregulation maintains the body temperature at the normal level despite the EM energy deposition or when thermoregulation is not challenged and there is no significant temperature change. In response to the first issue, investigations on the extremely low-level RFR have been established and some results confirmed but knowledge is yet inconclusive.

Regarding the second issue, a biological effect may have two meanings. It may mean an effect that occurs under circumstance of no evident change in temperature or the exposure level is low enough not to trigger thermoregulation in the biological body under irradiation, suggesting that physiological mechanisms maintain the exposed body at a constant temperature. Such case is related to nonthermal effect where the effect occurs through mechanisms other than those due to macroscopic heating. The second meaning is that RFR causes a biological effect, without the involvement of heat. This is sometimes referred to as *athermal effect*.

6.3.3. Toxicological Studies

Health effects are often the result of biological effects that accumulate over time and depend on exposure dose. For example, if an effect of EM exposure has been noticed on cultured cells, this does not essentially mean that the exposure will lead to adverse effect for the health of the organism as a whole. In general, the number of cellular and animal studies in the literature is large due to the large number of cellular processes and systems that may probably be affected by RFR.

A relationship between RFR and cancer would indicate that RFR somehow induces mutations in the DNA which, in turn, can disrupt cell growth and developing, leading to cancer. A number of laboratory experiments have been conducted to assess possible effects of RFR on genetic material. Investigations on different cell systems found no evidence for any direct genotoxic or mutagenic effects of continuous and pulsed RFR at different power densities. Tice et al. [49], as a part of comprehensive investigation of the potential genotoxicity of RF signals emitted by mobile phones, demonstrated that under extended exposure conditions, RFR from mobile phones at an average SAR of at least 5 W/kg are capable of inducing chromosomal damage in human lymphocytes. Similar findings were reported by d'Ambrosio et al. [50] while radiating human cells to 1748 MHz at 5 W/kg and by Mashevich et al. [51] when radiating human lymphocytes to continuous 830-MHz RF energy at SAR in the range 1.6–8.8 W/kg for 72 h. These results show that RFR has a genotoxic effect. Since the positive findings in the literature were consistently associated with hyperthermia, it will be concluded that RFR at low-intensity levels do not induce any genetic damage under nonthermal conditions.

Disturbance of normal cell cycle is a possible sign of uncontrolled cell growth, or cancer. Czerska et al. [52] reported an increased proliferation of cells exposed to 2.45-GHz RFR at SAR of 1 W/kg when the radiation was pulsed. Continuous wave (CW) RFR increased proliferation only when absorbed energy was high enough to induce heating. Other investigators reported increased and decreased cell proliferation rates after applying RFR of various SARs [53–55]. In contrast, d'Ambrosio et al. [50] found no significant changes in cell distribution or cell proliferation in cells exposed to 1748 MHz, either CW or phase only modulated wave (GMSK) for 15 min.

6.3.4. Noncancerous Effects on Animals

While most of experimental studies focus on carcinogenesis, tumor promotion, and mutagenic effects, other noncancer effects also need to be considered. RFR can induce morphological and physiological changes. According to Adey et al. [56], RF carriers sinusoidally modulated at ELF fields can induce changes to the CNS. However, Tsurita et al. [57] found no significant changes in the groups of rats exposed for 2–4 weeks to a 1439-MHz (2 W/kg) TDMA signal on the morphological changes of the brain. The exposure period was 2 or 4 weeks.

RFR can induce cataracts if the exposure intensity and the duration are sufficient. Many studies on the ocular effect of RFR on animals have reported no effects, despite the fact that most studies employed exposure levels greatly in excess of that seen with mobile phones [58,59].

Some changes in learning behavior occurred after RF exposure. Lai et al. [60] observed retarded learning of a task in rats exposed to 2.45 GHz. However, Bornhausen and Scheingraber [61] found that exposure in utero to the GSM (900 MHz, 217-Hz pulse-modulated RFR; 17.5 and 75 mW/kg) field did not induce any measurable cognitive deficits in exposed Wistar rats during pregnancy. Dubreuil et al. [62] noted that head-only exposure of rats to 900 MHz pulsed RFR (SAR of 1 or 3.5 W/kg) for 45 min had no effect on learning.

RFR-induced breakdown of the blood–brain barrier (BBB) have been studied either alone or in combination with magnetic fields. Many authors agree that exposure to RFR affects BBB in vivo [63–65]. However, other studies have not found RFR-induced disruption of the BBB [66,67].

6.3.5. Human Studies

Ocular Effects

The cornea and lens are the parts of the eye most exposed to RFR at high levels by their surface location and because heat produced by the RFR is more effectively removed from other eye regions by blood circulation.

One related modeling study of the human eye by Hirata et al. [68] showed that 5 mW/cm^2 caused a temperature change in the lens less than 0.3°C at frequencies from 0.6 to 6 GHz. This small temperature change is overestimated because the eye model was thermally isolated from the head and did not consider blood flow. Therefore, RF exposures much in excess of currently allowable exposure limits would be required to produce cataracts in human beings, and exposures below the cataractogenic level would be expected to cause other effects in other parts of the eye and face.

Reviews of the literature of RFR-induced cataracts [69,70] concluded that clinically significant ocular effects, including cataracts, have not been confirmed in human populations exposed for long periods of time to low-level RFR.

Brain Functions

The close placement of RFR sources such as mobile phones to the user's head has elevated possibilities of interference with brain activities. While many studies have addressed this issue, they have only investigated the short-term effects of RFR. The controversial findings in the literature suggest that some aspects of cognitive functions and measures of brain physiology may be affected without offering a uniform view. These include changes in memory tasks, response patterns, normal sleeping EEG patterns, and other brain functional changes. Subjective symptoms such as dizziness, disorientation, nausea, headache, and other unpleasing feelings such as a burning sentient or a faint pain might be a direct result of RFR although such symptoms are very general and may have many causes.

The actual outcomes from the majority of studies have no serious implications for human health since the effects were seen for just a few of many tests and they were far too small to have any serious functional significance.

In a review, Hossmann and Hermann [71] concluded that "Most of the reported effects are small as long as the radiation intensity remains in the nonthermal range. However, health risks may evolve from indirect consequences of mobile telephony, such as the sharply increased incidence rate of traffic accidents caused by telephony during driving, and possibly also by stress reactions which annoyed bystanders may experience when mobile phones are used in public places."

Cardiovascular System

Jauchem [72] reviewed cardiovascular changes in humans exposed to RFR. Both acute and long-term effects were investigated. The author reported that most studies showed no acute effect on blood pressure, heart rate, or electrocardiogram (ECG) waveform; others reported subtle effects on the heart rate.

6.3.6. Epidemiological Studies

Navy Personnel

Robinette et al. [73] conducted a study of mortality results on males who had served in the U.S. Navy during the Korean War. They selected 19,965 equipment-repair men who had occupational exposure to RFR. They also chose 20,726 naval equipment-operation men who, by their titles, had lower occupational exposure to RFR as a control group. The researchers studied mortality records for 1955–1974, in-service morbidity for 1950–1959, and morbidity for 1963–1976 in veterans administration hospitals. No difference on cancer mortality or morbidity was seen among the high-exposure and low-exposure groups.

Military Workers

Szmigielski [74] showed strong association between RF exposure and several types of cancer (including brain cancer and cancer of the alimentary canal) was reported in a cohort of about 120,000 Polish military personnel, of whom 3% had worked with RF heat sealers. Exposure was determined from assessments of field levels at various locations. The study did not consider the length of time at the location, the nature of the job, or the number of cases observed.

Traffic Radar Devices

Davis and Mostofi [75], in a brief communication, reported six cases of testicular cancer in police who used handheld radars between 1979 and 1991 among a cohort of 340 police officers employed at two police departments within contiguous counties in the north-central United States. The six cases had been employed as police officers as their primary lifetime occupation, and all had been exposed to traffic radar on a routine basis. The mean length of service prior to testicular-cancer diagnosis was 14.7 y, the mean age at diagnosis was 39 y, and all had used radar at least 4½ y before the diagnosis.

Finkelstein [76] presented the results of a retrospective cohort cancer study among 22,197 officers employed by 83 Ontario police departments. The standardized incidence ratio (SIR) for all tumor sites was 0.90. There was an increased incidence of testicular cancer (SIR = 1.3) and melanoma skin cancer (SIR = 1.45). No information about individual exposures to radar devices was provided.

6.3.7. RF Heat Sealers

Lagorio et al. [77] reported higher cancer mortality among Italian plastic ware workers exposed to RFR generated by dielectric heat sealers for the period 1962–1992. Six types of cancers were found in the exposed group. The standardized mortality ratio (SMR) analysis was applied to a small cohort of 481 women workers, representing 78% of the total person-years at risk. Mortality from malignant neoplasms was slightly elevated, and increased risks of leukemia and accidents were detected. The all-cancer SMR was higher among women employed in the sealing. Exposure assessment was based on the time assigned on jobs. Exposure to RFR was based on a previous survey, which showed that the radiation exceeded 1 mW/cm^2. The work area also included exposure to chemicals associated with cancer (solvents and vinyl chloride), which may have impact on the result.

Telecom Operators

In Norway, Tynes et al. [78] studied breast cancer incidence in female radio and telegraph operators with potential exposure to light at night, RFR (405 kHz–25 MHz), and ELF fields (50 Hz). The researchers linked the Norwegian Telecom cohort of female radio and telegraph operators working at sea to the Cancer Registry of Norway to conduct their study. The cohort consisted of 2619 women who were certified to work as radio and telegraph operators. The incidences of all cancers were not significant, but an excess risk was seen for breast cancer. They noted that these women were exposed to light at night, which is known to decrease melatonin levels, an expected risk factor for breast cancer.

Radio and Television Transmitters

An association between proximity of residences to TV towers and an increased incidence of childhood leukemia was found in an Australian study conducted by Hocking et al. [79]. The researchers studied the leukemia incidence among people living close to television towers (exposed group) and compared this to the incidence among those living further out from the towers (unexposed or control group). People were assigned to one of the two groups based on data from the New South Wales Cancer Registry and their accompanying address. The Hocking study concluded that there was a 95% increase in childhood leukemia associated with proximity to TV towers. No such association was found between RFR emitted by the TV towers and adult leukemia. McKenzie et al. [80] repeated the Hocking study, using more accurate estimates of the exposure to RFR. The researchers looked at the same area and at the same time period, but with more accurate estimates of the RF exposure that people received in various areas. They found increased childhood leukemia in one area near the TV antennas but not in other similar areas near the same TV antennas. They found no significant correlation between RF exposure and the rate of childhood leukemia. They also found that much of the "excess childhood leukemia" reported by the Hocking study occurred before high-power 24-h TV broadcasting had started.

In Italy, Michelozzi et al. [81] conducted a small area study to investigate a cluster of leukemia near a high-power radio transmitter in a peripheral area of Rome. The leukemia mortality within 3.5 km (5863 inhabitants) was higher than expected. The excess was due to a significant higher mortality among men (seven cases were observed). Also, the results showed a significant decline in risk with distance from the transmitter, only among men.

Mobile Phones

Most of the mobile phone studies (Table 6.1 [82–88]) reported no increased incidence of brain tumors among mobile phone users (analog or digital phones). Furthermore, there was no relationship between brain tumor incidence and duration of mobile phone use. Only one group of researchers in Sweden [82] has reported associations between analog phone use and brain tumors. Their results have found no support in the investigation of other researchers. It is also doubtful whether results for analog phone users can be extrapolated to digital phone users.

6.3.8. Safety Standards and Exposure Guidelines

How much RF energy is safe? This is a complex problem, comprising public health, life sciences, engineering, and social (including economic and legal) considerations. Currently, there are various safety standards established for RF exposure in most of the industrial world (Table 6.2 [43–48,89]).

Table 6.1 Summary of Epidemiological Studies of Cellular Phones and Cancer Risk

Investigator	Description[a]	Risk measure	Outcome
		Brain tumors	
Hardell et al., 1999 [82]	CC: Sweden (1994–1996); (GSM/NMT phones); 209 brain tumor cases; 425 controls.	OR = 0.98 (0.69–1.41); Same side of the head: OR = 2.42 (0.97–6.05)	Right brain tumors for users who used the phone at their right ear. Stronger for temporal or occipital localization of the tumor on right side (only for analog phones). Temporal or occipital localization of the tumor on the same side as phone use for the left side use.
Muscat et al., 2000, 2002 [83,84]	CC: USA (1994–1998); 469 brain cancer; 422 controls.	OR = 0.85 (0.6–1.2)	No significant association between primary brain cancer and years of mobile phone use, number of hours of use per month, or the cumulative number of hours of use.
Inskip et al., 2001 [85]	CC: USA (1994–1998); 489 Glioma; 197 Meningiomad; 96 Acoustic neuroma; 799 controls.	OR = 1.0 (0.6–1.5); Glioma: 0.9 (0.5–1.6); Meningioma: 0.2 (0.3–1.7); Acoustic neuroma: 1.4 (0.6–3.5).	The results do not support the existence of an association between mobile phone use and certain cancers (glioma, meningioma, or acoustic neuroma). There was no difference for side of head.
Johansen et al., 2001 [86]	CE: Denmark (1982–1995); 420,095 users from two operators; 3391 cancers; 3825 expected.	SIR = 0.89 (0.86–0.92); Brain: SIR = 0.95 (0.81–1.12); Salivary gland; SIR = 0.72 (0.29–1.49); Leukemia: SIR = 0.97 (0.78–1.21).	No relationship between brain tumor risk and RF dose compared by duration of phone use, date since first subscription, age at first subscription, or type of phone used.
Auvinen et al., 2002 [87]	CC: Finland (1996); 398 brain tumors; 198 gliomas; 34 salivary gland; 5 controls per case.	Brain tumor: OR = 1.3 (0.9–1.8); Salivary gland: OR = 1.3 (0.4–4.7); Gliomas: OR = 2.1 (1.3–3.4) (Analog); Gliomas; OR = 1.0(0.5–2.0) (Digital).	No clear association between use of mobile phones and risk of cancer has been provided. Gliomas were associated with the use of analog but not digital phones.
		Melanoma of the eye	
Stang et al., 2001 [88]	HBPBCC: Germany; (1994–1997); 118 case; 475 control.	OR = 3.0 (1.4–6.3)	Association between RF exposure from mobile phones and uveal melanoma.

[a]OR: Odds Ratio; CC: Case Control; CE: Case ecological; HBPBCC: Hospital-based population-based case control.

Table 6.2 Maximum Permissible Exposures to RFR

Standard	Frequency range	Whole body SAR (W/kg)		Local SAR in head (W/kg)		Local SAR in limbs (W/kg)	
		Public	Occupational	Public	Occupational	Public	Occupational
ARPANSA [48]	100 kHz–6 GHz	0.08 (6)[a]	0.4 (6)	2 [10][b] (6)	10 [10] (6)	4 [10] (6)	20 [10] (6)
Safety Code 6 [47]	100 kHz–10 GHz	0.08 (6)	0.4 (6)	1.6 [1] (6)	8 [1] (6)	4 [10] (6)	20 [10] (6)
ICNIRP [45]	100 kHz–6 GHz	0.08 (6)	0.4 (6)	2 [10] (6)	10 [10] (6)	4 [10] (6)	20 [10] (6)
FCC [89]	100 kHz–6 GHz	0.08 (30)	0.4 (6)	1.6 [1]	8 [1] (6)	4 [10]+[c]	20 [10] (6)+
NRPB [44]	100 kHz–6 GHz	0.4 (15)		10 [10] (6)		20 [100] (6)	
ANSI/IEEE[43]	100 kHz–6 GHz	0.08 (30)	0.4 (6)	1.6 [1] (30)	8 [1] (6)	4 [10] (30)+	20 [10] (6)+

[a]() Averaging time in minutes.
[b][] Averaging mass in grams.
[c]+ in hands, wrists, feet and ankles.

SAR is the rate at which RF energy is absorbed by the tissue and thus is a good predictor of thermal effects. SAR is defined as

$$\text{SAR} = \frac{\sigma |E|^2}{\rho} = c\frac{dT}{dt}$$

where E is the effective value of the electric field intensity (V/m), dT/dt is the time derivative of the temperature (K/s), σ is the electrical conductivity (S/m), ρ is the mass density (kg/m^3), and c is the specific heat (J/kg K). The unit of SAR is W/kg. The SAR is the dosimetric measure that is used for extrapolating across species.

SAR calculations and estimates usually use many EM properties of biological tissues (e.g., complex dielectric constants and conductivity of different tissues) whose accuracy depends on their acquisition techniques, which are mostly in vivo.

There are two major types of SAR: (1) a whole-body average SAR and (2) a local (spatial) peak SAR when the power absorption takes place in a confined body region, as in the case of the head exposed to mobile phone. Whole-body SAR measurements are significant to estimate elevations of the core body temperature. As SAR increases, the possibility for heating and, therefore, tissue damage also rises. The whole-body SAR for a given organism will be highest within a certain resonant frequency range, which is dependent on the size of the organism and its orientation relative to the electric and magnetic field vectors and the direction of wave propagation. For the average human the peak whole-body SAR occurs in a frequency range of 60–80 MHz, while the resonant frequency for a laboratory rat is about 600 MHz.

Both SARs are averaged over a specific period of time and tissue masses of 1 or 10 g (defined as a tissue volume in the shape of a cube). Averaging the absorption over a larger amount of body tissue gives a less reliable result. The 1-g SAR is a more precise representation of localized RF energy absorption and a better measure of SAR distribution. Local SAR is generally based on estimates from the whole-body average SAR. It incorporates substantial safety factors (for example, 20).

There are two local SAR safety limits applicable to mobile phones: 1.6 W/kg averaged over 1 g ($\text{SAR}_{1\,g}$) in North America, and 2 W/kg averaged over 10 g ($\text{SAR}_{10\,g}$) developed by the ICNIRP of the European Union and accepted for use in Australia, Japan, and other parts of the world. Whether 1.6 W/kg or 2 W/kg is a right limit for RF exposure remains controversial.

Exposure to RFR from mobile phones is in the region close to antenna, the near field. However, exposure from other sources is in the far field, which is often quantified in terms of power density, expressed in units of watts per square meter (W/m^2). At the lower frequencies, about 0.1 to 10 MHz, the energy absorbed is less important than current density and total current, which can affect the nervous system. There is an overlap region at the upper part of this range where either current density or energy absorption rate is the limiting quantity. The standards at the lower frequencies are concerned with preventing adverse effects on the central nervous system (CNS) and electric shock. Exposure limits at these lower frequencies also involve numerous technical issues as well, but are not the focus of this review.

6.4. DOSIMETRY

Dosimetry in this manner considers the measurement or determination by calculation of the internal fields, induced current density, specific absorption (SA), or SAR distributions in objects like models (phantoms), animals, humans, or even parts of human body exposed to RFR.

Internal dosimetry can be divided into two categories [90]: macroscopic and microscopic dosimetry. In macroscopic dosimetry, the EM fields are determined as an average over some volume of space, such as in mathematical cells that are small in size. For example, the electric field in a given mathematical cell of 1 mm is assumed to have the same value everywhere within 1 mm^3 volume of the cell. The same is applied for magnetic fields. While in microscopic dosimetry, the fields are determined at a microscopic (cellular) level. Or the other way, the mathematical cells over which the EM fields are determined are microscopic in size. Microscopic dosimetry is useful for studies at the cellular level, which may throw light on EM interaction mechanisms.

6.4.1. Theoretical Dosimetry

The internal field in any biological material irradiated by RFR is calculated by solving Maxwell's equations. Practically, this is a difficult task and may be done for only a few special cases. Because of the mathematical difficulties encountered in the process of calculation a combination of techniques is used to find SAR in any biological object. Each technique gives information over a limited range of parameters in such a way that suits the chosen model.

In general, computational methods for analyzing EM problems fall into three categories: analytical techniques, numerical techniques, and expert systems. Analytical techniques apply assumptions to simplify the geometry of the problem in order to apply a closed-form solution. Numerical techniques attempt to solve basic field equations directly, due to boundary conditions posed by the geometry. However, expert systems estimate, but do not calculate, the values of fields for the parameters of interest based on a rules database.

Finite difference time-domain (FDTD) method is one of the most popular modeling techniques currently being used for EM interactions and SAR analysis. Starting from the evaluated SAR distribution, the thermal responses as a function of time, until the steady-state reaches may also be calculated through the application of finite difference method (FDM) or other numerical techniques.

6.4.2. Dosimetry of Induced Electric Fields

Induced electric field and current density values in biological systems are used as dosimetric measures in quantifying interactions with EMFs. It is known that an electric field that is initially uniform becomes distorted in the immediate vicinity of any biological body (for example, human). Whether the human is electrically grounded or is standing on an insulating platform also will considerably affect the field distribution.

Professor Maria A. Stuchly and her team at the University of Victoria focus on development of efficient numerical EM modeling techniques and their applications to solving complex problems at low frequencies. A practical example of the electric fields and current densities induced in a human body in close proximity to a 60-Hz transmission line was evaluated by the group [91]. The total-scattered field formulation was employed, along with a quasistatic formulation of the finite difference time-domain (FDTD) method. The demonstrated induced fields and current densities were significantly higher than originally predicted for the uniform electric field exposure on a ground plane.

Dawson et al. [92], within the same group, employed the scalar potential finite difference (SPFD) method to estimate tissue- and site-specific electric fields and current densities due to contact currents in anatomically realistic models of an adult and a child. Three pathways of contact current were modeled: hand to opposite hand and both feet, hand to hand only, and hand to both feet. For a contact current of 1 mA, as the occupational reference level set by the ICNIRP, the current density in brain does not exceed the basic restriction of $10\,mA/m^2$. The restriction is exceeded slightly in the spine by a factor of more than 2 in the heart. For a contact current of 0.5 mA, as the general public reference level, the basic restriction of $2\,mA/m^2$ is exceeded several folds in the spine and heart. Several microamperes of contact current produced tens of mV/m within the child's lower arm bone marrow. The differences in induced electric field and current density values between child body and those of adult body are due to larger size of adult relative to the child. The above findings are supported by Akimasa et al. [93] of the same group.

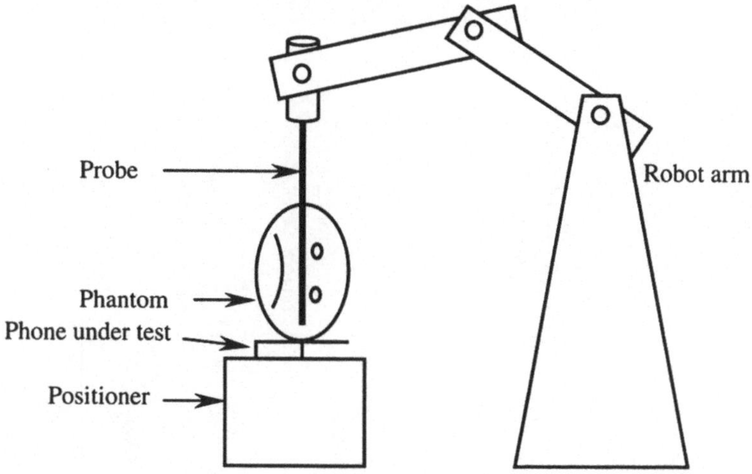

Figure 6.2 A schematic of a SAR measurement system.

6.4.3. Instrumentation

Since measuring actual SAR in the human body is difficult, SAR is estimated by measurements using phantom models. Phantoms are tissue-equivalent synthetic materials simulating biological bodies. They may be simple or complex depending on the tissue composition as well as the shape.

SAR distribution in the phantom is derived from the measurement of electric field strength inside the body with implantable isotropic electric field probes (small antennas). An electric field probe often consists of electrically short dipole with a diode sensor across its terminals and highly resistive lines to carry the detected signal for measurement. Sensors of the probe are designed to function as true square-law detectors where the output voltage is proportional to the square of the electric field.

A multiple-axis probe positioning system and additional instrumentation regarding data processing and calibration are needed while measuring SAR. Probe placement is conducted either manually or by a robot. A probe supported by nonmetallic robotic arm moves from one point to another in a homogeneous liquid simulating tissue. The liquid is contained in a manikin (a RF transparent shell for the phantom) simulating a human head or another part of the human body. The head model, for example, is usually placed on its side (left or right ear) that allows a handset to be placed underneath the head to facilitate field measurement. A SAR measurement system is illustrated in Fig. 6.2 [1].

A few major factors influence the results of SAR measurements including probe calibration in phantom, tissue properties, and data acquisition system.

6.4.4. In-Head Dosimetry of Mobile Phones

Dosimetry of mobile phones targets SAR generated in the human head due to RFR. The energy absorbed in the head is mainly due to electric fields induced by the magnetic fields generated by currents flowing through the feed point, along the antenna and the body of the phone. The RF energy is scattered and attenuated as it propagates through the tissues of the head, and maximum energy absorption is expected in the more absorptive high-water-content tissues near the surface of the head.

In-head dosimetry can be achieved by evaluating mobile devices with a dummy head model called *phantom*. A phantom is a device that simulates the size, contours, and electrical characteristics of human tissue at normal body temperature. It is composed of a mannequin (solid shell) cut in half and filled with tissue-equivalent synthetic material solution, which has electrical properties of tissues. The phantom is typically set up in relation to other SAR measurement equipment. Measured pieces of equipment for this setup include a robot arm and miniature isotropic electric

Table 6.3 Summary of SAR Levels and Temperature Rise in Human Head

Investigator	Description of source	SAR (W/kg)	Temperature rise
Dimbylow, 1994 [94]	900 MHz: $\lambda/4$; 600 mW; 1.8 GHz; $\lambda/4$; 125 mW; Calculated.	For 900 MHz $SAR_{1\,g} = 2.17$; $SAR_{10\,g} = 1.82$ For 1.8 GHz $SAR_{1\,g} = 0.7$; $SAR_{10\,g} = 0.48$	
Balzano et al., 1995 [95]	Motorola: 800–900 MHz; 600 mW and 2 W; Measured.	For analog (600 mW) Classic antenna: $SAR_{1\,g} = 0.2–0.4$; Flip antenna: $SAR_{1\,g} = 0.9–1.6$; Extended antenna: $SAR_{1\,g} = 0.6–0.8$. For GSM (2 W) Classic antenna: $SAR_{1\,g} = 0.09–0.2$; Flip antenna: $SAR_{1\,g} = 0.2–0.3$; Extended antenna: $SAR_{1\,g} = 0.1–0.2$.	
Anderson and Joyner, 1995 [96]	AMPS phones; 600 mW; 800/900 MHz.	SAR in the eye: 0.007–0.21; Metal-framed spectacles enhanced SARs in the eye by 9–29%; SAR in brain: 0.12–0.83.	Eye: 0.022°C due to SAR of 0.21 W/kg. Brain: 0.034°C due to SAR of 0.83 W/kg.
Okoniewski and Stuchly, 1996 [97]	Handset; 1 W; 915 MHz; $\lambda/4$; Calculated.	$SAR_{1\,g} = 1.9$; $SAR_{10\,g} = 1.4$	
Lazzi and Gandhi, 1998 [98]	Handset; Helical antenna 600 mW; 835 MHz. 125 mW; 1900 MHz; Calculated and measured.	$SAR_{1\,g} = 3.90$ (calculated); $SAR_{1\,g} = 4.02$ (measured). $SAR_{1\,g} = 0.15$ (calculated); $SAR_{1\,g} = 0.13$ (measured).	
Gandhi et al., 1999 [99]	AMPS phones; 600 mW; 800/900 MHz; Calculated and measured.	$SAR_{1\,g} > 1.6$ unless antennas are carefully designed and placed further away from the head.	
Van Leeuwen et al., 1999 [100]	Mobile phones; 250 mW; Calculated.	$SAR_{10g} = 1.6$	0.11°C
Wang and Fujiwara, 2000 [101]	Portable phone: 900 MHz; 600 mW; Helical antenna; Calculated.	$SAR_{1\,g} = 2.10$; $SAR_{10\,g} = 1.21$	
Bernardi et al., 2000 [102]	AMPS phones; 600 mW; 900 MHz; Calculated.	$SAR_{1\,g} = 2.2–3.7$	Ear: 0.22–0.43°C. Brain: 0.08°C to 0.19°C.
Van de Kamer and Lagendijk, 2002 [103]	Dipole antenna; 250 mW; 900 MHz; Calculated.	Cubic $SAR_{1\,g} = 1.72$; Arbitrary $SAR_{1\,g} = 2.55$; Cubic $SAR_{10\,g} = 0.98$; Arbitrary $SAR_{10\,g} = 1.73$.	

field probe. A phone is positioned against the mannequin operating at full power while the computer-controlled probe inserted into the tissue maps the electric fields inside. Computer algorithms determine the maximum electric field and then calculate a 1-g or 10-g average over a body to give a SAR value.

The local peak SARs differ depending on many factors such as the antenna type, antenna radiation efficiency, antenna inclination with the head, distance of antenna from head, effect of the hand holding the handset, and the structural accuracy and resolution of the head model. Therefore, values of SARs are a function of various conditions set by each investigator. In other words, SAR is a result of a complex physical phenomenon of reactive coupling of the whole radiating structure with the human tissue. A significant contributor to the uncertainty in estimating SAR is the absence of a standard tissue averaging technique of the local SAR values over 1 or 10 g.

In recent years, many dosimetrical studies have been performed for calculating or measuring power absorbed in phantoms simulating human heads exposed to RFR (Table 6.3 [94–103]). It is evident that many SAR values exceeded the safety limits. However, the temperature rise is far too small to have any lasting effects. Temperature measurements are significant only in case of high SARs. Increases in temperature (0.03–0.19°C) are much lower than the threshold temperature for neuron damage (4.5°C for more than 30 min), cataract induction (3–5°C), and physiological effects (1–2°C). Therefore, the temperature increases caused by mobile phone exposure have no effect on the temperature-controlling functions of the human brain. In fact, the thermostabilizing effect of brain perfusion often prevents temperature increase.

6.5. CONCLUSION AND RESEARCH NEEDS

In evaluating the significant amount of information and wide range of cases studied, the conclusion seems to be that the current studies indicate no evident pattern of increased health risk associated with EM fields. Many of the early studies are methodologically weak and the results not reliable.

In conclusion, effects of EM fields are only a threat if the dosage of exposure is very high. In the case of most EM sources, especially those from mobile phones, the dose is not very high but still detectable. The detection of biological responses to low-level EM exposure requires the design of sophisticated sensitive research procedure. The sensitivity creates a greater possibility of producing contradictory results. Such research depends critically on the skill and experience of the researcher and it is necessary that results be compared with prudent investigation in properly structured and independent research laboratories. Further research is required to narrow a gap of knowledge.

REFERENCES

1. Habash, R.W.Y. *Electromagnetic Fields and Radiation: Human Bioeffects and Safety;* Marcel Dekker: New York, NY, 2001.
2. Moulder, J.E. Biological studies of power-frequency fields and carcinogenesis. IEEE Engg. Med. Biol. **1996**, *15*, 31–40.
3. Foster, K.R. Electromagnetic field effects and mechanisms. IEEE Engg. Med. Biol. **1996**, *15*, 50–56.
4. Adair, P.K. Constraints on biological effects of weak extremely low-frequency electromagnetic fields. Phy. Rev. Lett. **1991**, *A43*, 1039–1048.
5. Eichwald, C.; Walleczek, J. Magnetic field perturbations as a tool for controlling enzyme-regulated and oscillatory biochemical reactions. Biophy. Chem. **1998**, *74*, 209–224.
6. Magnussen, T. *Electromagnetic Fields*; EMX Corporation: San Jose, CA, 1999.
7. Adey, W.R. Cell membranes: The electromagnetic environment and cancer promotion; Neurochem. Res. **1988**, *13*, 671–677.
8. Litovitz, T.A.; Krause, D.; Mullins, J.M. Effect of coherence time of the applied magnetic field on ornithine decarboxylase activity. Biochem. Biophys. Res. Comm. **1991**, *178*, 862–865.

9. Lai, H.; Singh, N.P. Acute exposure to a 60-Hz magnetic field increases DNA strand breaks in rat brain cells. *Bioelectromagnetics* **1997**, *18*, 156–165.
10. Wu, R.W.; Yang, H.; Chiang, H.; Shao, B.J.; Bao, J.L. The effects of low-frequency magnetic fields on DNA unscheduled synthesis induced by methylnitro-nitrosoguanidine in vitro. Electro Magnetobiol. **1998**, *17*, 57–65.
11. Tofani, S.; Barone, D.; Cintorino, M.; de Santi, M.M.; Ferrara, A.; Orlassino, R.; Ossola, P.; Peroglio, F.; Rolfo, K.; Ronchetto, F. Static and ELF magnetic fields induce tumor growth inhibition and apoptosis. *Bioelectromagnetics* **2001**, *22*, 419–428.
12. Cohen, M.M.; Kunska, A.; Astemborski, J.A.; McCulloch, D.; Paskewitz, D.A. The effect of low-level 60-Hz electromagnetic fields on human lymphoblastoid cells. II. Sister-chromatid exchanges in peripheral blood lymphocytes and lymphoblastod cell lines. Mutation Res. **1985**, *172*, 177–184.
13. Rosenthal, M.; Obe, G. Effects of 50-Hz Electromagnetic fields on proliferation and on chromosomal alterations in human peripheral lymphocytes untreated or pretreated with chemical mutagens. Mutation Res. **1989**, *210*, 329–335.
14. Scarfi, M.R.; Lioi, M.B.; Zeni, O.; Franceschetti, G.; Franceschi, C.; Bersani, F. Lack of chromosomal aberration and micronucleus induction in human lymphocytes exposed to pulsed magnetic fields. Mutation Res. **1994**, *306*, 129–133.
15. Paile, W.; Jokela, K.; Koivistoinen, A.; Salomaa, S. Effects of 50-Hz sinusoidal magnetic fields and spark discharges on human lymphocytes in vitro. Bioelectrochem. Bioenerg. **1995**, *36*, 15–22.
16. Sasser, L.B.; Morris, J.E.; Miller, D.L.; Rafferty, C.N.; Ebi, K.L.; Anderson, L.E. Lack of a co-promoting effect of a 60-Hz magnetic field on skin tumorigenesis in SENCAR mice. *Carcinogenesis* **1998**, *19*, 1617–1621.
17. Babbitt, J.T.; Kharazi, A.I.; Taylor, J.M.G.; Rafferty, C.N.; Kovatch, R.; Bonds, C.B.; Mirell, S.G.; Frumkin, E.; Dietrich, F.; Zhuang, D.; Hahn, T.J.M. Leukemia/lymphoma in mice exposed to 60-Hz magnetic fields. *Results of the Chronic Exposure Study* TR-110338, EPRI, Los Angeles, 1998.
18. Boorman, G.A.; McCormick, D.L.; Findlay, J.C.; Hailey, J.R.; Gauger, J.R.; Johnson, T.R.; Kovatch, R.M.; Sills, R.C.; Haseman, J.K. Chronic toxicity/oncogenicity evaluation of 60-Hz (power frequency) magnetic fields in F344/N Rats. Toxicologic Pathol. **1999**, *27*, 267–278.
19. Stuchly, M.A.; McLean, J.R.N.; Burnett, R.; Goddard, M.; Lecuyer, D.W.; Mitchel, R.E.J. Modification of tumor promotion in the mouse skin by exposure to an alternating magnetic field. Cancer Lett. **1992**, *65*, 1–7.
20. Liburdy, R.P.; Sloma, T.R.; Sokolic, R.; Yaswen, P. ELF magnetic fields, breast cancer and melatonin: 60-Hz fields block melatonin's oncostatic action on ER + breast cancer cell proliferation. J. Pineal Res. **1993**, *14*, 89–97.
21. Selmaoui, B.; Touitou, Y. Sinusoidal 50-Hz magnetic fields depress rat pineal NAT activity and serum melatonin role of duration and intensity of exposure. Life Sciences **1995**, *57*, 1351–1358.
22. Harland, J.D.; Liburdy, R.P. ELF inhibition of melatonin and tamoxifen action on MCF-7 cell proliferation: field parameters. *BEMS Meeting*, Victoria, British Columbia, Canada, 1996.
23. Rogers, W.R.; Reiter, R.J.; Barlow-Walden, L.; Smith, H.D.; Orr, J.L. Regularly scheduled, daytime, slow-onset 60-Hz electric and magnetic field exposure does not depress serum melatonin concentration in nonhuman primates. Bioelectromagnetics **1995**, Suppl. 3, 111–118.
24. Rogers, W.R.; Reiter, R.J.; Smith, H.D.; Barlow-Walden, L. Rapid-onset/offset, variably scheduled 60-Hz electric and magnetic field exposure reduces nocturnal serum melatonin concentration in nonhuman primates. Bioelectromagnetics **1995**, Suppl. 3, 119–122.
25. Graham, C.; Cook, M.R.; Sastre, A.; Riffle, D.W.; Gerkovich, M. Multi-night exposure to 60-Hz magnetic fields: Effects on melatonin and its enzymatic metabolite. J. Pineal Res. **2000**, *28*, 1–8.
26. Sazonova, T. A Physiological Assessment of the Work Conditions in 400 kV and 500 kV Open Switch Yards. In *Scientific Publications of the Institute of Labor Protection of the All-Union Central Council of Trade Unions* 46, Profizdat, USSR, 1967. (Available from IEEE, Piscataway, NJ, Special Issue Number 10.)
27. Zenon, S. Behavioural effects of EMFs mechanisms and consequences of power frequency electromagnetic field exposures, *Electromagnetics Meeting*, Bristol, UK, 24–25 September 1998.
28. Stuchly, M.A. Human exposure to static and time varying magnetic fields. Health Phy. **1986**, *51*, 215–225.

29. Korpinen, L.; Partanen, J.; Uusitalo, A. Influence of 50-Hz electric and magnetic fields on the human heart. Bioelectromagnetics **1993**, *14*, 329–340.
30. Wertheimer, N.; Leeper, E. Electrical wiring configurations and childhood cancer. Am. J. Epidemiol. **1979**, *109*, 273–284.
31. Savitz, D.A.; Wachtel, H.; Barnes, F.A.; John, E.M.; Tvrdik, J.G. Case-control study of childhood cancer and exposure to 60-Hz magnetic fields. Am. J. Epidemiol. **1988**, *128*, 21–38.
32. London, S.J.; Thomas, D.C.; Bowman, J.D.; Sobel, E.; Chen, T.S.; Peters, J.M. Exposure to residential electric and magnetic fields and risk of childhood leukemia. Am. J. Epidemiol. **1991**, *134*, 923–937.
33. Feychting, M.; Ahlbom, A. Magnetic fields and cancer in children residing near swedish high-voltage power lines. Am. J. Epidemiol. **1993**, *138*, 467–481.
34. Linet, M.S.; Hatch, E.E.; Kleinerman, R.A.; Robison, L.L.; Kaune, W.T.; Friedman, D.R.; Severson, R.K.; Haines, C.M.; Hartsock, C.T.; Niwa, S.; Wacholder, S.; Tarone, R.E. Residential exposure to magnetic fields and acute lymphoblastic leukemia in children. New England J. Med. **1997**, *337*, 1–7.
35. McBride, M.L.; Gallagher, R.P.; Theriault, G.; Armstrong, B.G.; Tamaro, S.; Spinelli, J.J.; Deadman, J.E.; Finchman, S.; Robson, D.; Choi, W. Power-frequency electric and magnetic fields and risk of childhood of leukemia in Canada. *Am. J. Epidemiol.* **1999**, *149*, 831–842.
36. Skinner, J.; Mee, T.J.; Blackwell, R.P.; Maslanyj, M.P.; Simpson, J.; Allen, S.G. Exposure to power frequency electric fields and the risk of childhood cancer in the UK. Br. J. Cancer **2002**, *87*, 1257–1266.
37. Sahl, J.D.; Kelsh, M.A.; Greenland, S. Cohort and nested case-control studies of hematopoietic cancers and brain cancer among electric utility workers. *Epidemiology* **1993**, *4*, 104–114.
38. Theriault, G.; Goldberg, M.; Miller, A.B.; Armstrong, B.; Guenel, P.; Deadman, J.; Imbernon, E.; To, T.; Chevalier, A.; Cyr, D.; Wall, C. Cancer risks associated with occupational exposure to magnetic fields among electric utility workers in Ontario and Quebec, Canada, and France: 1970–1989. Am. J. Epidemiol. **1994**, *139*, 550–572.
39. London, S.J.; Bowman, J.D.; Sobel, E.; Thomas, D.C.; Garabrant, D.H.; Pearce, N.; Bernstein, L.; Peters, J.M. Exposure to magnetic fields among electrical workers in relation to leukemia risk in Los Angeles County. Am. J. Indust. Med. **1994**, *26*, 47–60.
40. Johansen, C.; Olsen, J. Risk of Cancer among danish utility workers—A nationwide cohort study. Am. J. Epidemiology **1998**, *147*, 548–555.
41. Villeneuve, P.J.; Agnew, D.A.; Miller, A.B.; Corey, P.N.; Purdham, J.T. Leukemia in electric utility workers: The evaluation of alternative indices of exposure to 60-Hz electric and magnetic fields. Am. J. Indust. Med. **2000**, *37*, 607–617.
42. Villeneuve, P.J.; Agnew, D.A.; Johonson, K.C.; Mao, Y. Brain cancer and occupational exposure to magnetic fields among men: Results from a canadian population-based case-control study. Int. J. Epidemiol. *31*, 210–217.
43. IEEE C95.1–1991, Safety levels with respect to human exposure to radio-frequency electromagnetic fields, 3 kHz to 300 GHz, IEEE, Piscataway, NJ, 1992.
44. NRPB, Board Statement on Restrictions on Human Exposure to Static and Time-Varying Electromagnetic Fields. Documents of the PRPB, Vol. 4, No. 5, National Radiological Protection Board, Chilton, Didcot, Oxon, UK, 1993.
45. ICNIRP, Guidelines for limiting exposure to time-varying electric, magnetic, and electro-magnetic fields (up to 300 GHz). Health Phy. **1998**, *74*, 494–522.
46. TCO'99. Certification, Display (CRT), TCO Report No. 1, Stockholm, Sweden, 1999.
47. Safety Code 6, Limits of Human Exposure to Radiofrequency Electromagnetic Fields in the Frequency Range from 3 kHz to 300 GHz, Environmental Health Directorate, Health Protection Branch, Health Canada, Canada, 1999.
48. ARPANSA, Maximum exposure levels to Radio-frequency Fields, 3 kHz–300 GHz, Radiation Protection Series No. 3, Australian Radiation Protection and Nuclear Safety Agency, Australia, 2002.
49. Tice, R.R.; Hook, G.G.; Donner, M.; McRee, D.; Guy, A.W. Genotoxicity of radio-frequency signals. I. Investigation of DNA damage and micronuclei induction in cultured human blood cells. Bioelectromagnetics **2002**, *23*, 113–126.
50. D'Ambrosio, G.; Massa, R.; Rosaria, M.; Zeni, S.O. Cytogenetic damage in human lymphocytes following GMSK phase-modulated microwave exposure. Bioelectromagnetics **2002**, *23*, 7–13.

51. Mashevich, M.; Folkman, D.; Kesar, A.; Barbul, A.; Korenstein, R.; Jerby, E.; Avivi, E. Exposure of human peripheral blood lymphocytes to electromagnetic fields associated with cellular phones leads to chromosomal instability, Bioelectromagnetics **2003**, *24*, 82–90.
52. Czerska, E.M.; Elson, E.C.; Davis, C.C.; Swicord, M.L.; Czerski, P. Effects of continuous and pulsed 2450-MHz radiation on spontaneous lymphoblastoid transformation of human lymphocytes in vitro. Bioelectromagnetics **1992**, *13*, 247–259.
53. Cleary, S.F.; Du, Z.; Cao, G.; Liu, L.M.; McCrady, C. Effect of radio-frequency radiation on cytolytic T lymphocytes. Fed. Am. Soc. Experimental Biol. J. **1996**, *10*, 913–919.
54. Kwee, S.; Raskmark, P. Changes in cell proliferation due to environmental non-ionizing radiation: 2. Microwave radiation. Bioelectrochem. Bioenerget. **1998**, *44*, 251–255.
55. Velizarov, S.; Raskmark, P.; Kwee, S. The effects of radio-frequency fields on cell proliferation are non-thermal. Bioelectrochem. Bioenerget. **1999**, *48*, 177–180.
56. Adey, W.R., Bawin, S.M.; Lawrence, A.F. Effects of weak amplitude modulated microwave fields on calcium efflux from awake cat cerebral cortex. *Bioelectromagnetics* **1982**, *3*, 295–307.
57. Tsurita, G.; Nagawa, H.; Ueno, S.; Watanabe, S.; Taki, M. Biological and morphological effects on the brain after exposure of rats to a 1439-MHz TDMA field. Bioelectromagnetics **2000**, *21*, 364–371.
58. Kamimura, Y.; Saito, K.–I.; Saiga, T.; Amenyima, Y. Effect of 2.45 GHz microwave irradiation on monkey eyes. IEICE Trans. Comm. **1994**, *E77-B*, 762–765.
59. Lu S.–T.; Mathur, S.P.; Stuck, B.; Zwick, H.; D'Andrea, H.; Zeriax, J.M.; Merritt, J.H.; Lutty, G.; McLeod, D.S.; Johnson, M. Effects of high-peak-power microwaves on the retina of the rhesus monkey. Bioelectromagnetics **2000**, *21*, 439–454.
60. Lai, H.; Horita, A.; Guy, A.W. Microwave irradiation affects radial-arm maze performance in the rat. Bioelectromagnetics **1994**, *15*, 95–104.
61. Bornhausen, M.; Scheingraber, H. Prenatal exposure to 900-MHz, cell-phone electromagnetic fields had no effect on operant-behavior performances of adult rats. Bioelectromagnetics **2000**, *21*, 566–574.
62. Dubreuil, D.; Jay, T.; Edeline, J.M. Does head-only exposure to GSM-900 electromagnetic fields affect the performance of rats in spatial learning tasks? Behavioural Brain Res. **2002**, *129*, 203–210.
63. Lin, J.C.; Lin, M.F.; Microwave hyperthermia-induced blood–brain barrier alterations, Radiat. Res. **1982**, *89*, 77–87.
64. Neubauer, C.; Phelan, A.M.; Kues, H.; Lange, D.G. Microwave irradiation of rats at 2.45 GHz activates pinocytotic-like uptake of tracer by capillary endothelial cells of cerebral cortex. Bioelectromagnetics **1990**, *11*, 261–268.
65. Persson, B.R.R.; Salford, R.L.G.; Brun, A. Blood–brain barrier permeability in rats exposed to electromagnetic fields used in wireless communication. Wireless Network **1997**, *3*, 455–461.
66. Fritze, K.; Wiessner, C.; Kuster, N.; Sommer, C.; Gass, P.; Hermann, D.P.; Kiessling, M.; Hossmann, K.-A. Effect of GSM microwave exposure on the genomic response of the rat brain. Neuroscience **1997**, *81*, 627–639.
67. Finnie, J.W.; Blumbergs, P.C.; Manavis, J.; Utteridge, T.D.; Gebski, V.; Davies, R.A.; Vernon-Roberts, B.; Kuchel, T.R. Effect of long-term mobile communication microwave exposure on vascular permeability in mouse brain. Pathology **2002**, *34*, 344–347.
68. Hirata, A.; Matsuyama, S.; Shiozawa, T. Temperature rises in the human eye exposed to em waves in the frequency range 0.6–6 GHz. IEEE Trans. Electromagnet. Compatibility **2000**, *42*, 386–393.
69. Elder, J.A. Special senses: Cataractogenic effects. In *Biological Effects of Radio-frequency Radiation*; Elder, J.A., Cahill, D.F., Eds.; Washington, DC: Environmental Protection Agency, 1984; Environmental Protection Agency Report, EPA-600/8-83-026F: 5-64-5-68.
70. Elder, J.A. Occular Effects of Radio-Frequency Radiation. IEEE Subcommittee 28.4 White Paper, 2001.
71. Hossmann, K.-A.; Hermann, D.M. Effects of electromagnetic radiation of mobile phones on the central nervous system. Bioelectromagnetics **2003**, *24*, 49–62.
72. Jauchem, J.R.; Ryan, K.L.; Frei, M.R.; Dusch, S.J.; Lehnert, H.M.; Kovatch, R.M. Repeated exposure of C3H/HeJ mice to ultra-wideband electromagnetic pulses: Lack of effects on mammary tumors. Radiation Res. **2001**, *155*, 369–377.
73. Robinette, C.D.; Silverman, C. Causes of health following occupational exposure to microwave radiation (Radar) 1950–1974. In *Symposium on Biological Effects and Measurement of Radiofrequency/ Microwaves*; Hazzard, D.G. Ed.; Dept. of Health, Education, and Welfare, HEW Publication No. (FDA) 77–8026, Washington, DC, 1977.

74. Szmigielski, S. Cancer morbidity in subjects occupationally exposed to high-frequency (Radio frequency and microwave) Electromagnetic radiation. Science Total Environment **1996**, *180*, 9–17.
75. Davis, R.L.; Mostofi, F.K. Cluster of testicular cancer in police officers exposed to handheld radar. Am. J. Indust. Med. **1993**, *24*, 231–233.
76. Finkelstein, M.M. Cancer incidence among ontario police officers. Am. J. Indust. Med. **1998**, *34*, 157–162.
77. Lagorio, S.; Rossi, S.; Vecchia, P.; De Santis, M.; Bastianini, L.; Fusilli, M.; Ferrucci, A.; Desideri, E.; Comba, P. Mortality of plastic-ware workers exposed to radio frequencies. Bioelectromagnetics **1997**, *18*, 418–421.
78. Tynes, T.; Hannevik, M.; Andersen, A.; Vistnes, A.I.; Haldorsen, T. Incidence of breast cancer in Norwegian female radio and telegraph operators. Cancer Causes Control **1996**, *7*, 197–204.
79. Hocking, B.; Gordon, I.; Grain, H.; Hatfield, G. Cancer incidence and mortality and proximity to TV towers. Med. J. Australia **1996**, *65*, 601–605.
80. McKenzie, D.R.; Yin, Y.; Morrell, S. Childhood incidence of acute lymphoblastic leukemia and exposure to broadcast radiation in sydney—a second look. Australia and New Zealand J. Public Health **1998**, *22*, 360–367.
81. Michelozzi, P.; Ancona, C.; Fusco, D.; Forastiere, F.; Perucci, C.A. Risk of leukemia and residence near a radio transmitter in Italy. Epidemiology **1998**, *9* (Suppl), 354.
82. Hardell, L.; Nasman, A.; Pahlson, A.; Hallquist, A.; Mild, K.H. Use of cellular telephones and the risk for brain tumors: a case-control study. Int. J. Oncol. **1999**, *15*, 113–116.
83. Muscat, J.E.; Malkin, M.G.; Thompson, S.; Shore, R.E.; Stellman, S.D.; McRee, D.; Neugut, A.I.; Wynder, E.L. Handheld cellular telephone use and risk of brain cancer. J. AMA **2000**, *284*, 3001–3007.
84. Muscat, J.E.; Malikin, M.G.; Shore, R.E.; Thompson, S.; Neugut, A.I.; Stellman, S.D.; Bruce, J. Handheld cellular telephones and risk of acoustic neuroma. *Neurology* **2002**, *58*, 1304–1306.
85. Inskip, P.D.; Tarone, R.E.; Hatch, E.E.; Wilcosky, T.C.; Shapiro, W.R.; Selker, R.G. Cellular telephone use and brain tumors. New England J. Med. **2001**, *344*, 79–86.
86. Johansen, C.; Boice, Jr. J.D.; McLaughlin, J.K.; Olsen, J.H. Cellular telephones and cancer—A nationwide cohort study in denmark. J. Nat. Cancer Inst. **2001**, *93*, 203–207.
87. Auvinen, A.; Hietanen, M.; Luukkonen, R.; Koskela, R.S. Brain tumors and salivary and cancers among cellular telephone users. Epidemiology **2002**, *13*, 356–359.
88. Stang, A.; Anastassiou, G.; Wolfgang, A.; Bromen, K.; Bornfeld, N.; Jöckel, K.-H. The possible role of radio-frequency radiation in the development of uveal melanoma. Epidemiology **2001**, *12*, 7–12.
89. FCC, Guidelines for Evaluating the Environmental Effects of Radio Frequency Radiation, Federal Communications Commission, 96–326, Washington, DC, 1996.
90. Durney, C.H.; Christensen, D.A. *Basic Introduction to Bioelectromagnetics*; CRC Press: Boca Raton, FL, 1999.
91. Potter, M. E.; Okoniewski, M.; Stuchly, M.A. low-frequency finite difference time-domain (FDTD) for modeling of induced fields in humans close to line sources. J. Computational Phy. **2000**, *162*, 82–103.
92. Dawson, T.W.; Caputa, K.; Stuchly, M.A.; Kavet, R. Induced electric fields in the human body associated with 60-Hz contact currents. IEEE Trans. Biomed. Engg. **2001**, *48*, 1020–1026.
93. Akimasa, H.; Caputa, K.; Dawson, T.W.; Stuchly, M.A. Dosimetry in models of child and adult for low-frequency electric fields. IEEE Trans. Biomed. Engg. **2001**, *48*, 1007–1011.
94. Dimbylow, P.J.; Mann, S.M. SAR calculations in an anatomically realistic model of the head for mobile communications transceivers at 835 MHz and 1.8 GHz. Phys. Med. Biol. **1994**, *39*, 1537–1553.
95. Balzano, Q.; Garay, O.; Manning, T.J.; Electromagnetic energy exposure of simulated users of portable cellular telephones. IEEE Trans. Vehicular Technol. **1995**, *44*, 390–403.
96. Anderson, V.; Joyner, K.H. Specific absorption rate levels measured in a phantom head exposed to radio-frequency transmissions from analog hand-held mobile phones. Bioelectromagnetics **1995**, *16*, 60–69.
97. Okoniewski, M.; Stuchly. M.A. A study of the handset antenna and human body interaction. IEEE Trans. Microwave Theory Techni. **1996**, *44*, 1855–1864.

98. Lazzi, G.; Gandhi, O.P. On modeling and personal dosimetry of cellular telephone helical antennas with the FDTD code. IEEE Trans. Antennas Propagat. **1998**, *46*, 525–529.
99. Gandhi, Om, P.; Gianluca, L.; Tinniswood, A.; Yu, Q.-S. Comparison of numerical and experimental methods for determination of SAR and radiation patterns of handheld wireless telephones. Bioelectromagnetics **1999**, *20*, 93–101.
100. Van Leeuwen, G.M.; Lagendijk, J.J.; Van Leersum, B.J.; Zwamborn, A.P.; Hornsleth, S.N.; Kotte, A.N. Calculation of change in brain temperatures due to exposure to a mobile phone. Phy. Med. Biol. **1999**, *44*, 2367–2379.
101. Wang, J.; Fujiwara, O. FDTD analysis of dosimetry in human model for a helical antenna portable telephone. Electronics Communications in Japan **2000**, *E83-B*, 549–554.
102. Bernardi, P.; Cavagnaro, M.; Pisa, S.; Piuzzi, E. Specific absorption rate and temperature increases in the head of a cellular-phone user. IEEE Trans. Microwave Techniq **2000**, *48*, 1118–1125.
103. Van de Kamer, J.B.; Lagendijk, J.J.W. Computation of high-resolution SAR distributions in a head due to a radiating dipole antenna representing a hand held mobile phone. Phy. Med. Biol. **2002**, *47*, 1827–1835.

7

Biomedical Applications of Electromagnetic Engineering

James C. Lin
University of Illinois at Chicago
Chicago, Illinois

7.1. INTRODUCTION

Electromagnetic energy in the frequency region below 300 GHz is nonionizing and has wavelengths in air longer than 1 mm. Energy with wavelengths longer than 10 m (frequencies lower than 30 MHz) has conduction and propagation properties that differ greatly from those of wavelengths that approximate the human body's physical dimensions. Since its interaction with biological media differs according to the specific spectral band, these properties can give rise to a wide range of applications. Moreover, physiological responses can often vary at different frequencies. For example, thermal therapies avoid frequencies lower than 10 kHz to prevent stimulation of excitable muscular and cardiac tissues. Thus, advances in the use of electromagnetic technology for biomedical research and practice would be enhanced by a thorough understanding of biophysical interactions.

Short-wave and microwave diathermy have been used to heat muscle masses to relieve stress and strain, to stimulate blood circulation, and to reduce inflammation for more than 50 y [1,2]. Although nuclear magnetic resonance imaging or MRI has been developed only during the past 20 y, it has become one of the most extensively used diagnostic radiological imaging procedures. This chapter reviews recent biomedical applications that involve electromagnetic technology. Specifically, it describes applications in neuromagnetic imaging and stimulation, physiological monitoring, elimination of hypothermia, thermal ablation therapy, and hyperthermia treatment of cancer. Some of the applications have already found their niche in clinical practice.

7.2. THERMAL ABLATION THERAPIES

Percutaneous catheter ablation of arrhythmogenic foci has become an important therapy for selected patients with drug refractory, symptomatic tachyarrhythmias [3]. The increased interest stems, in part, from the nonpharmacological approach and minimally invasive nature of the procedure [3–5]. Microwave catheter antennas and radio-frequency (RF) electrodes are used to deliver electromagnetic (EM) fields and waves into the surrounding heart tissue. The frequencies used for the RF band are 500 to 750 kHz [6,7] and that for the microwave band are 915 and 2450 MHz [5,8]. Endocardiac conducting tissues responsible for causing arrhythmia or abnormal heart rhythm are destroyed by thermal energy applied through a catheter to the tissue.

7.2.1. EM Energy Propagation in Tissue

When RF energy is used, the applied voltage induces a current to flow between a small electrode inside or on the surface of the body to a large grounded, dispersive electrode on the surface (Fig. 7.1). In cauterizing tissue, a train of short RF pulses at high voltage is delivered through a pair of electrodes to create cutting and coagulation. For catheter ablation, RF energy is applied as a sinusoidal current through a small endocardial electrode to provide effective tissue heating [4].

A characteristic of RF frequency is that the associated wavelength is at least an order of magnitude longer than the dimensions of the human body. Its propagation behavior is therefore quasi-static and can be approximated using Laplace's formulation in electromagnetic field theory [7].

The absorption of EM energy in tissues is governed by the dielectric permittivity and conductivity. At the RF frequencies used for ablation, the conductive energy dissipation is considerably higher than dielectric energy dissipation. Accordingly, a reasonable approximation is obtained by neglecting dielectric permittivity and considering only the tissue conductivity such that Laplace's equation becomes

$$\nabla \cdot \sigma \nabla V = 0 \tag{7.1}$$

where σ is the tissue electrical conductivity and V is the electrical potential. The density of current ($\mathbf{J}$) flowing at any point in the tissue is given from the Ohm's law,

$$\mathbf{J} = -\sigma \nabla V \tag{7.2}$$

The current flow is impeded by tissue resistance (which is inversely proportional to conductivity), and RF energy is extracted or transferred to the tissue. The transferred or absorbed energy is

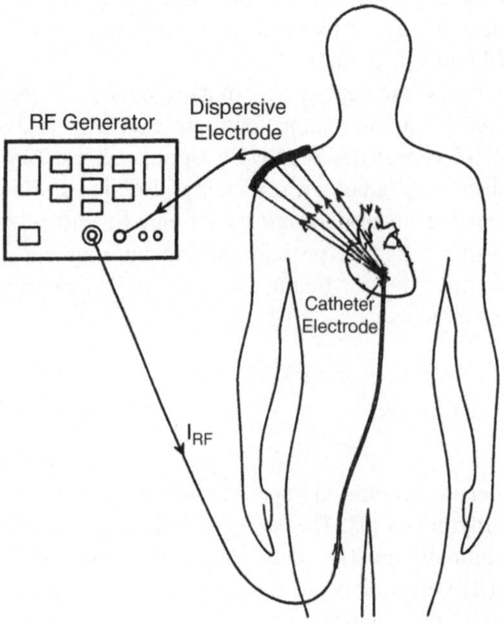

Figure 7.1 Schematic diagram of a RF ablation system. The applied RF voltage induces a current to flow between a small electrode inside or on the surface of the body to a large grounded, dispersive electrode on the surface.

converted to heat in accordance with Joule's law, which states

$$\mathbf{W} = \frac{\mathbf{J}^2}{\sigma} = \sigma(\nabla V)^2 \tag{7.3}$$

where **W** denotes the rate of energy absorption or the heating potential generated by RF energy as applied through the catheter electrode in a unit volume of tissue. The SI unit of **W** is watts per cubic meter (W/m^3).

A solution to Eq. (7.1) requires that V be specified at all points of the boundary throughout the region of interest. A pertinent set of boundary conditions is the voltage on the surface of the electrodes. In the case of unipolar ablation, RF energy is delivered from the electrode at the catheter tip inside the heart to a large, flat dispersive electrode on the skin surface. The voltage distribution between the active and dispersive electrodes in a homogeneous tissue is approximately given by

$$V(r) \sim \frac{V_o}{r} \tag{7.4}$$

where V_o is the voltage on the surface of the spherical unipolar electrode and r the radial distance from the active electrode. The inverse proportion of voltage distribution to radial distance indicates a lower risk of cardiac stimulation or muscle contraction associated RF energy from cardiac ablation at 100 V or less, which is the usual voltage used in RF ablation. The current density and time rate of heat generation are given, respectively, by

$$\mathbf{J} \sim \sigma \frac{V_o}{r^2} \tag{7.5}$$

$$\mathbf{W} \sim \sigma \frac{V_o^2}{r^4} \tag{7.6}$$

Clearly, RF energy absorption decreases as the fourth power of distance from the active electrode. The rapid decrease suggests that applied RF energy diverges from the small electrode. Consequently, active tissue heating is localized to a very short distance from the electrode–tissue interface. For effective cardiac ablation, it is essential to maintain direct contact between the RF electrode and cardiac tissue. Slight pressure exerted on the myocardium by the catheter is useful. If the density of tissue (ρ) is known, W can also be expressed as a specific absorption rate (SAR) quantity in units of W/kg, such that

$$\mathrm{SAR} = \frac{\mathrm{W}}{\rho} \tag{7.7}$$

Note that SAR is a measure of the rate of RF energy deposition at points surrounding the catheter electrode. This distribution of SAR serves as the source of lesion formation in cardiac ablation following thermalization of absorbed RF energy. Figure 7.2 illustrates the SAR measured in a tissue-equivalent model for a catheter electrode operating at 500 kHz. It can be seen that the drop in SAR is about 7.5 dB/mm away from the electrode. At 3 mm the decrease is about 20 dB or 100 times [9]. The size of lesions produced would be the combined result of SAR, duration of RF application, and heat conduction in tissue.

RF heating would quickly become insignificant beyond a few millimeters from the active electrode. Heating would fall well below the thermal noise floor at the dispersive electrode on the body

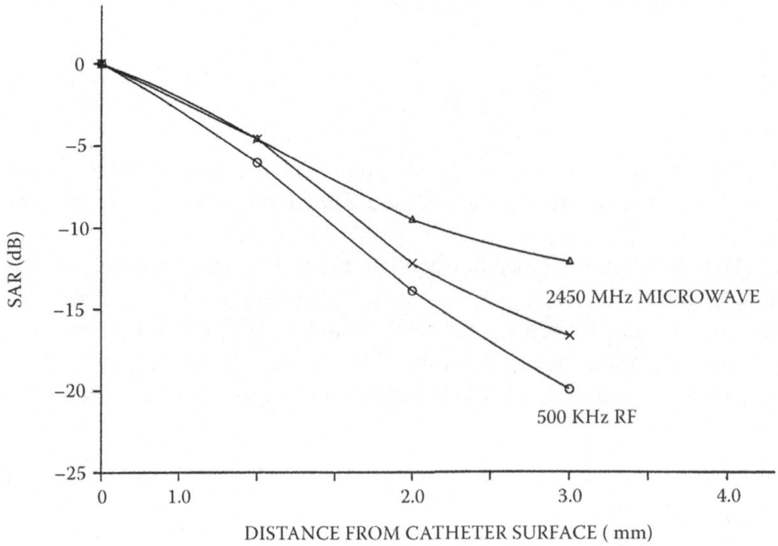

Figure 7.2 A comparison of SAR distributions produced by microwave catheter antennas and RF electrode as a function of radial distance from the catheter surface.

surface, provided that the dispersive electrode is large and in good contact everywhere. Increasing RF power delivery influences the magnitude of active heating (SAR) and its subsequent passive spread in lesion formation by RF ablation, but it has less influence on the region of active heat generation.

Desiccation and coagulation of tissue close to the electrode would decrease the tissue's electrical conductivity and raise the resistance to current flow. This, in turn, would further impede effective tissue heating and limit the size of RF-induced lesions. Lesions beyond the immediate vicinity of the electrode–tissue interface occur as a result of passive heat transfer from the shallow high temperature region. Indeed, studies have shown that RF-induced lesions increase rapidly in size during the initial period of power application. Subsequently, the rate of increases diminishes rapidly as the resistance at the electrode–tissue interface rises, and the current flow falls [10–13]. This inevitable phenomenon of thermal lesion production may be assessed through changes in the impedance of the catheter electrode. It is noteworthy that measures to maintain good electrode–tissue contact such as increasing the contact pressure can enhance RF coupling to the tissue [12].

Thermal microwave energy has been investigated for its potential in producing larger and deeper lesions. Unlike RF ablation, microwave energy is delivered through a radiating antenna mounted to the tip of a catheter (Fig. 7.3). A dispersive electrode at the body surface is not needed. Tissue heating is produced exclusively by absorption of radiated microwave energy in the biological dielectric [13]. Endocardiac microwave antennas should increase the volume of direct heating as compared to RF ablation since the lesion size is determined by the antenna radiation pattern, microwave power, and duration of power delivery. Comparison of phantom and in vivo results from RF and microwave ablation catheters showed that the volume of direct heating is indeed larger (Fig. 7.2) and that microwave energy is suitable for transcatheter ablation procedures [9,14–19]. Several microwave catheter antennas have been developed with efficient energy transfer into the myocardium [20–23].

The propagation and radiation of microwaves in biological tissue are governed by frequency, power, and the antenna radiation pattern, as well as by tissue composition and dielectric permittivity. The biological tissues of interest in microwave cardiac ablation are blood, muscle, and tissues with low water content, such as fat, bone, or desiccated tissue. Some typical values of microwave dielectric constant and conductivity at 37°C are given in Table 7.1 for 915 and 2450 MHz—two frequencies of most interest [13,24]. There is a modest change in dielectric constant and conductivity

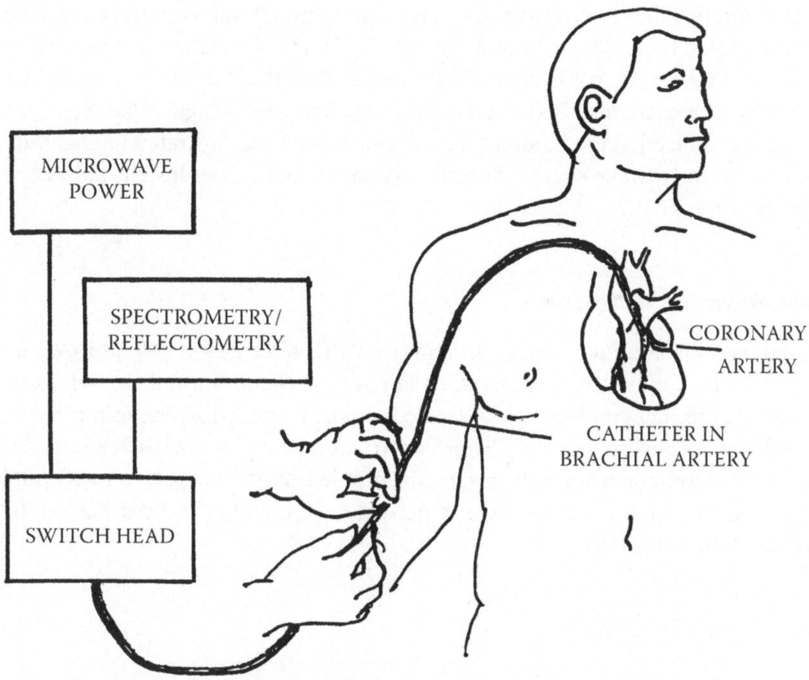

Figure 7.3 A microwave cardiac ablation system microwave energy is delivered through a radiating antenna mounted to the tip of a catheter.

Table 7.1 Dielectric Constant and Conductivity of Biological Tissues at 37°C.

	Dielectric constant			Conductivity (S/m)		
Frequency (MHz)	Blood	Muscle	Fat[a]	Blood	Muscle	Fat[a]
915	60	51	5.6	1.4	1.6	0.10
2450	58	47	5.5	2.1	2.2	0.16

[a]Fat, bone, or desiccated tissue.

as a function of frequency for all three types of tissues. However, differences among the three types of tissues are quite large.

As microwave fields propagate in the tissue medium, energy is extracted from the field and absorbed by the medium. This absorption results in a progressive reduction of the microwave power intensity as it advances into the tissue. The portion of energy extracted from the propagating microwave field is converted into heat production. The reduction is quantified by the depth of penetration. At 2.45 GHz, the depths of plane-wave penetration for blood, muscle, and fat are 19, 17, and 79 mm, respectively. For microwave-catheter antennas that do not have plane wavefronts, the penetration depth is reduced according to the specific antenna design. Nevertheless, these values clearly suggest that microwave energy can deposit energy directly into tissue at a distance, through radiative interaction of microwaves with cardiac tissues. Furthermore, the differences in the dielectric permittivity yield a depth of penetration for tissues with low water content that is about four times deeper than for muscle (or higher water content tissue) at 2.45 GHz. This means that a microwave field can propagate more readily through (is absorbed less by) low-water-content tissues than in tissues of high water content. It also implies that microwaves can propagate through intervening desiccated tissue to deposit energy directly in more deeply lying tissue. At 2.45 GHz, the dielectric constant for muscle is 20% lower than that for blood, but is about 800% higher than that for fat. While conductivities for blood and muscle are approximately the same, they are about 300% higher than those for tissues with low water content. Indeed, these inherent features have been demonstrated in phantom, animal, and human subjects for microwave energy. Specifically, larger and deeper lesions have been produced by microwave radiation [9,14].

7.2.2. Temperature Elevation in Tissue

In thermal therapeutic applications, the final temperature may be affected by tissue blood flow and thermal conduction. Specifically, the dynamic temperature variation in tissue is a function of tissue composition, blood perfusion, thermal conductivity of tissue, and heat generation due to metabolic processes in addition to RF energy absorption. The Pennes approximation to biological heat transfer via diffused conduction and blood convection in tissue can provide useful insights into RF and microwave ablation under the condition of uniform tissue perfusion by blood [25]. Specifically, the Pennes bioheat transfer equation states that

$$\frac{d(\rho c T)}{dt} = \mathbf{W} + \eta \nabla^2 T - V_s (T - T_o) \tag{7.8}$$

where T is the temperature in tissue (°C), T_o is the temperature in blood and tissue (°C), ρ is the density of tissue (kg/m^3), c is the specific heat of tissue (J/kg °C), **W** is the heating potential generated by RF power deposition (W/m^3), η is the coefficient of heat conduction of tissue (W/(°C m^3)), and V_s is the product of flow rate and heat capacity of blood (W/(°C m^3)). The numerical values of these parameters for muscle tissue are $\rho = 1000$ (kg/m^3), $c = 3500$ (J/kg °C), and $\eta = 0.508$ (W/°C m^3). However, heat transfer by blood convection is considerably faster in well perfused tissue, as is seen by a high $V_s = 7.780 \times 10^6$(W/°C m^3). Note that this equation neglects metabolic heat production since the usual period of RF application is 30 to 60 s, a period insufficient for significant metabolic heat contribution to tissue ablation. Note that for applications of short duration, the time rate of heating and the spatial distribution of radiated microwave energy are function of power deposition (the rate of energy absorption) and the antenna radiation patterns, respectively.

7.2.3. Cardiac Ablation

Radiofrequency ablation has become the preferred treatment modality for a variety of cardiac arrhythmias [26,27]. When RF energy is used, the applied voltage induces a current to flow between a small electrode inside or on the surface of the body to a large grounded, dispersive electrode on the surface. An important precaution is that the dispersive electrode must be large and in uniform contact with the skin to safely guard against high current density occurring at the edges. Considerations of the biophysical aspects of RF cardiac ablation indicate that the dissipation or absorption of RF energy is the source of tissue temperature elevation and the cause for subsequent lesion formation in RF ablation [7]. The divergent nature of RF current flow from the catheter electrode and the rapid dissipation of RF energy by tissue resistance combine to limit the depth and size of lesions produced. Several electronic [10–12,28,29] and mechanical techniques [30–37] can be invoked to enhance catheter ablation, produce larger and deeper lesions, and to overcome some of the aforementioned limitations. However, all of these techniques are baffled fundamentally by the effective current density required for ablation through resistive heating. Treatment of certain subepicardial arrhythmogenic substrates remains a challenge for RF ablation.

Several microwave catheter antennas have been developed to deliver ablating energy to the target tissue substrate. A drawback of some catheter antennas is that a considerable amount of microwave energy is reflected by the antennas to the skin surface and is deposited at the point of antenna insertion into the blood vessel [38–41]. The problem was addressed by several catheter antennas with efficient energy transfer into the myocardium [20–23]. A particular design feature of these catheter antennas is that there is good dielectric and impedance matching and a minimal amount of power is reflected from the antenna or flowing up the transmission line. Thus, they minimize heating of the coaxial cable or at the insertion point of the catheter into the body. These catheter antennas also serve as bipolar electrodes for endocardiac electrogram recording. Studies in dogs both during cardiopulmonary bypass and closed-chest operations have shown microwave energy greater than 200 J (joules) delivered to the heart through a split-tip dipole catheter antenna can produce an irreversible block of the heart rhythms [22,42,45]. This energy was achieved either by increasing the delivered power from 20 to 40 W or by increasing the treatment duration from 7 to 11 s (210 to 330 J per application). It produced a myocardial temperature of 65°C. These results show that microwave catheter ablation is a safe and suitable procedure for treatment of cardiac arrhythmias, including arrhythmias due to conduction pathways located deep in the myocardium.

7.2.4. Angioplasty

The primary goal of angioplasties is in dilation of atherosclerotic arteries to achieve maximum function for a prolonged period of time [45–49]. Under sterile condition, a percutaneous microwave catheter antenna or RF catheter electrode is guided through the arterial system and positioned in the appropriate coronary artery. A moderate dose (10–20 W) of 2450-MHz microwave or 500-kHz RF energy is delivered to the atherosclerotic plaque to reduce arterial narrowing by microwave-induced heating, softening and spreading of the plaque. In the microwave case, the catheter antenna consists of either a simple junction or sleeved slot radiator terminating a flexible coaxial cable 1 to 2 mm in diameter. Alternatively, the microwave thermal angioplasty is used in combination with conventional balloon angioplasties. A microwave catheter antenna is used to apply energy to heat the plaque through a water-filled balloon in remodeling the atherosclerotic lesion. Peak spatial temperatures in dog and rabbit myocardium have reached 60 to 80°C in 10 to 60-s, respectively, for a net power of 10 to 40 W to the catheter antenna. These techniques have been used successfully in animals and are presently undergoing clinical trials.

7.2.5. Benign Prostate Hyperplasia

Benign prostatic hyperplasia or hypertrophy is a major cause of morbidity in the adult male. At present, open surgery and transurethral resections of the prostate are the gold standards for treatment of benign prostatic hypertrophy. They can provide immediate relief of obstructive symptoms that remain fairly to extremely durable [50]. A new, less invasive procedure uses thermal energy delivered by microwaves (Fig. 7.4) [51]. In particular, an early report of thermal microwave technique from 1985 employed a transurethral microwave applicator [52]. It showed coagulation of the prostate in mongrel dogs and some salutary effects in an initial six patients treated with this device. An ensuing study used 2450 MHz microwave energy to treat 35 patients, and it compared a transurethral resection alone to preliminary microwave coagulation followed by transurethral resection of the gland [53]. Significant reduction in blood loss by initial treatment with the microwave thermal therapy was observed.

Numerous reports have appeared since that time on various aspects of both transrectal and transurethral microwave therapy of the prostate using 915 and 2450 MHz energy [54–59]. Most of the research in human subjects to date has focused on methods of delivery. Initial attempts to deliver the energy transrectally have not been effective and injury to the rectal mucosa has occurred due to

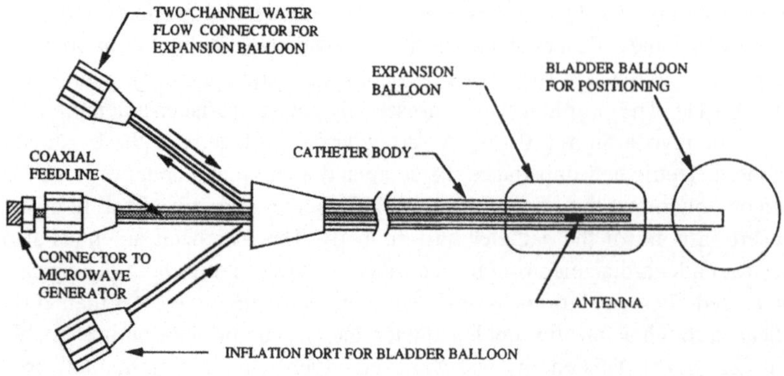

Figure 7.4 Schematic diagram of transurethral microwave balloon catheter used in hyperthermia treatment for benign prostatic hyperplasia (from Ref. 51).

the difficulty of interface cooling of this organ. Recent investigations have focused on transurethral delivery of the energy with cooling systems within the catheter to ensure urethral preservation [56–61]. Sensors placed in the microwave antenna maintain temperature on the urethral surface between 43 and 45°C. It is noted that while the number of treatment sessions and the temperature attained are extremely important predictors of response, sufficient hyperthermia volume is crucial for enhanced efficacy.

Virtually no data clearly demonstrating reduction in prostate volume in human subjects has been reported, although most investigators have shown improvement in measured urinary flow rates compared to preoperative studies. Randomized studies comparing microwave thermotherapy to transurethral resections conclude that microwave hyperthermia treatment had a definite therapeutic effect on symptomatic prostatic hypertrophy [62–69]. Thus, microwave thermal ablation of prostatic tissue and enlargement of the urethral with minimal clinical complications offers a therapeutic alternative to surgery in select patients with benign prostatic hyperplasia. Indeed, transurethral microwave thermotherapy is on its way to become a standard in minimally invasive treatment.

7.3. HYPERTHERMIA CANCER TREATMENT

Hyperthermia cancer therapy is a treatment procedure in which tumor temperatures are elevated to the range of 42–45°C. An important aspect of this development is the production of adequate temperature distribution in superficial and deep-seated tumors. A large number of antennas and applicators have been designed to produce therapeutic heating of a tumor of different volumes in a variety of anatomic sites [70–73]. For superficial tumors, single-contact applicators operating between 433 and 2450 MHz have been used. Because of the limited depth of energy penetration, these antennas have been applied to the heating of well-localized tumors extending to depths of up to a few centimeters. Techniques devised to provide noninvasive heating of deep-seated tumors include capacitive plates, helical coils and multiapplicator arrays. Capacitive plates operate at low frequencies so that the wavelengths are long compared to typical body dimensions. With appropriate design, this simple applicator can provide fairly uniform heating of tissue between the plates of two or three-electrode systems. The helical coil applicator gives rise to a power deposition pattern that varies slowly with radial distance and gives good penetration. The multiapplicator array concept is rapidly gaining utility in the clinic. Commercially available annular phased array systems operating at 60–90 MHz are designed to heat large anatomical regions such as the thorax, abdomen, and pelvic areas. The most promising region for this system appears to be the pelvis, since heating of the upper body is frequently limited by systemic hyperthermia and hemodynamic compensation, and by excessive heating of adjacent normal tissue structures. A major advantage of multiapplicator array systems is the ability to steer the hot spot electronically by varying the phase of each element, thereby allowing phased arrays operating at microwave frequencies to be used to effect selective heating of deep-seated tumors in a variety of anatomic sites.

Under certain conditions invasive method of power deposition such as thermoseed implants and interstitial antennas may be preferable for local hyperthermia of deep-seated tumors. Magnetic induction heating of arrays of thermoseed implants holds great promise for localized hyperthermia in deep-seated tumors. Quantitative study of power deposition at the implant and elsewhere inside the head has shown at least two orders of magnitude greater energy absorption by the implant than by the rest of the head. For tumors of large volume, interstitial techniques have been employed to generate the desired hyperthermic field (Fig. 7.2). RF electrodes operating in the frequency range of 0.5 to 10 MHz and microwave antennas operating between 300 and 2450 MHz have been employed for treatment of breast, head, and neck tumors. In most cases, it was necessary to implant an array of microwave antennas in order to produce the desired volume heating (Fig. 7.5). Interstitial techniques are attractive because, in combination with radiotherapy, interstitial hyperthermia renders a new modality of treatment for these malignancies with little additional risk to the patient (Fig. 7.6). Moreover, hyperthermia has been shown to be especially effective against hypoxic tumor cells that are low in pH and are resistant to ionizing radiation. Thus, the synergistic effect of this combined mode of treatment may portend reduced therapeutic dosages of drugs and/or ionizing radiation. This may not only lessen the extent of unacceptable radiation necrosis of normal tissue, but also enhance the mean survival rate.

While clinical and laboratory results have indicated a promising future for hyperthermia [74–80], its efficacy critically depends on the induction of a sufficient temperature rise throughout the tumor volume. Since each modality has its own liabilities there is no universally preferred modality. It is often necessary to use a wide range of external and implanted antennas and applicators to produce therapeutic heating of localized and regional tumors of different volumes at a variety of

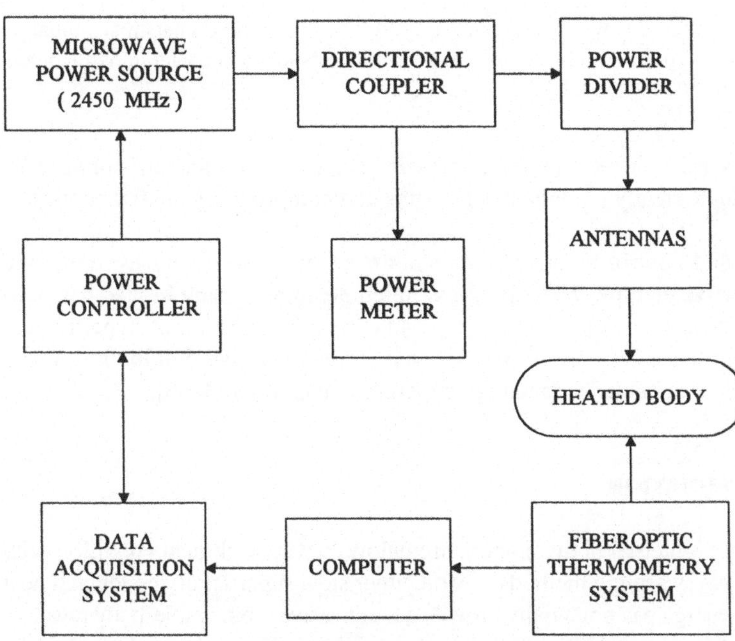

Figure 7.5 A system block diagram for interstitial hyperthermia.

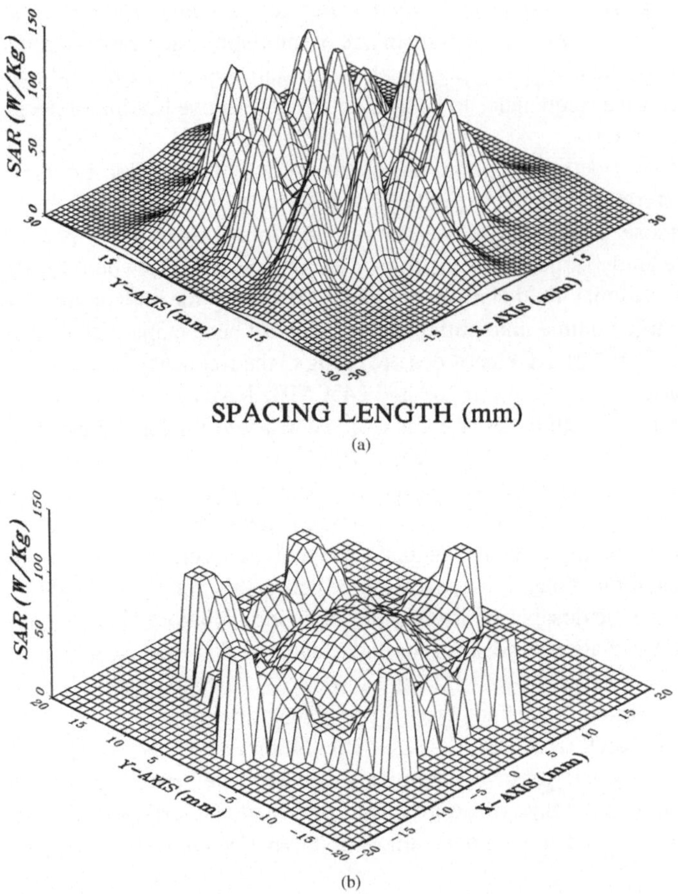

Figure 7.6 SAR distributions produced by an hexagonal array of six 2450-MHz interstitial miniature microwave catheter antennas: (a) computer simulation and (b) measurements made in phantom tissue model (additional details are given in Ref. 161).

anatomical sites. The type of applicator must therefore be selected, ad hoc, on the basis of specific site and type of the tumor which greatly complicates logistics and compromises quality control of the treatment.

Moreover, monitoring and control of tumor temperature in real time during hyperthermia treatment is essential for effective therapy. While progress in temperature sensing in vivo has been dramatic, considerable advance is needed prior to widespread clinical application of hyperthermia for cancer. Among approaches that may impact this outcome include invasive multipoint sensing [81,82], and noninvasive magnetic resonance imaging temperature mapping [83–86].

7.4. HYPOTHERMIA ELIMINATION

Severe hypothermia occurring at internal body temperature below 32°C is a clinical condition with a high rate of mortality. Current rewarming methods rely on either slow superficially conducted heat or invasive core perfusion. RF energy can potentially provide a noninvasive rapid whole-body rewarming that is safe and effective under most conditions. Likewise, extreme cold can seriously impair the performance and threaten the well-being of persons engaged in certain occupations. Conventional passive cold protection is inadequate because it necessitates the use of bulky gloves and precludes full use of hands and fingers, for example. Gloves with resistive wires through which a direct electrical current is passed to apply heat are limited by surface burns and excessive heat loss to the environment when used in water. In contrast, RF energy is known to be capable of bulk heating at depth in tissues and therefore has been studied to assess its feasibility both in elimination of deep hypothermia and in warming of the extremities exposed to severe environmental cold. In one study, a 13-turn helical coil RF rewarming system measured 34 cm long and 22 cm in diameter produced an average energy deposition rate of 5.5 W/kg in 10-kg rhesus monkeys with 60 W of 13.56-MHz power. The rectal temperature of the primate was raised from 28.5 to 34.5°C in 1 h compared to more than 2 h for conventional surface rewarming [87–90]. A study in human volunteers compared the rewarming effectiveness of a 13.56-MHz RF coil at a specific absorption rate of 2.5 W/kg with warm water immersion (40°C) and a mummy-type insulating sack under simulated mildly hypothermic conditions (35°C). It found that for mildly hypothermic individuals, active rewarming with RF at an SAR of 2.5 W/kg is about equivalent to passive rewarming with insulating sack, but less effective than warm water immersion [91].

Similar designs of helical coils operating at resonating frequency of 27.12 MHz have been used to efficiently warm hands and feet of subjects in cold air and water. The hand warming system consisted of a series of smaller insulated copper coils for each finger. Net powers of 15 and 25 W were delivered to the hands and feet, respectively. Subjects exposed to cold air at 10°C promptly reported warming sensations and gradual decrease of cold discomfort in tests where recorded finger tip temperatures were as low as 15°C [92]. These preliminary results from human volunteers demonstrate that RF technology can be applied to maintain hand and foot temperatures to within comfort ranges, and higher levels of the same energy can be used to rewarm whole-body hypothermia.

7.5. TISSUE IMAGING

A majority of imaging devices used by physicians to facilitate the diagnosis of diseases are based on sources that emit ionizing electromagnetic energy. Magnetic resonance imaging relies on the nonionizing static and RF magnetic fields and has been shown to offer a distinct advantage in a multitude of disease processes when compared to ionizing modalities. The current generation of MRI machines produces magnetic fields of 1.5 T (tesla) or less and therefore for imaging of hydrogen protons in tissue the frequency of applied RF field must be 63.9 MHz or lower according to Larmor resonance. MRI is the preferred imaging technique in the diagnosis of soft tissue disorders of the head, neck and spinal regions (see Fig. 7.7 for MRI images of brain anatomy). Also, it has been shown to be important in the evaluation of disease progression [93]. The use of MRI is increasing for diagnostic evaluation of cardiovascular diseases and abdominal disorders in the method of MR angiography. While cardiovascular evaluation may soon become the next major domain of application for MRI after neurological studies [94,95], the problem of motion artifact remains as a significant challenge for MRI of the abdomen. An instrumentation diagram for magnetic resonance imaging is shown in Fig. 7.8.

MRI is based on the nuclear magnetic resonance of water molecules. An applied static magnetic field induces the magnetic dipole moments associated with hydrogen protons in tissue to be aligned in the direction of the applied field. Since it takes time to reach complete alignment, the magnetization of the dipole moments, M is described by,

$$M = M_o[1 - e^{(-t/T_1)}] \tag{7.9}$$

where M_o is the magnitude of the instantaneous magnetization and T_1 is the longitudinal relaxation time. In the absence of an applied magnetic field, the spatial orientations of the magnetic dipole

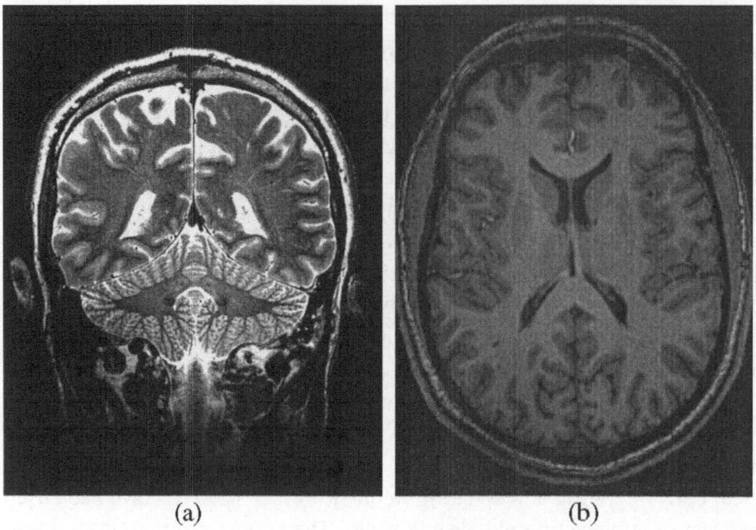

Figure 7.7 High-resolution MRI images of brain anatomy with different MR techniques: (a) coronal T2-weighted image obtained with a spin echo technique; (b) axial T1-weighted image obtained with a gradient echo technique. (Courtesy of Noam Alperin, Department of Radiology, University of Illinois, Chicago.)

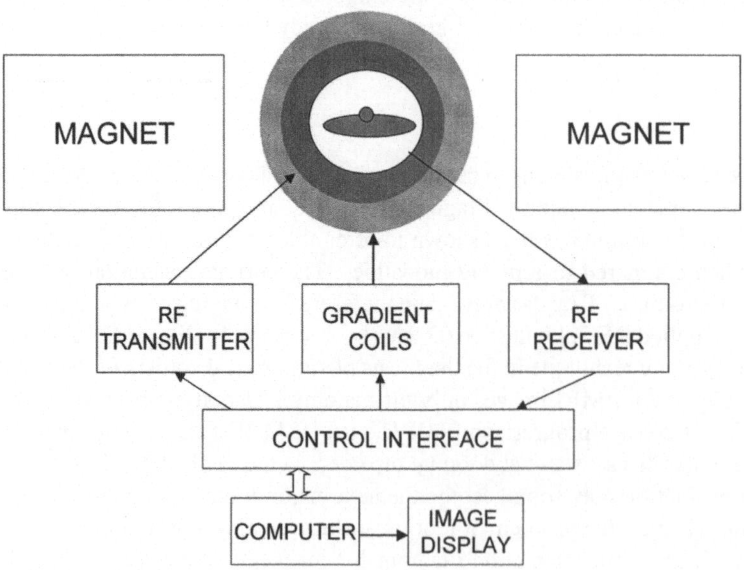

Figure 7.8 Instrumentation diagram for a typical magnetic resonance imaging system.

moments are random, $M = 0$. Each protons or dipole moment precesses about the static field, $\boldsymbol{B}_o$, in the transverse plane, at the Larmor frequency,

$$\omega_0 = \gamma \boldsymbol{B}_o \tag{7.10}$$

where γ is the gyromagnetic ratio. For protons, $\gamma/2\pi = 42.6$ MHz/T. A time-varying magnetic field, $\boldsymbol{B}_1$ that oscillates at ω_0, is applied in the direction transverse to $\boldsymbol{B}_o$ and then by choosing the duration of the excitation, the magnetic dipole moments can be tipped into the transverse direction and precess about $\boldsymbol{B}_1$. The decaying signal following the excitation is detected using a radio-frequency (RF) receiver.

To form an image, the signals originating from dipole moments in the body must be spatially resolved. This is accomplished through spatial encoding by using a gradient magnetic field. Specifically, a linear magnetic field gradient, G_x, in the x direction, is introduced so that,

$$\boldsymbol{B} = \boldsymbol{B}_o + G_x \boldsymbol{x} \tag{7.11}$$

The magnetic resonance signal coming from each anatomic region is spatially encoded such that each region contributes a signal whose frequency is proportional to,

$$\omega = \gamma \mathbf{B} = \gamma[\mathbf{B}_o + G_x x] \tag{7.12}$$

Thus the spatial location of the dipole moment giving rise to the MRI signal is obtained by a simple inversion of the Fourier transform of the received signal. The formation and display of the MR image is done efficiently, with the aid of a computer.

There are several other tissue imaging techniques currently under development that utilize electromagnetic technology [96–113]. The wide range of dielectric property variations offers a potential for higher contrast and better tissue characterization. Microwave tomography has been explored to reconstruct images associated with dielectric property variations in body cross section [96–101]. Microwave thermoelastic imaging uses microwave pulse-induced thermoelastic pressure waves to form planar or tomographic images [102–113]. Since the generation of thermoelastic or thermoacoustic pressure wave depends on permittivity, specific heat, and acoustic properties of tissue, microwave thermoelastic imaging possesses some unique features that are being explored as an imaging modality for noninvasive characterization of tissues.

7.6. NONINVASIVE AND REMOTE SENSING

Microwaves provide a convenient approach to detect and monitor physiological movements without compromising the integrity of these physiological events. In this case, microwave energy is directed to the target and the reflected signal is processed to yield information on the organ of interest or the physiological event under interrogation (Fig. 7.9). This noninvasive technique provides a capability for continuous monitoring as well as quantifying time-dependent changes in the cardiovascular and respiratory systems [114].

Currently, there are several areas in which noninvasive microwave contact or noncontact, close range, or remote approaches hold promise. These include heart rate, respiration, ventricular movement, pressure pulse sensing, and monitoring of superficial arterial circulation [115–126]. Low-frequency displacement of the precordium overlying the apex of the heart is related to movement in the left ventricle, and it echoes the hemodynamic events within the left ventricle. Microwave apexcardiograms obtained using 2.45 GHz showed close correlation to the hemodynamic events occurring within the left ventricle [118]. They involve detecting the reflected Doppler signal using an antenna located a few centimeters over the apex of the heart (Fig. 7.10). This approach has several advantages over more conventional techniques because it does not require any physical contact with the subject. Problems such as skin irritation, restriction of breathing and electrode connections are easily eliminated.

Doppler microwaves have been employed to interrogate the wall properties and pressure pulse characteristics at a variety of arterial sites, including the carotid, brachial, radial, and femoral arteries [119–121]. Microwave-sensed carotid pulse waveforms have been obtained in patients using

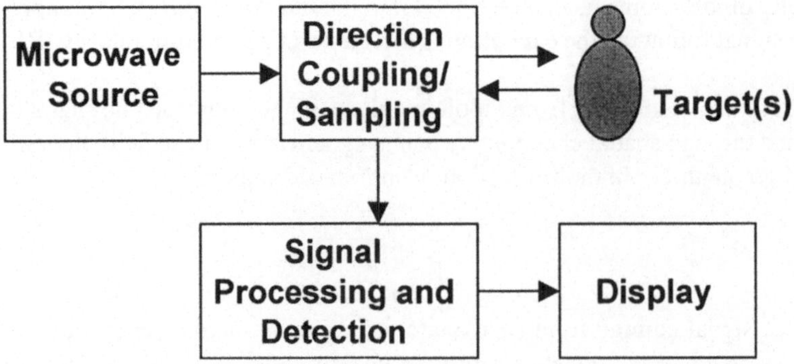

Figure 7.9 A microwave system for noninvasive sensing of physiological movements.

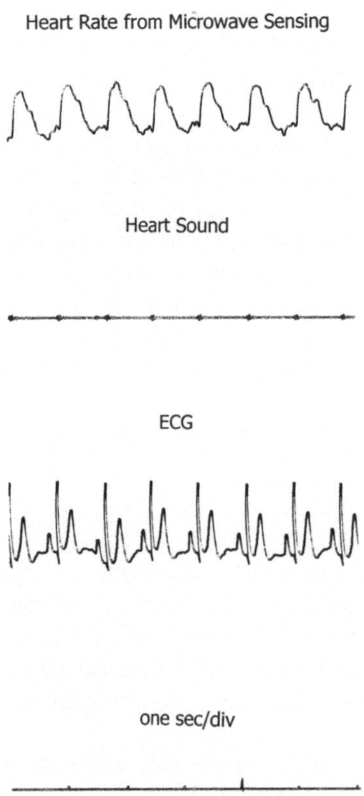

Figure 7.10 Human heart signal from remote, noncontract sensing using microwave radiation, showing simultaneous heart sound and electrocardiogram (ECG) recordings.

contact application of 25-GHz energy along with simultaneously recorded intra-aortic pressure waves. The resemblance of the microwave-sensed arterial pulse and the invasively recorded pressure wave is remarkable. These results confirm that a noninvasive Doppler microwave sensor can successfully and reproducibly detect pressure pulse waveforms of diagnostic quality. Because of its basis on motion detection, this continuous wave device will detect other movements as well. Other interesting applications are the use of microwaves for sensing cerebral edemas [122,123] and for speech articulator measurement [127].

The ability to detect remotely such vital signs as heart beat and respiration rates are particularly useful in situations where direct contact with the subject is either impossible or undesirable. Indeed, heart beat and respiration have been detected at distances of a few to tens of meters, with or without intervening nonmetallic barriers [124–126]. This approach has several advantages over more conventional techniques because it does not require any physical contact with the subject. Similar approaches may conceivably find uses in a variety of rescue related operations where direct physical contact with the subject is either impossible or undesirable.

7.7. MAGNETIC IMAGING AND STIMULATION

Magnetic fields are emerging as functional imaging modalities in the form of magnetocardiography (MCG) and magnetoencephalography (MEG) [128,129]. Magnetic stimulation for the diagnosis of neurological disorders represents yet another noncontact application of electromagnetic energy that is gaining in popularity [130,131]. MCG is a promising noninvasive modality for obtaining functional information on the electrical activity of the heart. The small biomagnetic signals (1 pT or less) are recorded using multichannel superconducting quantum interference devices (SQUID) with subjects in the supine position and the SQUID sensors placed directly over the thoracic surface [132]. It has been explored as a mapping tool to localize noninvasively cardiac electrical sources responsible for fetal atrial flutter and ventricular fibrillation [133–135]. In a like manner, high-resolution MEG measures the minute magnetic fields generated by the ionic currents in the brain [136]. MEG correlates well with abnormal electrical activity measured by electroencephalography (EEG), and sometimes it shows abnormalities not seen on an EEG. MEG is more accurate in spatial localization, as the magnetic field passes through the scalp and skull unimpeded, while the electrical signal of an EEG is attenuated and dispersed. It has been successfully used to localize cerebral sources of electrical activity [137–140].

In magnetic stimulation of the nervous system, a time-varying magnetic field, produced by passing a current through a wire coil, gives rise to an induced electric field in proximity to excitable tissues of the central and peripheral nervous system. Currents are delivered as 50–200 μs pulses with peak values of several thousand amperes. The technique is used to study the somatosensory or neuromuscular systems in humans [141–149]. The principal advantages of magnetic stimulation are that it is noninvasive and less painful than applying electrical currents through surface electrodes, and it has the ability to reach nerves lying well below the skin surface. The main disadvantages are that magnetic stimulation is not selective enough to restrict the region of excitation and it has relatively poor controllability and reproducibility.

7.8. BONE AND SOFT TISSUE HEALING

The therapeutic effects of low-frequency electric and magnetic fields have been studied extensively for their promotion of connective tissue repair. These studies concern bone repair and deal with acceleration of the healing of fresh fractures, delayed and nonunion, incorporation of bone grafts, osteoporosis, and osteonecrosis [150–152]. The most commonly used techniques included inductive and capacitive coupling, and implanted electrodes that induce voltages and currents similar to those produced, normally, during dynamic mechanical deformation of connective tissues. Indeed, there are three FDA-approved technologies:

1. Pulsed electromagnetic field (PEMF) technology using noncontact coils
2. Sinusoidal electric field (SEF) technology using skin contact electrodes
3. Direct-current technology using surgically implanted electrodes

Most studies indicate a dose-response relationship with current densities of 5 to 100 μA/cm^2, peak or 0.3 to 10 μA/cm^2, average in the bone and surrounding soft tissue. Sinusoidal and specialized waveforms with frequency contents extending from 10 Hz to 60 kHz have been employed. Although the majority of patients were treated with PEMF technology, more than 250,000 of them have been treated with these three technologies [150,152]. Success of treatment ranges from 65 to 90% depending upon such variables as infection, prior operation, and patient compliance. The total elapse times since initial trauma also influence rates of success; higher success rates are associated with short times since initial trauma. Moreover, several experimental and clinical investigations [153–160] involving tibial, scaphoid and other fractures have concluded that pulsed electromagnetic field is a safe and effective treatment for nonunion of bone fractures without discomfort, or the high costs of surgical repair.

As the specific requirements for field parameters are being defined, the range of treatable ills has been broadened to include nerve regeneration and wound healing, Specifically, the noninvasive treatment techniques of PEMF and SEF are being adapted for treatment of soft tissue injuries to reduce swelling and to accelerate wound healing [119–121]. While an acceleration of extracellular matrix synthesis and tissue healing has been observed in all these experimental systems and clinical applications, the underlying cellular mechanism of interaction of these fields upon the repair of cartilage and soft fibrous tissues is presently obscure.

REFERENCES

1. Lehmann, J.F. Diathermy. In *Handbook of Physical Medicine and Rehabilitation*; Krusen, F.H., Kottke, F.J., Elwood, P.M., Eds.; Saunders: Philadelphia, 1971, 273–345.
2. Lehmann, J.F. (Ed.). *Therapeutic Heat and Cold*; Williams & Wilkins: Baltimore, 1990.
3. Huang, S.K.S. (Ed.). *Radio-frequency Catheter Ablation of Cardiac Arrhythmias*; Futura: Armond, NY, 1995.
4. Wagshal, A.B.; Huang, S.K.S. Application of radiofrequency energy as an energy source for ablation of cardiac arrhythmias. In *Advances in Electromagnetic Fields in Living Systems*; Lin, J.C., Ed.; Plenum: New York, 1997; Vol. 2, 205–254.
5. Lin, J.C. Catheter microwave ablation therapy for cardiac arrhythmias. Bioelectromagnetics. **1999**, *20*, 120–132, Supplement 4.
6. Huang, S.K.; Bharati, S.; Graham, A.R.; Lev, M.; Marcus, F.I.; Odall, R.S. Closed chest catheter desiccation of the atrioventricular junction using radio-frequency energy—a new method of catheter ablation. J. Am. Coll. Cardiol. **1987**, *9*, 349–358.
7. Lin, J.C. Biophysics of radio-frequency ablation. In *Radio-frequency Catheter Ablation of Cardiac Arrhythmias: Basic Concepts and Clinical Applications*, 2nd Ed.; Huang, S.K.S., Wilber, D.J. Eds.; Futura: Armonk, New York, 2000; 13–24.
8. Beckman, K.J.; Lin, J.C.; Wang, Y.; Illes, R.W.; Papp, M.A.; Hariman, R.J. Production of reversible and irreversible atrio-ventricular block by microwave energy. Circulation **1987**, *76*, 1612.
9. Lin, J.C.; Wang, Y.J.; Hariman, R.J. Comparison of power deposition patterns produced by microwave and radio-frequency cardiac ablation catheters. Electronics Lett. **1994**, *30*, 922–923.
10. Blouin, L.T.; Marcus, F.I. The effect of electrode design on the efficiency of delivery of RF energy to cardiac tissue in vitro. PACE **1989**, *12*, 136–143.
11. Wittkampf, F.H.M.; Hauer, R.N.W.; EO Robles de Medina. Control of RF lesions size by power regulation. Circulation **1989**, *80*, 962–968.
12. Hoyt, R.H.; Huang, S.K.S.; Marcus, A.I.; Odell, R.S. Factors influencing transcatheter radio-frequency ablation of the myocardium. J. Appl. Cardiol. **1986**, *1*, 469–486.
13. Lin, J.C. Engineering and biophysical aspects of microwave and radio-frequency radiation. In *Hyperthermia*, Watmough, D.J., Ross, W.M., Eds.; Blackie: Glasgow, 1986; 42–75
14. Wonnell, T.L.; Stauffer, P.R.; Langberg, J.J. Evaluation of microwave and radio-frequency catheter ablation in a myocardium-equivalent phantom model. IEEE Trans. Biomed. Engg. **1992**, *39*, 1086–1095.

15. Langberg, J.J.; Wonnell, T.; Chin, M.C.; Finkbeiner, W.; Scheinman, M.; Stauffer, P. Catheter ablation of the atrioventricular junction using a helical microwave antenna: a novel means of coupling energy to the endocardium. PACE **1991**, *14*, 2105–2113.
16. Liem, L.B.; Mead, R.H.; Shenasa, M.; Chun, S.; Hayase, M.; Kernoff, R. Microwave catheter ablation using a clinical prototype system with a lateral firing antenna design. Pacing Clin. Electrophysiol. **1998**, *21*, 714–721.
17. Shetty, S.; Ishii, T.K.; Krum, D.P.; Hare, J.; Mughal, K.; Akhtar, M.; Jazeyeri, M.R. Microwave applicator design for cardiac tissue ablations. J. Microwave Power and Electromagnetic Energy **1996**, *31*, 59–66.
18. Haugh, C.; Davidson, E.S.; Estes 3rd, N.A.; Wang, P.J. Pulsing microwave energy: a method to create more uniform myocardial temperature gradients. J. Interv. Card. Electrophysiol. **1997**, *1*, 57–65.
19. Nevels, R.D.; Arndt, G.D.; Raffoul, G.W.; Carl, J.R.; Pacifico, A. Microwave catheter design. IEEE Trans. Biomed. Engg. **1998**, *45*, 885–890.
20. Lin, J.C.; Wang, Y.J. A catheter antenna for percutaneous microwave therapy. Microwave Optical Technol. Lett. **1995**, *8*, 70–72.
21. Lin, J.C.; Wang, Y.J. The cap-choke catheter antenna for microwave ablation treatment. IEEE Trans. Biomed. Engg. **1996**, *43*, 657–660.
22. Lin, J.C.; Hariman, R.J.; Wang, Y.J.; Wang, Y.G. Microwave catheter ablation of the atrioventricular junction in closed-chest dogs. Med. Biolog. Engg. Comput. **1996**, *34*, 295–298.
23. Pisa, S.; Cavagnaro, M.; Bernardi, P.; Lin, J.C. A 915-MHz Antenna for microwave thermal ablation treatment: physical design, computer modeling, and experimental measurement. IEEE Trans. Biomed. Engg. **2001**, *48*, 599–601.
24. Michaelson, S.M.; Lin, J.C. *Biological Effects and Health Implications of Radio-Frequency Radiation*; Plenum: New York, 1987.
25. Pennes, H.H. Analysis of tissue and arterial blood temperatures in the resting human forearm, J. Appl. Physiol. **1948**, *1*, 93–122.
26. Huang, S.K.S.; Wilber, D.J. (Eds.). *Radio-Frequency Catheter Ablation of Cardiac Arrhythmias: Basic Concepts and Clinical Applications*, 2nd Ed.; Futura: Armonk, New York, 2000.
27. Wagshal, A.B.; Huang, S.K.S. Application of radiofrequency energy as an energy source for ablation of cardiac arrhythmias. In *Advances in Electromagnetic Fields in Living Systems*; Lin, J.C., Ed.; Plenum: New York, 1997; Vol. 2, 205–254.
28. Wagshal, A.B.; Pires, L.A.; Bonavita, G.J.; Mittleman, R.S.; Huang, S.K.S. Does the baseline impedance measurement during radio-frequency catheter ablation influence the likelihood of an impedance rise. Cardiology **1996**, *87*, 42–45.
29. Strickberger, S.A.; Weiss, R.; Knight, B.P.; Bahu, M.; Bogun, F.; Brinkman, K.; Harvey, M.; Goyal, R. Randomized comparison of two techniques for titrating power during radio-frequency ablation of accessory pathways. J. Cardiovascular Electrophysiology **1996**, *7*, 795–801.
30. Langberg, J.J.; Gallagher, M.; Strickberegere, A.S.; Amirana, O. Temperature-guided radio-frequency catheter ablation with very large distal electrode. Circulation **1993**, *88*, 245–249.
31. Dinerman, J.L.; Berger, R.D.; Calkins, H. Temperature monitoring during radiofrequency ablation. J. Cardiovascular Electrophysiol **1996**, *7*, 163–173.
32. Pires, L.A.; Huang, S.K.S.; Wagshal, A.B.; Mittleman, R.S.; Rittman, W.J. Temperature-guided radio-frequency catheter ablation of closed-chest ventricular myocardium with a noval thermister-tipped catheter. Am. Heart J. **1994**, *127*, 1614–1618.
33. Wen, Z.C.; Chen, S.A.; Chiang, C.E.; Tai, C.T.; Lee, S.H.; Chen, Y.Z.; Yu, W.C.; Huang, J.L.; Chang, M.S. Temperature and impedance monitoring during radio-frequency catheter ablation of slow av node pathway in patients with atrioventricular node reentrant tachycardia. International J. Cardiol. **1996**, *57*, 257–263.
34. Mackey, S.; Thornton, L.; He, S.; Marcus, F.I.; Lampe, L.F. Simultaneous multipolar radio-frequency ablation in the monopolar mode increases lesion size. PACE **1996**, *19*, 1042–1048.
35. Langberg, J.J.; Lee, N.A.; Chin, M.C.; Rosenqvist, M. Radio-frequency catheter ablation: The effect of electrode size on lesion volume in vivo. PACE **1990**, *13*, 1242–1248.
36. Mittleman, R.S.; Huang, S.K.S.; Deguzman, W.T.; Cuenoud, H.; Wagshal, A.B.; Pires, L.A. Use of the saline infusion electrode catheter for improved energy delivery and increased lesion size in radio-frequency catheter ablation. PACE **1995**, *18*, 1022–1027.
37. Simmons, W.N.; Mackey, S.; He, D.S.; Marcus, F.I. Comparison of gold versus platinum electrodes on myocardial lesion size using radio-frequency energy. PACE **1996**, *19*, 398–402.

38. Langberg, J.J.; Wonnell, T.; Chin, M.C.; Finkbeiner, W.; Scheinman, M.; Stauffer, P. Catheter ablation of the atrioventricular junction using a helical microwave antenna: a novel means of coupling energy to the endocardium. PACE **1991**, *14*, 2105–2113.
39. Wonnell, T.L.; Stauffer, P.R.; Langberg, J.J. Evaluation of microwave and radio-frequency catheter ablation in a myocardium-equivalent phantom model. IEEE Trans. Biomed. Engg. **1992**, *39*, 1086–1095.
40. Liem, L.B.; Mead, R.H.; Shenasa, M.; Kernoff, R. In vitro and in vivo results of transcatheter microwave ablation using forward-firing tip antenna design. Pacing Clin. Electrophysiol. **1996**, *19*: 2004–2008.
41. Liem, L.B.; Mead, R.H.; Shenasa, M.; Chun, S.; Hayase, M.; Kernoff, R. Microwave catheter ablation using a clinical prototype system with a lateral firing antenna design. Pacing Clin. Electrophysiol. **1998**, *21*, 714–721.
42. Lin, J.C.; Beckman, K.J.; Hariman, R.J. Microwave ablation for tachycardia. Proc. IEEE/ EMBS International Conf. 1989, pp. 1141–1142.
43. Lin, J.C.; Beckman, K.J.; Hariman, R.J.; Bharati, S.; Lev, M.; Wang, Y.J. Microwave ablation of the atrioventricular junction in open heart dogs. Bioelectromagnetics **1995**, *16*, 97–105.
44. Lin, J.C. Catheter microwave ablation therapy for cardiac arrhythmias. Bioelectromagnetics. **1999**, *20*, Supplement 4, 120–132.
45. Lin, J.C. Transcatheter microwave technology for treatment of cardiovascular diseases. In *Emerging Electromagnetic Medicine*; O'Connor, M.E., Bentall, R.H.C., Monahan, J.C., Eds.; Springer-Verlag: New York, 1990; 125–132.
46. Rosen, A.; Wallinsky, P.; Smith, D.; Shi, Y.; Kosman, Z.; Martinez-Hernandez, A.; Rosen, H.; Sterzer, F.; Mawhinney, D.; Presser, A.; Chou, J.S.; Goth, P.; Lowery, G. Percutaneous transluminal microwave balloon angioplasty. IEEE. Trans. Microwave Theory Tech. **1990**, *38*, 90–93.
47. Lin, J.C. Microwave technology for minimally invasive interventional procedures. Chinese J. Med. Biolog. Engg. **1993**, *13*, 293–304.
48. Nardone, D.T.; Smith, D.L.; Martinez-Hernandez, A.; Consigny, P.M.; Kosman, Z.; Rosen, A.; Walinsky, P. Microwave thermal balloon angioplasty in the atherosclerotic rabbit. Am. Heart J. **1994**, *27*, 198–203.
49. Landau, C.; Currier, J.W.; Haudenschild, C.C.; Minihan, A.C.; Heymann, D.; Faxon, D.P. Microwave balloon angioplasty effectively seals arterial dissections in an atherosclerotic rabbit model. J. Am. Coll. Cardiol. **1994**, *23*, 1700–1707.
50. Aagaard, J.; Jonler, M.; Fuglsig, S.; Christensen, L.L.; Jorgensen, H.S.; Noorgaard, J.P. Total transurethral resection vs minimal transurethral resection of the prostate—A 10-year followup study of urinary symptoms, uroflowmetry, and residual volume. Bri. J. Urol. **1994**, *74*, 333–336.
51. Sterzer, F.; Mendecki, J.; Mawhinney, D.D.; Friedenthal, E.; Melman, A. Microwave treatments for prostate disease. IEEE Trans. Microwave Theory Tech. **2000**, *48*, 1885–1891.
52. Harada, T.; Etori, K.; Nishizawa, O.; Noto, H.; Tsuchida, S. Microwave surgical treatment of diseases of the prostate. Urology **1985**, *26*, 572–576.
53. Harada, T.; Tsuchida, S.; Nishizawa, O.; Kigure, T.; Noto, H.; Etori, K.; Kumazaki, T.; Koh, D.; Shimoda, J. Microwave surgical treatment of diseases of the prostate: Clinical application of microwave surgery as a tool for improved prostatic electroresection. Urologia Internationalis **1987**, *42*, 127–131.
54. Montorsi, F.; Galli, L.; Guazzoni, G.; Colombo, R.; Bulfamante, G.; Barbieri, L.; Grazioli, V.; Rogatti, P. Transrectal microwave hyperthermia for benign prostatic hyperplasia—long-term clinical, pathological and ultrastructural patterns. J. Urology **1992**, *148*, 321–325.
55. Debicki, P.; Astrahan, M.A.; Ameye, F.; Oyen, R.; Baert, L.; Haczewski, A.; Petrovich, Z. Temperature steering in prostate by simultaneous transurethral and transrectal hyperthermia. Urology **1992**, *40*, 300–307.
56. Astrahan, M.A.; Sapozink, M.D.; Cohen, D.; Luxton, G.; Kampp, T.D.; Boyd, S.; Petrovich, Z. Microwave applicator of transurethral hyperthermia of benign prostatic hyperplasia. Intern. J. Hyperthermia **1989**, *5*, 383–396.
57. Strohmaier, W.L.; Bichler, K.H.; Fruchter, S.H.; Wilbert, D.M. Local microwave hyperthermia of benign prostatic hyperplasia. J. Urol. **1990**, *144*, 913–917.
58. Lindner, A.; Braf, Z.; Lev, A.; Golomb, J.; Lieb, Z.; Seigel, Y.; Servadio, C. Local hyperthermia of the prostatic gland for the treatment of benign prostate hypertrophy and urinary retention. Br. J. Urol. **1990**, *65*, 201–203.
59. Carter, SStC.; Patel, A.; Reddy, P.; Royer, P.; Ramsay, J.W.A. Single-session transurethral microwave thermotherapy for the treatment of benign prostate obstruction. J. Endourol. **1991**, *5*, 137–143.

60. Baert, L.; Willemen, P.; Ameye, F.; Astrahan, M.A.; Lanholz, B.; Petrovich, Z. Transurethral microwave hyperthermia: An alternative treatment for prostdynia. Prostate **1991**, *19*, 113–119.
61. Bostwick, D.G.; Larson, T.R. Transurethral microwave thermal therapy—pathologic findings in the canine prostate. Prostate **1995**, *26*, 116–122.
62. Zerbib, M.; Steg, A.; Conquy, S.; Martinache, P.R.; Flam, T.A.; Debre, B. Localized hyperthermia vs the sham procedure in obstructive benign hyperplasia of the prostate—a prospective randomized study. J. Urology **1992**, *147*, 1048–1052.
63. Dahlstrand, C.; Walden, M.; Geirsson, G.; Pettersson, S. Transurethral microwave thermotherapy versus transurethral resection for symptomatic benign prostatic obstruction: a prospective randomized study with a 2-year follow-up. Br. J. Urol. **1995**, *76*, 614–618.
64. Larson, T.R.; Collins, J.M.; Corica, A. Detailed interstitial temperature mapping during treatment with a novel transurethral microwave thermoablation system in patients with benign prostatic hyperplasia. J. Urology **1998**, *159*, 258–264.
65. Jepsen, J.V.; Bruskewitz, R.C. Recent developments in the surgical management of benign prostatic hyperplasia. Urology **1998**, *51*, 23–31.
66. Ramsey, E.W.; Miller, P.D.; Parsons, K. A novel transurethral microwave thermal ablation system to treat benign prostatic hyperplasia: results of a prospective multicenter clinical trial. J. Urol. **1997**, *158*, 112–119.
67. D'Ancona, F.C.; Francisca, E.A.; Hendriks, J.C.; Debruyne, F.M.; De La Rosette, J.J. High-energy transurethral thermotherapy in the treatment of benign prostatic hyperplasia: criteria to predict treatment outcome. Prostate Cancer Prostatic Dis. **1999**, *2*, 98–105.
68. Wagrell, L.; Schelin, S.; Nordling, J.; Richthoff, J.; Magnusson, B.; Schain, M.; Larson, T.; Boyle, E.; Duelund, J.; Kroyer, K.; Ageheim, H.; Mattiasson, H.A. Feedback microwave thermotherapy versus TURP for clinical BPH—a randomized controlled multicenter study. Urology **2002**, *60*, 292–299.
69. Norby, B.; Nielsen, H.V.; Frimodt-Moller, P.C. Transurethral interstitial laser coagulation of the prostate and transurethral microwave thermotherapy vs. transurethral resection or incision of the prostate: results of a randomized, controlled study in patients with symptomatic benign prostatic hyperplasia. Br. J. Urol. Int. **2002**, *90*, 853–862.
70. Lin, J.C. (Ed.). Special issue on phased arrays for hyperthermia treatment of cancer. Trans. IEEE Microwave Theory Tech. **1986**, *34*, 481–482.
71. Fessenden, P.; Hand, J.W. Hyperthermia therapy physics. In *Medical Radiology—Radiation Therapy Physics*; Smith, A.R., Ed.; Springer-Verlag: Berlin, 1995; 315–363.
72. Lin, J.C. Hyperthermia therapy. In *Encyclopedia of Electrical and Electronics Engineering*; Webster, J.G., Ed.; Wiley: New York, 1999; Vol. 9. 450–460.
73. Rosen, A.; Vender Vosrt, A.; Kotsuka, Y. (Eds.). Special issue on medical applications in medicine. IEEE Trans. Microwave Theory Tech. **2000**, *48*, 1885–1891.
74. Vernon, C.C.; Hand, J.W.; Field, S.B.; Machin, D.; Whaley, J.B.; Vanderzee, J.; Vanputten, W.L.J.; Vanrhoon, G.C.; Vandijk, J.D.P.; Gonzalez, D.G.; Liu, F.F.; Goodman, P.; Sherar, M. Radiotherapy with or without hyperthermia in the treatment of superficial localized breast cancer—results from five randomized controlled trials. Int. J. Radiation Oncol. Biol. Phy. **1996**, *35*, 731–744.
75. Kuwano, H.; Sumiyoshi, K.; Watanabe, M.; Sadanaga, N.; Nozoe, T.; Yasuda, M.; Sugimachi, K. Preoperative hyperthermia combined with chemotherapy and irradiation for the treatment of patients with esophageal carcinoma. Tumori. **1995**, *81*, 18–22.
76. Emami, B.; Scott, C.; Perez, C.A.; Asbell, S.; Swift, P.; Grigsby, P.; Montesano, A.; Rubin, P.; Curran, W.; Delrowe, J.; Arastu, H.; Fu, K.; Moros, E. Phase III study of interstitial thermoradiotherapy compared with interstitial radiotherapy alone in the treatment of recurrent or persistent human tumors—a prospectively controlled randomized study by the Radiation Therapy Oncology Group. Int. J. Radiation Oncol. Biol. Phy. **1996**, *34*, 1097–1104.
77. Matsuda, T. The present status of hyperthermia in Japan. Ann. Acad. Med. Singapore **1996**, *25*, 420–424.
78. Overgaard, J.; Gonzalez, D.G.; Hulshof, M.C.C.M.; Arcangeli, G.; Dahl, O.; Mella, O.; Bentzen, S.M. Hyperthermia as an adjuvant to radiation therapy of recurrent or metastatic malignant melanoma—a multicenter randomized trial by the European Society for Hyperthermic Oncology. Int. J. Hyperthermia **1996**, *12*, 3–20.
79. Falk, M.H.; Issels, R.D. Hyperthermia in oncology. Int. J. Hyperthermia, **2001**, *17*, 1–18.
80. Moroz, P.; Jones, S.K.; Gray, B.N. Status of hyperthermia in the treatment of advanced liver cancer. J. Surg. Oncol. **2001**, *77*, 259–269.

81. Vanbaren, P.; Ebbini, E.S. Multipoint temperature control during hyperthermia treatments—theory and simulation. IEEE Trans. Biomed. Engg. **1995**, *42*, 818–827.
82. Qi, C.; Li, D.J. Thermometric analysis of intra-cavitary hyperthermia for esophageal cancer. Int. J. Hyperthermia **1999**, *15*, 399–407.
83. Lewa, C.J.; de Certaines, J.D. Body temperature mapping by magnetic resonance imaging. Spectroscopy Lett. **1994**, *27*, 1369–1419.
84. Young, I.R.; Hand, J.W.; Oatridge, A.; Prior, M.V. Modeling and observation of temperature changes in vivo using MRI. Magnetic Resonance Med. **1994**, *32*, 358–369.
85. Macfall, J.R.; Prescott, D.M.; Charles, H.C.; Samulski, T.V. H-1 MRI phase thermometry in vivo in canine brain, muscle, and tumor tissue. Med. Phys. **1996**, *23*, 1775–1782.
86. Kowalski, M.E.; Behnia, B.; Webb, A.G.; Jin, J.M. Optimization of electromagnetic phased-arrays for hyperthermia via magnetic resonance temperature estimation. IEEE Trans. Biomed. Engg. **2002**, *49*, 1229–1241.
87. Olsen, R.G.; David, T.D. Hypothermia and electromagnetic rewarming in the rhesus monkey. Aviation, Space, and Environmental Medicine **1984**, *55*, 1111–1117.
88. Olsen, R.G.; Ballinger, M.B.; David, T.D.; Lotz, W.G. Rewarming of the hypothermic rhesus monkey with electromagnetic radiation. Bioelectromagnetics **1987**, *8*:183–193.
89. Olsen, R.G. Reduced temperature afterdrop in rhesus monkeys with RF rewarming. Aviat Space Environ Medicine **1988**, *59*, 78–80.
90. Hesslink, Jr. R.L.; Pepper, S.; Olsen, R.G.; Lewis, S.B.; Homer, L.D. Radio frequency (13.56 MHz) energy enhances recovery from mild hypothermia. J. Appl. Physiol. **1989**, *67*, 1208–1212.
91. Kaufman, J.W.; Hamilton, R.; Dejneka, K.Y.; Askew, G.K. Comparative effectiveness of hypothermia rewarming techniques: radio-frequency energy vs. warm water. Resuscitation. **1995**, *29*, 203–214.
92. Lloyd, J.R.; Olsen, R.G. Radiofrequency energy for rewarming of cold extremities. Undersea Biomed. Res. **1992**, *19*, 199–207.
93. Stark, D.D.; Bradley, Jr. W.G. *Magnetic Resonance Imaging*, 3rd Ed.; Mosby-Year Book: St. Louis, 1999.
94. Nield, L.E.; Qi, X.; Yoo, S.J.; Valsangiacomo, E.R.; Hornberger, L.K.; Wright, G.A. MRI-based blood oxygen saturation measurements in infants and children with congenital heart disease. Pediatr. Radiol. **2002**, *32*, 518–522.
95. Plein, S.; Ridgway, J.P.; Jones, T.R.; Bloomer, T.N.; Sivananthan, M.U. Coronary artery disease: assessment with a comprehensive MR imaging protocol—initial results. Radiology **2002**, *225*, 300–307.
96. Larsen, L.E.; Jacobi, J.H. Microwave scattering parameter imagery of an isolated canine kidney. Med. Phys. **1979**, *6*, 394–403.
97. Guerquin-Kern, J.L.; Gautherie, M.; Peronnet, G.; Jofre, L.; Bolomey, J.C. Active microwave tomographic imaging of isolated, perfused animal organs. Bioelectromagnetics **1985**, *6*, 145–156.
98. Meaney, P.M.; Paulsen, K.D.; Hartov, A.; Crane, R.K. Microwave imaging for tissue assessment: initial evaluation in multitarget tissue-equivalent phantoms. IEEE Trans. Biomed. Engg. **1996**, *43*, 878–890.
99. Semenov, S.Y.; Svenson, R.H.; Boulyshev, A.E.; Souvorov, A.E.; Borisov, V.Y.; Sizov, Y.; Starostin, A.N.; Dezern, K.R.; Tatsis, G.P.; Baranov, V.Y. Microwave tomography—two-dimensional system for biological imaging. IEEE Transactions on Biomedical Engineering **1996**, *43*, 869–877.
100. Franchois, A.; Joisel, A.; Pichot, C.; Bolomey, J.C. Quantitative microwave imaging with a 2.45-GHz planar microwave camera. IEEE Trans. Med. Imag. **1998**, *17*, 550–561.
101. Semenov, S.Y.; Svenson, R.H.; Bulyshev, A.E.; Souvorov, A.E.; Nazarov, A.G.; Sizov, Y.E.; Posukh, V.G.; Pavlovsky, A.; Repin, P.N.; Starostin, A.N.; Voinov, B.A.; Taran, M.; Tatsis, G.P.; Baranov, V.Y. Three-dimensional microwave tomography: initial experimental imaging of animals. IEEE Trans. Biomed. Eng. **2002**, *49*, 55–63.
102. Olsen, R.G.; Lin, J.C. Acoustical imaging of a model of a human hand using pulsed microwave irradiation. Bioelectromagnetics **1983**, *4*, 397–400.
103. Lin, J.C.; Chan, K.H. Microwave thermoelastic tissue imaging—system design. IEEE Trans. Microwave Theory Tech. **1984**, *32*, 854–860.
104. Chan, K.H. Microwave-induced thermoelastic tissue imaging. PhD dissertation, University of Illinois, Chicago, 1988.
105. Chan, K.H., Lin, J.C. Microwave-induced thermoelastic tissue imaging. Proc. IEEE/EMBS Annual International Conference, New Orleans, 1988, pp. 445–446.

106. Su, J.L. Computer-assisted tomography using microwave-induced thermoelastic waves. Thesis PhD dissertation, University of Illinois, Chicago, 1988.
107. Su, J.L.; Lin, J.C. Computerized Thermoelastic Wave Tomography, World Cong. Med Phys Biomed Engg, Kyoto, Japan, July, 1991.
108. Lin, J.C. Auditory perception of pulsed microwave radiation. In *Biological Effects and Medical Applications of Electromagnetic Fields*; Gandhi, O.P. Ed.; Prentice-Hall: New York, 1990, Chapter 12, 277–318.
109. Dajani, N.F. 3D finite difference time-domain scattering and computed tomography in microwave medical imaging. PhD dissertation, University of Illinois, Chicago, 2001.
110. Kruger, R.A.; Reinecke, D.R.; Kruger, G.A. Thermoacoustic computed tomography—technical considerations. Med. Phys. **1999**, *26*, 1832–1837.
111. Kruger, R.A.; Miller, K.D.; Reynolds, H.E.; Kiser, W.L.; Reinecke, D.R.; Kruger, G.A. Breast cancer in vivo: Contrast enhancement with thermoacoustic CT at 434 MHz—Feasibility study. Radiology **2000**, *216*, 279–283.
112. Ku, G.; Wang, L.V. Scanning thermoacoustic tomography in biological tissue. Med. Phys. **2000**, *27*, 1195–1202.
113. Xu, M.; Wang, L.V. Pulsed-microwave-induced thermoacoustic tomography: filtered back-projection in a circular measurement configuration. Med. Phys. **2002**, *29*, 1661–1669.
114. Lin, J.C. Microwave sensing of physiological movement and volume change. Bioelectromagnetics **1992**, *13*, 557–565.
115. Lohman, B.; Boric-Lubecke, O.; Lubecke, V.M.; Ong, P.W.; MM Sondhi. A digital signal processor for Doppler radar sensing of vital signs. IEEE Eng. Med. Biol. Mag. **2002**, *21*, 161–164.
116. Lin, J.C. Noninvasive Microwave measurement of respiration, *Proc. IEEE* **1975**, *63*, 1530.
117. Lin, J.C.; Dawe, E.; Majcherek, J. A noninvasive microwave apnea detector, Proceedings 1977 San Diego Biomedical Symposium, Academic Press, 1977, pp. 441–443.
118. Lin, J.C.; Kiernicki, J.; Kiernicki, M.; Wollschlaeger, P.B. Microwave apexcardiography. IEEE Trans. Microwave Theory Tech. *27*, 618–620, 1979.
119. Lee, J.Y.; Lin, J.C. A microprocessor based noninvasive pulse wave analyzer. IEEE Trans. Biomed. Engg. **1985**, *32*, 451–455.
120. Papp, M.A.; Hughes, C.; Lin, J.C.; Pouget, J.M. Doppler microwave: a clinical assessment of its efficacy as an arterial pulse sensing technique. Invest. Radiol. **1987**, *22*, 569–573.
121. Thansandote, A.; Stuchly, S.S.; Smith, A.M. Monitoring variations of biological impedances using microwave Doppler radar. Phys. Med. Biol. **1983**, *28*, 983–990.
122. Lin, J.C.; Clarke, M.J. Microwave imaging of cerebral edema, *Proc. IEEE*, **1982**, *70*, 523–524.
123. Clarke, M.J.; Lin, J.C. Microwave sensing of increased intracranial water content. Invest. Radiol. **1983**, *18*, 245–248.
124. Lin, J.C. Microwave propagation in biological dielectrics with application to cardiopulmonary interrogation. In *Medical Applications of Microwave Imaging*; Larsen, L.E., Jacobi, J.H. Eds.; IEEE Press: New York, 1986, 47–58.
125. Chen, K.M.; Misra, D.; Wang, H.; Chuang, H.R.; Postow, E. An X-band microwave life-detection system. IEEE Trans. Biomed. Eng. **1986**, *33*, 697–701.
126. Chan, K.H.; Lin, J.C. Microprocessor based cardiopulmonary rate monitor, Med. Biol. Engg. and Comput. **1987**, *25*, 41–44.
127. Holzrichter, J.F.; Burnett, G.C.; Ng, L.C.; Lea, W.A. Speech articulator measurements using low power EM-wave sensors. J. Acoust. Soc. Am. **1998**, *103*, 622–625.
128. Hukkinen, K.; Kariniemi, V.; Katila, T.E.; Laine, H.; Lukander, R.; Makipaa, P. Instantaneous fetal heart rate monitoring by electromagnetic methods. Am. J. Obstet. Gynecol. **1976**, *125*, 1115–1120.
129. Tesche, C.D. Noninvasive detection of ongoing neuronal population activity in normal human hippocampus. Brain Res. **1997**, *749*, 53–60.
130. Ueno, S.; Tashiro, T.; Harada, K. Localized stimulation of neural tissues the brain by means of a paired configuration of time-varying magnetic fields. J. Appl. Phys. **1988**, *64*, 5862–5864.
131. Ueno, S.; Matsuda, T.; Hiwaki, O. Localized stimulation of the human brain and spinal cord by a pair of opposing pulsed magnetic fields. J. Appl. Phys. **1990**, *67*, 5838–5840.
132. Stroink, G. Principles of cardiomagnetism. In *Advances in Biomagnetism*; Williamson, S.J. Ed.; Plenum Press: New York, 1989; 47–56.

133. Peters, M.J.; Stinstra, J.G.; van den Broek, S.P.; Huirne, J.A.F.; Quartero, H.W.F.; ter Brake, H.J.M.; Rogalla, H. On the fetal magnetocardiogram. Bioelectrochem. Bioenerget. **1998**, *47*, 273–281.
134. Wakai, R.T.; Leuthold, A.C.; Martin, C.B. Atrial and ventricular fetal heart rate patterns in isolated congenital complete heart block detected by magnetocardiography. Am. J. Obstet. Gynecol. **1998**, *179*, 258–260.
135. Stinstra, J.; Golbach, E.; van Leeuwen, P.; Lange, S.; Menendez, T.; Moshage, W.; Schleussner, E.; Kaehler, C.; Horigome, H.; Shigemitsu, S.; Peters, M.J. Multicenter study of fetal cardiac time intervals using magnetocardiography. BJOG (an international journal of obstetrics and gynaecology). **2002**, *109*, 1235–1243.
136. Barth, D.S. Magnetoencephalography. In *The Treatment of Epilepsy: Principles and Practice*; Wyllie, E., Ed.; Lea & Febiger: Philadelphia, 1993, 285–297.
137. Sekihara, K.; Abraham-Fuchs, K.; Stefan, H.; Hellstrandt, E. Multichannel biomagnetic system for study of electrical activity in the brain and heart. Radiology **1990**, *176*, 825–830.
138. Pantev, C.; Gallen, C.; Hampson, S.; Buchanan, S.; Sobel, D. Reproducibility and validity of neuromagnetic source localization using a large array biomagnetometer, Am. J. EEG Technol. **1991**, *31*, 83–101.
139. Babb, C.W.; Coon, D.R.; Rechnitz, G.A. Biomagnetic neurosensors. 3. Noninvasive sensors using magnetic stimulation and biomagnetic detection. Anal. Chem. **1995,** *67*, 763–769.
140. Roberts, T.P.; Rowley, H.A. Magnetic source imaging as a tool for presurgical functional brain mapping. Neurosurg. Clin. N. Am. **1997**, *8*, 421–438.
141. Amassian, V.E.; Cracco, R.Q.; Maccabe, J.P. Focal stimulation of human cerebral cortex with the magnetic coil: A comparison with electrical stimulation, Electroencephal Clin. Neurophysiol. **1989**, *74*, 401–416.
142. Chokroverty, S. *Magnetic Stimulation in Clinical Neurophysiology*; Butterworth: Boston, 1990.
143. Ueno, S.; Tashiro, T.; Harada, K. Localized stimulation of neural tissues the brain by means of a paired configuration of time varying magnetic fields. J. Appl. Phys. **1988,** *64*, 5862–5864.
144. Ueno, S.; Masuda, T.; Fujiki, M. Functional mapping of the human motor cortex obtained by focal and vectorial magnetic stimulation of the brain. IEEE Trans. Magn. **1990**, 26, 1539–1544.
145. Kobayashi, M.; Ueno, S.; Kurokawa, T. Importance of soft tissue inhomogeneity in magnetic peripheral nerve stimulation. Electroencephalog. Clin. Neurophysiol.: Electromyog. Motor Control **1997**, *105*, 406–413.
146. Evans, B.A. Magnetic stimulation of the peripheral nervous system, J. Clin. Neurophysiol. **1991**, *8*, 77–84.
147. McMillan, A.S.; Watson, C.; Walshaw, D. Transcranial magnetic-stimulation mapping of the cortical topography of the human masseter muscle. Arch. Oral Biol. **1998**, *43*, 925–931.
148. Ishii, R.; Schulz, M.; Xiang, J.; Takeda, M.; Shinosaki, K.; Stuss, D.T.; Pantev, C. MEG study of long-term cortical reorganization of sensorimotor areas with respect to using chopsticks. NeuroReport **2002**, *13*, 2155–2159.
149. Kanno, A.; Nakasato, N.; Hatanaka, K.; Yoshimoto, T. Ipsilateral area 3b responses to median nerve somatosensory stimulation. Neuroimage **2003**, *18*, 169–177.
150. Bassett, C.A. Beneficial effects of electromagnetic fields. J. Cell Biochem. **1993**, *51*, 387–393.
151. Aaron, R.K.; Ciombor, D.M. Therapeutic effects of electromagnetic fields in the stimulation of connective tissue repair. J. Cell Biochem. **1993**, *51*, 42–46.
152. Polk, C. Therapeutic applications of low frequency electric and magnetic fields. In *Advances in Electromagnetic Fields in Living Systems*; Lin, J.C., Ed.; Plenum Press: New York, 1994; Vol. 1, 129–153.
153. Pienkowski, D.; Pollack, S.R.; Brighton, C.T.; Griffith, N.J. Low-power electromagnetic stimulation of osteotomized rabbit fibulae. A randomized, blinded study. J. Bone Joint Surg. Am. **1994**, *76*, 489–501.
154. Grace, K.; Revell, W.; Brookes, M. The effects of pulsed electromagnetism on fresh fracture healing: osteochondral repair in the rat femoral groove. Orthopedics **1998**, *21*, 297–302.
155. Godley, D. Nonunited carpal scaphoid fracture in a child: treatment with pulsed electromagnetic field stimulation. Orthopedics **1997**, *20*, 718–719.
156. Kenkre, J.E.; Hobbs, F.D.; Carter, Y.H.; Holder, R.L.; Holmes, E.P. A randomized controlled trial of electromagnetic therapy in the primary care management of venous leg ulceration. Fam. Pract. **1996**, *13*, 236–241.
157. Patino, O.; Grana, D.; Bolgiani, A.; Prezzavento, G.; Mino, J.; Merlo, A.; Benaim, F. Pulsed electromagnetic fields in experimental cutaneous wound healing in rats. J. Burn Care Rehabil. **1996**, *17*, 528–531.

158. Scardino, M.S.; Swaim, S.F.; Sartin, E.A.; Steiss, J.E.; Spano, J.S.; Hoffman, C.E.; Coolman, S.L.; Peppin. B.L. Evaluation of treatment with a pulsed electromagnetic field on wound healing, clinicopathologic variables, and central nervous system activity of dogs. Am. J. Vet. Res. **1998**, *59*, 1177–1181.
159. Ito, H.; Shirai, Y. The efficacy of ununited tibial fracture treatment using pulsing electromagnetic fields: relation to biological activity on nonunion bone ends. J. Nippon Med. Sch. **2001**, *68*, 149–153.
160. Inoue, N.; Ohnishi, I.; Chen, D.; Deitz, L.W.; Schwardt, J.D.; Chao, E.Y. Effect of pulsed electromagnetic fields (PEMF) on late-phase osteotomy gap healing in a canine tibial model. J. Orthop. Res. **2002**, *20*, 1106–1114.
161. Lin, J.C.; Hirai, S.; Chiang, C.L.; Hsu, W.L.; Su, J.L.; Wang, Y.J. Computer simulation and experimental studies of SAR distributions of interstitial arrays of sleeved-slot microwave antennas for hyperthermia treatment of brain tumors. IEEE Trans. Microwave Theory Techniq. **2000**, *48*, 2191–2197.

8

Measurement Techniques for the Electromagnetic Characterization of Biological Materials

Mohammad-Reza Tofighi and Afshin Daryoush
Drexel University
Philadelphia, Pennsylvania

8.1. INTRODUCTION

The knowledge of material parameters at RF frequencies and above is important in various industrial [1,2], medical, and regulatory applications [3,4]. Data are available for a variety of materials ranging from PCB substrates and laminates [1,2] to biological samples [5].

Complex permittivity measurements of biological materials are of particular importance since interest exists to understand the interaction of electromagnetic energy with biological tissues up to millimeter wave frequencies. Some of this interest stems from the potential health hazards due to the advent of wireless communications [4], therapeutic applications of microwaves [3], and microwave imaging [6].

The therapeutic [3], dosimetric [4], or other applications [7–10] require a knowledge of dielectric properties of the tissues. From the electromagnetic engineering point of view, studying the bulk dielectric properties remains the most direct way of characterizing any substance. With the knowledge of these properties as they appear in Maxwell's equation, the absorption of energy and the field distribution which are the results of the solution to a boundary value problem can be obtained. Today, varieties of techniques exist to numerically solve Maxwell's equation. On the other hand, at the microscopic level, the underlying physics is much more complicated compared with the existing formulations for the bulk effects and not well understood yet. However, observations made at the macroscopic level can greatly contribute to an understanding of the microscopic phenomena.

There are many good references dedicated to the issue of the interaction of RF/microwaves with biological systems and medical applications of microwaves [3,4,6–10]. Most popular applied publications are in regards to EM energy absorption in human body as a result of cell phone radiation [4,11–16], applications where heating is needed [2,17,18], hyperthermia treatment of cancer tumors [19–21], microwave catheters for ablation [3,22], and microwave imaging [6,23–25].

This chapter is primarily focused on biological materials. Section 8.2 presents the theory of relaxation and the relaxation phenomena in biological substances. Various techniques for permittivity measurement are addressed in Sec. 8.3. Section 8.4 highlights the technical issues related to the characterization of biological materials using the commonly used open-ended coaxial probe. Moreover, a two-port measurement test fixture is presented that provides the accurate complex permittivity of brain tissue up to 50 GHz.

8.2. COMPLEX PERMITTIVITY OF MATERIALS

The dielectric theory literature, mathematical models (i.e., Debye and Cole-Cole), and physical mechanisms involved are briefly reviewed in this section. Techniques for permittivity measurement are reviewed in the rest of the chapter.

8.2.1. Applications and Significance

From a macroscopic view, in electromagnetic problems dielectric properties of materials are quantified by their bulk complex permittivity. This complex permittivity is represented by:

$$\varepsilon = \varepsilon_0(\varepsilon' - j\varepsilon'') = \varepsilon_0\left(\varepsilon' - \frac{j\sigma}{\omega\varepsilon_0}\right) = \varepsilon_0\varepsilon'(1 - j\tan\delta) \tag{8.1}$$

where $\varepsilon_0 = 8.854 \times 10^{-12}$ (F/m) is the free space permittivity, σ is the conductivity (S/m), and tan $\delta = \varepsilon''/\varepsilon'$ is the loss tangent. The imaginary part is the term associated with the absorption of electrical energy. Since a field in linear systems can be represented by a summation of plane waves, the plane wave propagation parameters bear physically meaningful information. Three important wave parameters are the phase constant (β), the attenuation constant (α), and the wavelength (λ). For a plane wave, these parameters are related to the medium constitutive parameters and at frequency of (ω are given as [24]:

$$\alpha = \frac{\omega}{c}\sqrt{\frac{\varepsilon'}{2}}\sqrt{\sqrt{1+\tan\delta^2} - 1}(\text{rad/m}) \tag{8.2}$$

$$\beta = \frac{\omega}{c}\sqrt{\frac{\varepsilon'}{2}}\sqrt{\sqrt{1+\tan\delta^2} + 1}(\text{Np/m}) \tag{8.3}$$

$$\lambda = \frac{\lambda_0}{\sqrt{\varepsilon'/2}\sqrt{\sqrt{1+\tan\delta^2} + 1}}(\text{m}) \tag{8.4}$$

where c is the speed of the light in vacuum, and λ_0 is the wavelength in free space. The inverse of attenuation, i.e., the penetration depth, is also an important parameter describing the wave penetration in lossy materials.

8.2.2. Relaxation Theory

The extent to which the fields interact with the materials depends on the dielectric parameters of those materials. The classical theory of dielectrics can be found in books by Frohlich [26] and Daniel [27]. Frohlich's book [26] covers the basic macroscopic theory, static and dynamic properties. That book reviews important topics in dielectric theory such as dipolar interaction, dipolar molecules in gases and dilute solutions, Debye theory, and resonance absorption. Classical relaxation theory is presented comprehensively in Daniel's book [27].

Due to the complexity of biological substances, a complete understanding of bioelectrical interactions requires at least a sufficient knowledge of biochemistry, biology, electrical engineering, and physics [28–32]. The book by Pethig [28] presents the electronic and dielectric properties of biological material by bringing together the relevant issues from all these disciplines. He reviews not only the basic dielectric theory but also the dielectric properties of biopolymers (which are amino acids, polypeptides, and side chains), the role of water in biological systems, heterogeneous material (Maxwell-Wagner and counterion theory), and quantum mechanical aspects.

Works by Schwan and coworkers, started as early as 1950s [30–32], have provided significant enhancement in relating the macroscopic theory to the microscopic phenomena and in the interpretation of different relaxation mechanisms, from dc up to the microwave region.

The electrical properties of a material exposed to electromagnetic fields are in general frequency dependent. A material, which demonstrates significant permittivity changes in the frequency range of interest is referred to as a dispersive one in that range.

In order to understand the physics of dielectrics, the first step is developing a theory for the static properties of dielectric materials, i.e., when the electric field has no time variation. A good review of these properties in sufficient details relevant to biological tissues can be found in Ref. 28.

A phenomenological approach for the mathematical modeling of dispersion is the Debye theory [27,28]. The theory suggests a first-order differential equation system, similar to charge of a linear RC circuit. The complex permittivity in the frequency domain reduces to the well-known Debye equation

$$\varepsilon = \varepsilon' - j\varepsilon'' = \varepsilon_\infty + \frac{\varepsilon_s - \varepsilon_\infty}{1 + j\omega\tau} \tag{8.5}$$

where ε_s and ε_∞ are the static and optical dielectric constants and τ (i.e., the relaxation time) is a time constant of this first-order system. To study the Debye relaxation phenomena at microwave frequency range, ε_∞ is considered as the permittivity value at sufficiently high frequencies, where the orientational effects have disappeared. For small and relatively simple molecular structures (e.g., water), there is often only a single relaxation process (cf. Fig. 8.1). In contrast, for polymers and biological tissues, the dielectric dispersion can consist of several components associated with small side chain movements and the whole macromolecular movement (cf. Fig. 8.2). From Eq. (8.5), the real and imaginary parts of complex permittivity, ε, are

$$\varepsilon' = \varepsilon_\infty + \frac{\varepsilon_s - \varepsilon_\infty}{1 + (\omega\tau)^2} \tag{8.6}$$

$$\varepsilon'' = \frac{(\varepsilon_s - \varepsilon_\infty)\omega\tau}{1 + (\omega\tau)^2} \tag{8.7}$$

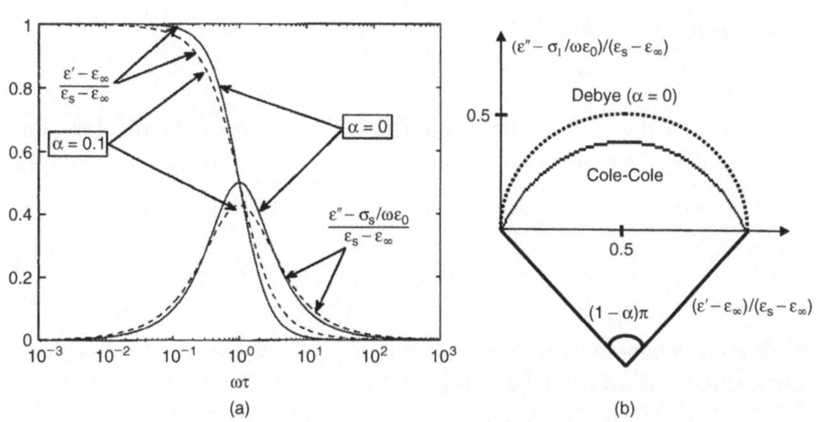

Figure 8.1 Cole-Cole plot for ε' and ε'' as (a) a function of ω and (b) arc plot in complex plane. $\alpha = 0$ corresponds to Debye equation.

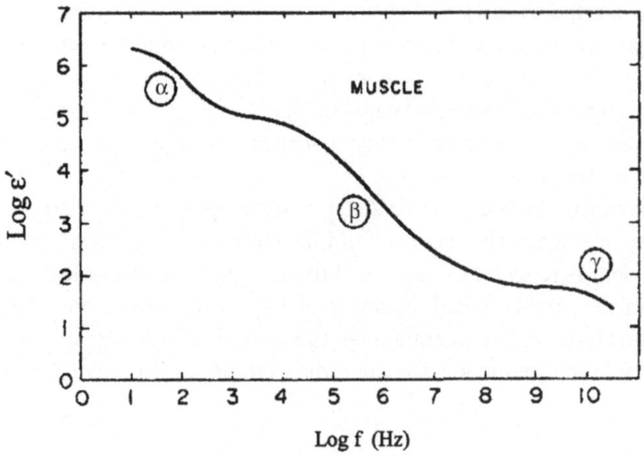

Figure 8.2 The real part of permittivity of muscle tissues. α, β, and γ dispersions are identified [31] (with permission from IEEE).

For Debye relaxation the characteristic frequency is defined as $f_c = 1/2\pi\tau$, which is half way between its low and high frequency values. Dielectric relaxation behavior can be best represented by Argand diagrams [27] on the complex plane

$$\left(\varepsilon' - \frac{\varepsilon_s + \varepsilon_\infty}{2}\right)^2 + \left(\varepsilon'' - \frac{\sigma_I}{\omega\varepsilon_0}\right)^2 = \left(\frac{\varepsilon_s - \varepsilon_\infty}{2}\right)^2 \tag{8.8}$$

where σ_I is the ionic conductivity of the medium. The plot of $(\varepsilon'' - \sigma_I/\omega\varepsilon_0)$ vs. ε' will lead to a semicircle centered at $(\varepsilon_s + \varepsilon_\infty)/2$. Debye theory is the basis of relaxation models proposed for interpretation of the observed dispersion of real materials.

8.2.3. Models for Relaxation

Most real materials do not exhibit single time constant relaxation behavior. In concentrated systems, the electrical interaction between the relaxing species will usually lead to a distribution of relaxation time, $p(\tau)$, and with the help of this distribution, the following relation is used [27,28].

$$\varepsilon = \varepsilon_\infty + (\varepsilon_s - \varepsilon_\infty)\int_0^\infty \frac{p(\tau)}{1 + j\omega\tau}\, d\tau - \frac{j\sigma_I}{\omega\varepsilon_0} \tag{8.9}$$

Gaussian, Cole-Cole, Fuoss-Kirkwood, and Davidson-Cole are some of the distribution function introduced in the literature [27]. The most useful distribution was first introduced by Cole and Cole [33], which leads to

$$\varepsilon = \varepsilon_\infty + \frac{(\varepsilon_s - \varepsilon_\infty)}{1 + (j\omega\tau)^{1-\alpha}} \tag{8.10}$$

In this expression, the ionic conductivity σ_I is ignored. From Eq. (8.10) the real and imaginary parts of complex permittivity are obtained from the following relations $(0 < \alpha < 1)$:

$$\varepsilon' - \varepsilon_\infty = \frac{(\varepsilon_s - \varepsilon_\infty)[1 + (\omega\tau)^{1-\alpha}\sin(\alpha\pi/2)]}{1 + 2(\omega\tau)^{1-\alpha}\sin(\alpha\pi/2) + (\omega\tau)^{2(1-\alpha)}} \tag{8.11}$$

$$\varepsilon'' = \frac{(\varepsilon_s - \varepsilon_\infty)(\omega\tau)^{1-\alpha}\cos(\alpha\pi/2)}{1 + 2(\omega\tau)^{1-\alpha}\sin(\alpha\pi/2) + (\omega\tau)^{2(1-\alpha)}} \tag{8.12}$$

Figure 8.1 compares a Cole-Cole plot for $\alpha = 0.1$ with the Debye plot (i.e., Cole-Cole plot with $\alpha = 0$). Cole-Cole plot for ε' and ε'' as a function of ω is illustrated in Fig. 8.1a. It is seen that the Cole-Cole distribution yields a broader spectrum. A plot of real versus imaginary parts of permittivity reveals that the locus is a semicircle in the case of the Debye equation and is an arc with a subtended angle of $(1 - \alpha)\pi$ in the case of the Cole-Cole model (Figure 8.1b).

Both Debye and Cole-Cole models are examples of physically realizable system. The complex permittivity of a physically realizable system follows the Kramers-Kronig relations [27–29].

$$\varepsilon'(f) - \varepsilon_\infty = \frac{2}{\pi}\int_0^\infty \frac{x\varepsilon''(x)}{x^2 - f^2}\,dx \tag{8.13}$$

$$\varepsilon''(f) = \frac{-2f}{\pi}\int_0^\infty \frac{\varepsilon'(f) - \varepsilon_\infty}{x^2 - f^2}\,dx \tag{8.14}$$

8.2.4. Relaxation Mechanisms in Biological Substances

Three different relaxation mechanisms are defined by Schwan and Foster [31]. These mechanisms are interfacial polarization (Maxwell-Wagner effect), dipolar orientation, and ionic diffusion.

Biological tissues are electrically heterogeneous and hence composed of different entities. In a heterogeneous medium, a charge accumulation exists between the interfaces. This charge accumulation is a consequence of the boundary conditions that the internal electric fields must satisfy at interfaces between different media. Consequently, a dielectric relaxation is observed in bulk properties. A useful example of this theory, known as Maxwell-Wagner theory, is a simple model for the cell suspension as analyzed by Schwan [30–32], where spheres covered by shells with different permittivity values are suspended in a third medium.

On the other hand, the partial orientation of permanent dipoles is responsible for dipolar relaxation. The time constant for dipolar relaxation ranges from microseconds for large globular proteins, to picoseconds for smaller polar molecules such as water. Consequently, the center frequency of the dispersion will be in the MHz to GHz region [31].

Dipolar relaxation effects are major contributors to the permittivity of tissues in the MHz to GHz region. In contrast to the tissues, water as a pure liquid exhibits a nearly single relaxation time with a characteristic frequency close to 20 GHz at room temperature and 25 GHz at 37°C [28].

The third major class of polarization mechanisms arises from ionic diffusion in the electrical double layers adjacent to charged surfaces [28,29]. This phenomenon is called the counterion effect. In contrast to the Maxwell-Wagner effect, which is a macroscopic phenomenon, this effect is a surface phenomenon. Counterion effect is responsible for the dispersion in the KHz frequency range of the solutions of biological particles and long chain macromolecules such as DNA, which show dielectric constants in the order of 10^4 below 1 KHz [28,29]. The reason for the observation of this dispersion is assumed to be the formation of a double layer of charge around a particle with surface charges.

Figure 8.2 illustrates a typical dielectric relaxation behavior exhibited by all tissues. Two significant features of the plot are highlighted by Schwan [32]. These features are the very large dielectric constant at low frequencies and three distinct relaxation regions at low, medium, and very high frequencies. These regions are called α, β, and γ, respectively [32]. Each of these relaxation regions is in its simplest form characterized by a Cole-Cole relation [cf. Eq. (8.10)].

The Maxwell-Wagner effect is responsible for β dispersion, around 50 KHz. Among the three dispersion regions (i.e., α, β, γ), the α dispersion (about 80 Hz) is the least understood one [31]. One

Table 8.1 Range of Characteristic Frequencies Observed for Biological Materials for α, β, γ, and δ Dispersions [32]

Dispersion	Frequency range (Hz)
α	1–10^4
β	10^4–10^8
δ	10^8–10^9
γ	2×10^{10}

Source: Reprinted with permission from [32]. Copyright (2003) American Chemical Society.

possibility for α dispersion is the counterion effect. Frequency dependent impedance of intracellular structures, such as tubular apparatus in muscle cells, is assumed as the other possibility [31].

γ dispersion has a relaxation frequency near 20 GHz. The temperature dependence of the relaxation frequency for γ dispersion in tissues is equal to that of water and is about 2%/°C [31]. This dispersion is primarily due to the presence of water. A minor additional relaxation between β and γ dispersion was first observed by Schwan [33] (for Hemoglobin) and then reaffirmed by Grant et al. [34]. It is called δ dispersion and is observed in a broad frequency range from some 200 to 3000 MHz and is mainly due to proteins bound water [31,32]. The variability of the characteristic frequencies for the various mechanisms i.e., α, β, γ, and δ from one biological tissue to another is given at Table 8.1 [32].

It is believed that the tissue water sets the γ-dispersion relaxation frequency in a similar manner to the pure water [35]. Therefore, the dielectric property of pure water which is well established [36–38], from dc up to the infrared, becomes important behavior of these tissues.

8.3. TECHNIQUES FOR PERMITTIVITY MEASUREMENT

Since the 1940s, many techniques have been introduced for measuring the complex permittivity of materials at RF and microwave frequencies. Reviews of some of early techniques are available in book chapters by Westphal [39], Fox and Sucher [40], and a survey by Bussey [41]. State of the art techniques of the 1970s, especially for biological tissues and liquids, are illustrated in a book by Grant et al. [42]. A review of the coaxial line reflection method, the most widely used technique for biological materials today, is provided by Stuchly and Stuchly [43]. Afsar et al. [44] review a variety of methods available by 1986, for complex permittivity measurement of both lossy and low-loss materials in a broad range of frequencies from 1 MHz to 1500 GHz.

Recent advances in processing speeds and improvement in functionality and accuracy of test equipment, have been major factors for the development of automated complex permittivity measurement systems in recent years. Today, dielectric probe measurement systems, such as the Agilent 85070 family [45], exist that can measure the complex permittivity of materials conveniently and quickly and are compatible with a variety of network analyzers. Moreover, with the emergence of numerical techniques capable of solving Maxwell's equations in reasonable time for complex structures, we are witnessing the arrival of new methodologies that can increase the measurement accuracy, specifically at millimeter wave range where traditional methods face some limitations.

In what follows we review some of the technique identified in above mentioned references and elsewhere. We are particularly interested in techniques historically used in characterization of dielectrics from a few hundred MHz up to millimeter wave frequencies. We would also like to emphasize methods that are widely used for biological and lossy materials. In this regard, this section will address those works, which are extensively referenced or are unique in their nature.

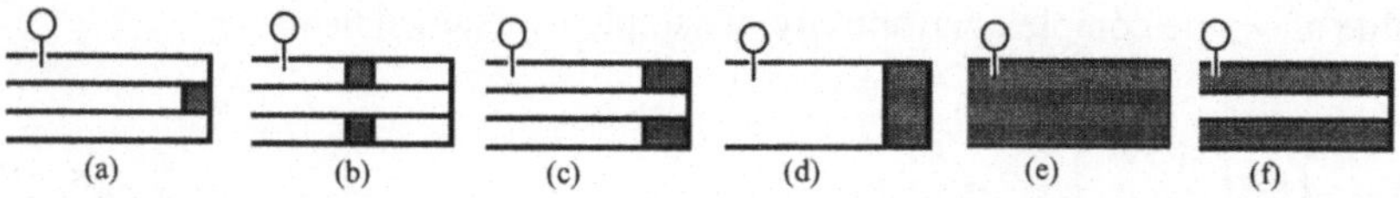

Figure 8.3 Various transmission-line techniques for complex permittivity measurement.

8.3.1. Basic Methods

Westphal [39] provides the formulation of some transmission line techniques. These methods are the basis of measurement instrumentation later developed and used by many researchers [30–32, 34–35,46–47]. Figure 8.3 illustrates some of these techniques, in which an air-filled coaxial line or a hollow waveguide, short terminated at the end, is partially filled with a disk shape sample (a–d) or is filled with a liquid (e, f).

Measurement of the standing pattern using a slotted line and a detector yield the complex permittivity of the sample. The process is as follows: If $\beta_a(= 2\pi/\lambda_{ga}$, where λ_{ga} is the air-filled transmission line wavelength) and Z_a are the propagation constant and characteristic impedance of the air field transmission line, $S\ (=E_{max}/E_{min})$ is the standing wave ratio, and x_a is the location of the first minimum with respect to the sample surface, the input impedance for configurations (b), (c), and (d), at the sample surface, is given by [39]

$$Z_{in} = Z_a \frac{1/S - j\tan\beta_a x_a}{1 - j(1/S)\tan\beta_a x_a} \tag{8.15}$$

For (d) and (c),

$$Z_{in} = Z_s \tanh\gamma_s d \tag{8.16}$$

where index s refers to the sample region and d is the sample thickness. The characteristic impedance of the TEM (coaxial case) or dominant TE_{10} mode (waveguide case) for a nonmagnetic medium is

$$Z_i = \frac{j\omega\mu_0}{\gamma_i} \quad (i = a, s) \tag{8.17}$$

The unknown value of γ_s is obtained by solving the following equation

$$\frac{\tanh\gamma_s d}{\gamma_s d} = -\frac{jZ_{in}}{\beta_a d Z_a} \tag{8.18}$$

The desired equation for configuration (b), which is a preferred method for a thin sample by placing it at the region of high electric field and setting the distance between the sample and end wall as quarter wavelength, is

$$\frac{\coth\gamma_s d}{\gamma_s d} = -\frac{jZ_{in}}{\beta_a d Z_a} \tag{8.19}$$

The above equations are solved for γ_s. Ambiguity in the solution can be resolved if an approximate knowledge of its value is available, or two consecutive measurements for (b) and (c) are performed, and Eqs. (8.18) and (8.19) are multiplied.

Knowing the value of γ_s, the complex permittivity of sample is obtained using

$$\varepsilon' - j\varepsilon'' = \lambda_0^2\left[\left(\frac{1}{\lambda_c}\right)^2 - \left(\frac{\gamma_s}{2\pi}\right)^2\right] \tag{8.20}$$

where λ_c is the cutoff wavelength (infinity for coaxial line). For low-loss sample, the loss of waveguide cannot be ignored. Westphal provides correction for the waveguide loss [39].

A small sample at the end of inner conductor in configuration (a) is a capacitively terminated coaxial line. This structure resembles a reentrant structure where a capacitance $C = \varepsilon C_0$ at the end is introduced, where C_0 is the capacitance with no sample (i.e., the sample is air). From a knowledge of C_0, by neglecting the fringing field, the relations for obtaining the complex permittivity is straightforward:

$$\varepsilon' - j\varepsilon'' = \frac{1}{(j\omega\varepsilon_0 C_0 Z_{\text{in}})} \tag{8.21}$$

Configurations (e) and (f) are ideal for liquid samples [42]. Direct measurement of γ_s can be made from the standing wave pattern quite easily if the liquid is low loss. For high-loss liquids, the standing wave pattern diminishes rapidly and a traveling wave technique using a microwave bridge is suggested. The bridge has two arms, one containing the liquid cell and one containing an attenuator and a phase shifter. By balancing the bridge, an estimate of $\gamma_s = \alpha_s + j\beta_s$ is obtained [42].

For liquids, Steel and Sheppard [46,47] and Grant et al. [38] change the sample thickness by a movable short. The amplitude of the signal as a function of sample thickness is recorded and is used for parameter extraction. In some cases, a combination of the above-mentioned techniques is more appropriate [37].

8.3.2. Resonant Cavity Measurement

Measurement of the complex permittivity using a resonant cavity coupled to the sample provides more accuracy than the previous methods. However, this method suffers from the fact that it is applicable only at resonance frequencies of the cavity in the lowest order mode (or a specified mode in the case of waveguide), which are distinctly away from each other or from other modes. The basic idea in using the resonant method is that the resonance frequency (f_0) and quality factor (Q) of the cavity will change as a result of coupling by the sample. The characteristic parameters of the sample, i.e., permittivity and permeability, can be measured by monitoring the changes in these two parameters.

Perturbation Technique

Perturbing the fields inside the cavity by a small sample has been a popular method for decades, where the perturbation theory [48] can be applied to find the electric or magnetic properties of the sample [49–55].

The basic formula for this method relates the change in the resonance frequency of the cavity to the field distribution inside, before and after placing the sample, which is [49,56]

$$\frac{\Delta\omega}{\omega_0} = \frac{\iiint_{V_1}[(E_1D_0 - E_0D_1) - (H_1B_0 - H_0B_1)]dv}{\iiint_V (E_0D_0 - H_0B_0)dv} \tag{8.22}$$

where, V_1 and V are the corresponding volumes of the sample and the cavity respectively, $\Delta\omega\ (= \omega - \omega_0)$ is the change in the resonance frequency of the cavity, and field indices 0 and 1 refer

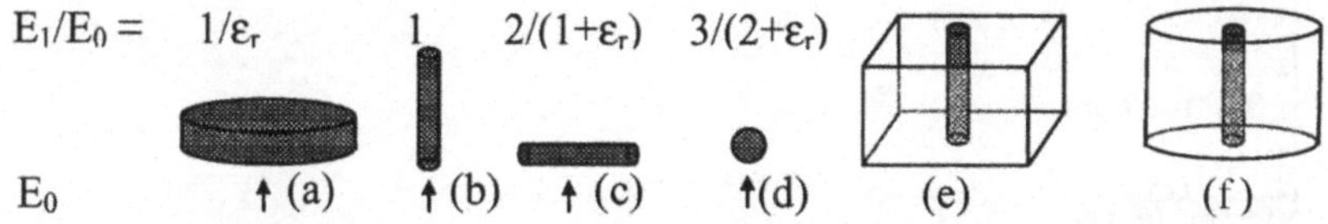

Figure 8.4 (a–d) Useful relations for applying perturbation theory. (e, f) Samples embedded in rectangular and circular resonators.

to the cavity fields before and after placing the sample. Assuming that the perturbation is small and an isotropic and homogeneous sample [56] is placed in an air filled cavity,

$$\frac{\Delta\omega}{\omega_0} = \frac{\iiint_{V_1} [\varepsilon_0(1-\varepsilon_r)E_0 \cdot E_1 - \mu_0(1-\mu_r)H_0 \cdot H_1]dv}{\iiint_V (\varepsilon_0 E_0^2 - \mu H_0^2)dv} \tag{8.23}$$

The quantity in the denominator is proportional to the stored energy in the cavity. The above relation can be further simplified if it is assumed that the fields outside the sample (volume V_1) are unchanged, and the field inside the sample can be related to the field outside by a quasistatic approximation. Some of these approximated relations between E and E_1 are given by Harrington [48] as shown in Fig. 8.4.

The dielectric samples are usually placed at the location of the maximum electric field intensity, where as magnetic samples are measured when inserted in a high magnetic field region. In any case, applying the perturbation theory by further simplification of Eq. (8.23) requires no appreciable field variation within the sample. For a nonmagnetic material, by satisfying this requirement, the numerator of Eq. (8.23) reduces to $\varepsilon_0(1 - \varepsilon_r)E_0E_1V_1$, in which E_0 and E_1 are the fields at the location of sample placement. On the other hand, the denominator depends on the cavity shape and modal distribution and is four times the energy stored in the cavity.

In the case that sample is lossy, ε_r is a complex quantity and $\Delta\omega/\omega_0$ in the above relations is replaced by [48,49,51]

$$\frac{\Delta\omega}{\omega_0} = \frac{\omega-\omega_0}{\omega_0} + \frac{j}{2}\left(\frac{1}{Q} - \frac{1}{Q_0}\right) \tag{8.24}$$

where Q_0 and Q are the unloaded quality factors of the cavity before and after placing the sample. Still by replacing ε_r by ε' in Eqs. (8.22) and (8.23), these equations can be used for obtaining ε' from the change in real frequency, and ε'' is obtained by the following relation [48].

$$\frac{\varepsilon'-1}{\varepsilon''} = 2\left(\frac{Q_0Q}{Q_0-Q}\right)\left(\frac{\omega_0-\omega}{\omega_0}\right) \tag{8.25}$$

As an example, by placing a thin sample (Figure 8.4b) in the middle of a TE_{101} rectangular resonant cavity (Figure 8.4e), along the direction of the E field, complex permittivity is obtained by [41,54]

$$\begin{aligned} \varepsilon' &= 1 + \frac{f_0 - f}{f}\frac{V}{2V_1} \\ \varepsilon'' &= \frac{Q_0 - Q}{Q_0Q}\frac{V}{4V_1} \end{aligned} \tag{8.26}$$

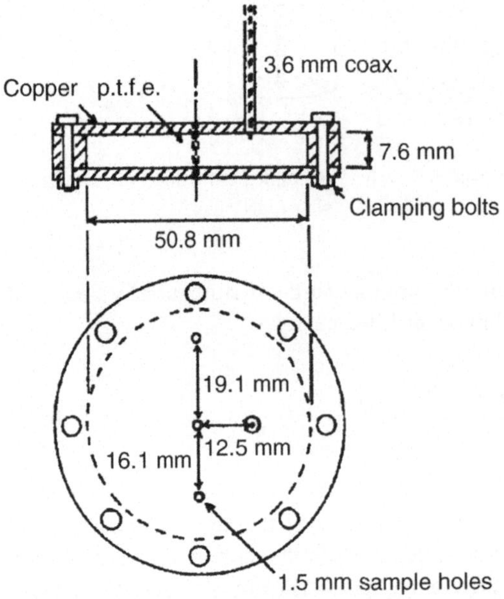

Figure 8.5 The cylindrical cavity used by Land and Campbell [55] for measurement of biological samples at 3.2 GHz (with permission from IOP Publishing Ltd.).

A very popular technique is using a cylindrical cavity resonating at TM_{010} mode (Figure 8.5). In this case as used by Land and Campbell [55],

$$\varepsilon' = 1 + 2\frac{f_0 - f}{f}C \tag{8.27}$$

$$\varepsilon'' = \frac{Q_0 - Q}{Q_0 Q}C \tag{8.28}$$

where, for a sample of radius s, inside a dielectric field cavity with the radius a and relative permittivity ε_c', C is given by

$$C = \varepsilon_c' \left(\frac{a}{s}\right)^2 J_1^2(ka) \tag{8.29}$$

In this relation k is the wavenumber in the medium filling the cavity and $J_1()$ is the first order Bessel function of the first kind. Land and Campbell [55] propose a cavity method, where the sample is placed in three sample container holes made inside a PTFE filled cylindrical resonant cavity (cf. Fig. 8.5). The cavity resonates at 3.2 GHz for the TM_{010} mode. It has a length of 7.6 mm and a diameter of 50.8 mm. The sample is so small that the perturbation technique can be used to find the complex permittivity from the measurement of Q and f_0. They use this technique to measure variety of liquids and tissues such as water, saline, fat, and breast. The accuracy is reported to be $\pm 2.5\%$ for the dielectric constant and $\pm 3.5\%$ for the material loss factor.

Another practical system is the strip line cavity suggested by Waldron [49] and developed by NIST [52,53] (Fig. 8.6).

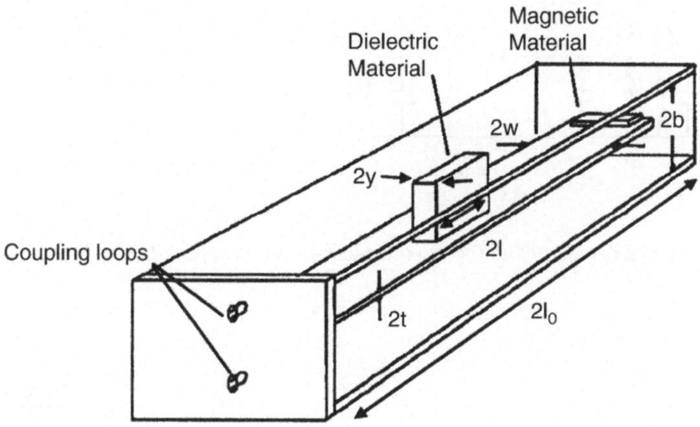

Figure 8.6 NIST strip-line cavity [53] (with permission from IEEE).

The formulas for a cavity with height $2b$, length $2l_0$, with a strip with thickness $2t$ and width $2w$, holding a sample with the height $b - t$, length $2l$, and width $2y$ are as follows [49,52].

$$\frac{\omega - \omega_0}{\omega} + \frac{j}{2}\left(\frac{1}{Q} - \frac{1}{Q_0}\right) = -A(\varepsilon - 1)\frac{2yl}{l_0} \tag{8.30}$$

with ε being the relative complex permittivity and

$$A = \frac{\pi\alpha}{2(\alpha + \beta)K(1/\alpha)}\sqrt{\frac{\alpha^2 - \beta^2}{\alpha - 1}} \tag{8.31}$$

α and β are parameters depending on the cavity dimensions (w, t, b) [49,52] and K is the complete elliptic integral of the first kind.

Sample-Terminated Coaxial Cavity

In practice, the resonance method can be used in all the cases shown in Fig. 8.3 by terminating the left side of transmission lines to a short wall as explained by Westphal [39]. A more recent approach is the one used by Tanabe and Joines [57] and Xu et al. [58], in which a coaxial line is terminated to a sample (cf. Fig. 8.7). Resonator is constructed by placing a capacitive gap at the input. C_r, L_r, and G_r are the equivalent components associated with the coaxial line length constituting the resonator. C_f is the fringing field inside due to the higher order mode, and C and G are the capacitance and conductance outside the interface due to the radiation and capacitive effects.

In the method discussed by Tanabe and Joines [57], the fringing and radiation effects are ignored and the admittance at the interface is considered to be a result of the open end static capacitor $C(\varepsilon')$ given by

$$C(\varepsilon') = \varepsilon'\varepsilon_0 h(\varepsilon') \tag{8.32}$$

h is a function of inner and outer radii of the coaxial lines ($a = 0.14364$ cm and $b = 0.47250$ cm) and the dielectric of the coaxial line ($\varepsilon_c = 2.05$). It is obtained by static methods and is represented by an experimental relation as a function of ε'. For a time varying case, a conductance G can be recognized that is

$$G = \omega C(\varepsilon') \tan\delta \tag{8.33}$$

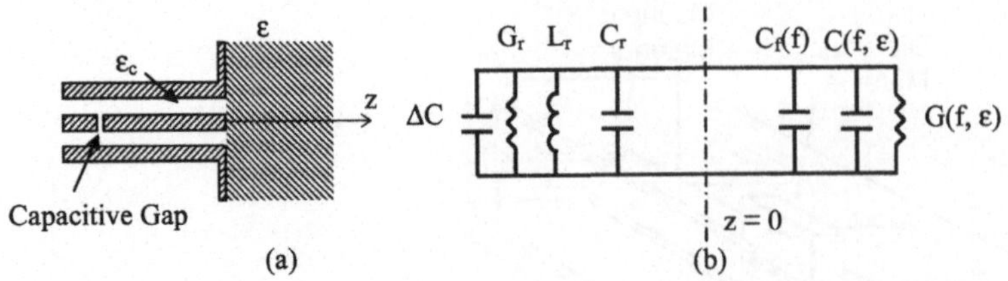

Figure 8.7 (a) Sample-terminated coaxial resonator and (b) its equivalent circuit [58] (© 2003 IEEE).

Note that

$$\frac{G}{C} = \omega \tan \delta \tag{8.34}$$

Tanabe and Joines [57] use the equivalent transmission line methods where the capacitive gap and the open end are treated as the extension of the coaxial cavity with the same characteristic impedance as the coaxial line. The overall cavity will have an effective resonance length of Le $(= nc/2f)$, where n is the mode order of TEM_n mode and c $(= 2.998 \times 10^8\,\text{m/s})$ is the speed of light in free space.

The mathematical derivation of the end results are somewhat tedious and are not presented. The end formulas are

$$\varepsilon' = \frac{Y_0}{\omega\varepsilon_0 h(\varepsilon')} \frac{A_1(1 - A_2^2)}{1 + A_1^2 A_2^2} \tag{8.35}$$

$$\tan \delta = \frac{C_e}{CQ}\left(1 - \frac{Qf_0 C_{e0}}{Q_0 f C_e}\right) \tag{8.36}$$

where

$$A_1 = \tan\left[\frac{f}{f_0}\tan^{-1}\left(\frac{\omega_0 C_0}{Y_0}\right) + n\pi\frac{\Delta f}{f_0}\right] \tag{8.37}$$

$$A_2 = \tanh\left[\left(\frac{\sqrt{1+\tan^2\delta} - 1}{\sqrt{1+\tan^2\delta} + 1}\right)^{1/2} \tan^{-1} A_1\right] \tag{8.38}$$

$$C_e = \frac{1}{2}\left[C + \frac{Y_0 l}{v_1}\left(1 + \frac{\omega^2 C^2}{Y_0}\right)\right] \tag{8.39}$$

$$C_{e0} = \frac{1}{2}\left[C_0 + \frac{Y_0 l}{v_1}\left(1 + \frac{\omega_0^2 C_0^2}{Y_0}\right)\right] \tag{8.40}$$

In the above relations C_e and C_{e0} are the total equivalent capacitance of the resonator, $Y_0(=1/Z_0)$ is the characteristic admittance of the line, v_1 is the wave velocity in the coaxial line,

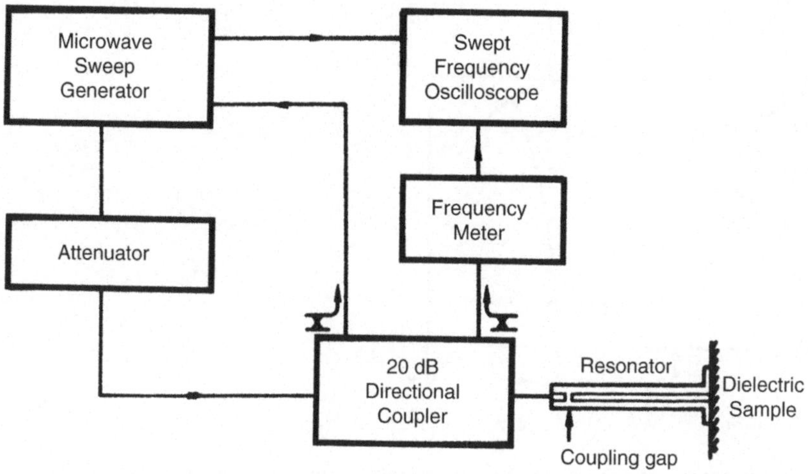

Figure 8.8 Schematic block diagram of the dielectric measurement technique reported by Tanabe and Joines [57] (with permission from IEEE).

$C_0 = C\,(\varepsilon' = 1)$, and l is the equivalent length of the coaxial cavity plus the gap capacitor. This latter quantity can be obtained from the air terminated measurement and C_0. Measurement of Q and f, and Q_0 and f_0 (cf. Fig. 8.8) provide enough information to calculate the complex permittivity through Eqs. (8.35) and (8.37) by successive iterations.

Tanabe and Joines [57] apply their technique for measuring the permittivity of water, methanol, and skin, as well as loss less dielectric such as polyethylene. They claim an accuracy of 5% for real part of permittivity and 25% on loss tangent within the frequency range of 1 to 4 GHz range.

The above method is further modified by Xu et al. [58]. They introduce the effect of fringing capacitance and the radiation at the sample side. They employ the method of Marcuvitz [59] for radiation admittance of the probe and end up with a series presentation for both G and C as functions of ε and f. The method is similar to the previous one and a pair of nonlinear equations for ε' and $\tan\delta$ is obtained. The coefficients of the capacitance series, truncated to the first two terms, and the fringing capacitance are obtained by resonance measurement of open, short, and a known dielectric-terminated cavity. The coefficient for the conductance series are explicitly given as a function of coaxial line parameters. They employ this method for a variety of tissues from 0.1 to 11 GHz.

Open Resonators

At frequencies above 30 GHz, the size and the quality factor of closed cavities decrease. Yet resonance methods are still applicable by employing open and Fabry-Perot resonators [60–65]. These structures can be used for measurements up to 300 GHz [61].

Developing a theory for his method, Jones [60] uses an open resonator that is constructed by a flat and a concave mirror and operating at 35 GHz. The sample sheet is placed on top of the flat mirror. In his procedure, the resonance for the empty cavity is obtained by moving the mirror using a micrometer. Then the quality factor of the empty cavity is measured by changing the frequency. After inserting the sample, the resonance for the same mode is restored by reducing the resonator length. Finally, Q of the loaded resonator is measured.

Jones' technique has been used by researchers over time by modifying it to be automated and applied with a fixed-frequency source (Afsar et al. [64–65]) or used in conjunction with a network analyzer [62].

Afsar and others [63–65] introduce an automated system based on a fixed frequency of 60 GHz and changing the cavity length (cf. Fig. 8.9).

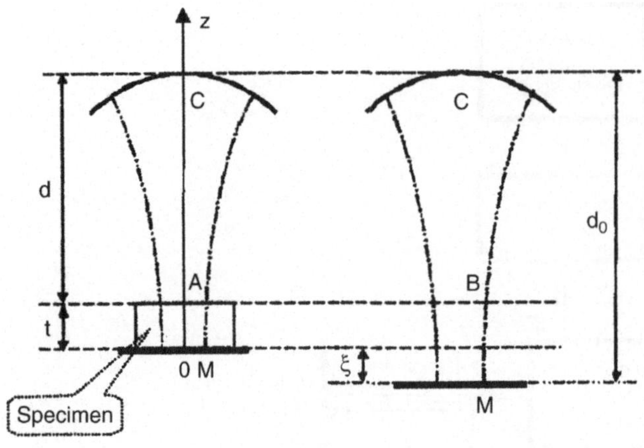

Figure 8.9 Diagram of a hemispherical open resonator with and without the specimen (from Afsar et al. [65], with permission from IEEE).

The transcendental equation for the refractive index, obtained by applying a wave impedance matching condition of gaussian beam boundary condition to the fields at the air sample interface, is

$$\frac{1}{n}\tan(nkt - \Phi_{AM}) = -\tan(kd - \Phi_{AC}) \tag{8.41}$$

where n is the refractive index of the specimen, k is the free-space wave number. Furthermore, Φ_{AM} and Φ_{AC} are expressed as

$$\Phi_{AM} = \tan^{-1}\frac{t}{nZ_0} - \tan^{-1}\frac{1}{nkR_1(t)} \tag{8.42}$$

$$\Phi_{AC} = \tan^{-1}\frac{d'}{Z_0} - \tan^{-1}\frac{1}{kR} - \tan^{-1}\frac{t}{nZ_0} + \tan^{-1}\frac{1}{kR_2(t)} \tag{8.43}$$

R is the radius of the curvature of the mirror and also

$$R_1(t) = t + \frac{n^2 Z_0^2}{t} \qquad R_2(t) = \frac{R_1(t)}{n} \tag{8.44}$$

Finally, $Z_0 = \sqrt{d'(R-d')}$, with $d' = d + t/n$. For an empty cavity ($t = 0$, $n = 1$), the above relations reduce to

$$kd_0 = q\pi + \tan^{-1}\sqrt{\frac{d_0}{R - d_0}} - \tan^{-1}\frac{1}{kR} \tag{8.45}$$

where q is the mode number of $TEM_{0,0,q}$ mode that yields d_0, the resonance length of the empty cavity. A correction for mismatching between the wave front of the gaussian beam and the surface of the spherical mirror, as well as the upper surface of the sample, is

$$d = d_0 - t - \xi + \frac{t(n-\Delta)}{n^2k^2w_t^2} + \frac{3}{4k^2R} \tag{8.46}$$

$$w_t^2 = \frac{2Z_0(1+t^2/n^2Z_0^2)}{k} \tag{8.47}$$

$$\Delta = \frac{n^2}{n^2\cos^2(nkt - \Phi_{AM}) + \sin^2(nkt - \Phi_{AM})} \tag{8.48}$$

where ξ is the shift length to restore the resonance with and without specimen. Equations (8.41) and (8.46) can be solved to find n and d. The loss tangent is found through the following relation:

$$\tan\delta = \frac{1}{Q_\varepsilon}\frac{2nk(d+t\Delta)}{2nkt\Delta - \Delta\sin 2(nkt - \Phi_{AM})} \tag{8.49}$$

where

$$\frac{1}{Q_\varepsilon} = \frac{1}{Q_L} - \frac{1}{Q'_L} \quad \text{and} \quad Q'_L = Q_0\frac{2(t\Delta + d)}{d_0(\Delta+1)} \tag{8.50}$$

in which Q_L and Q_0 are the quality factor of the cavity with and without the sample.

Using this technique, Afsar et al. [64,65] have measured the dielectric properties of various dielectrics at 60 GHz. For instance, $\varepsilon' = 2.063 \pm 0.004$ and $\tan\delta = 0.00029 \pm 0.00003$ is calculated for Teflon.

8.3.3. Open-Ended Transmission Line

The advent of accurate Automatic Network Analyzers (ANA) and new calibration techniques has revolutionized the measurement of microwave device and networks, since the late 1970s. In the last two decades, the extraction of complex permittivity from the measured reflection coefficient (input admittance) of a coaxial line terminated to a specimen (cf. Fig. 8.10) has been the most widely used method of the tissue permittivity extraction [43,45,66–89].

Stuchly and Stuchly [43] review a variety of coaxial line terminated structure including those already shown in Fig. 8.3. The basic idea involves the reflection coefficient (S_{11}) measurement at the input of the coaxial line. First, the values for the equivalent circuit models (cf. Fig. 8.10) are analytically or numerically calculated as a function of frequency and permittivity. Then, the unknown real and

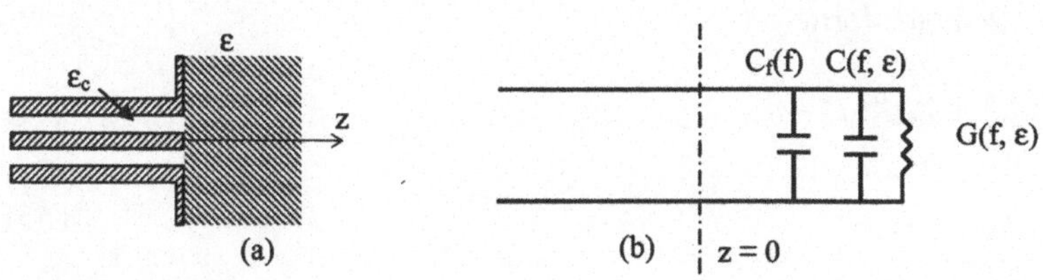

Figure 8.10 A coaxial probe terminated to a lossy medium with complex permittivity ε: (a) the schematic of the probe and (b) the equivalent circuit model.

imaginary parts of the complex permittivity are obtained. A variety of approaches exist in the literature to formulate the terminating admittance of the coaxial line. Some of them are briefly reviewed here.

Lumped-Capacitance Method

This approach yields the same complex permittivity extraction relations for structures in Figs. 8.3a (i.e., reentrant coaxial line) and 8.10 (i.e., open-ended coaxial line) [43,66,67]. The former has the advantage of requiring a small sample and the latter is useful for a sufficiently big sample having a flat surface to contact with the probe. The fringing capacitance C_f is due to higher order, nonpropagating coaxial modes on the left side of the coaxial–sample interface and is assumed not to change by changing the sample permittivity. G and C are the conductance and capacitance on the right side of the interface. The lumped-element approach assumes that these two quantities are directly proportional to their corresponding complex permittivity components (real or imaginary) as if they are created as a result of filling the static fringing capacitance in air (C_0), in the right side of the interface, with a medium with the permittivity of ε.

$$C = \varepsilon' \varepsilon_0 C_0 \qquad G = \varepsilon'' \varepsilon_0 \omega C_0 \tag{8.51}$$

Note that their relation is the same as Eq. (8.34). If $S_{11} = |S_{11}| \exp j\phi$ is obtained by a measurement, ε' and ε'' can be obtained by

$$\varepsilon' = \frac{2|S_{11}|\sin(-\phi)}{\omega Z_0 C_0 (1 + 2|S_{11}|\cos\phi + |S_{11}|^2)} - \frac{C_f}{C_0} \tag{8.52}$$

$$\varepsilon'' = \frac{1 - |S_{11}|^2}{\omega Z_0 C_0 (1 + 2|S_{11}|\cos\phi + |S_{11}|^2)} \tag{8.53}$$

Note that C_f and C_0 should be known separately by analytical or numerical methods or from the measurements of two known samples. The latter is usually a more practical method. However, a numerical solution for these quantities using the method of moment (MOM) [90] or finite element technique (FEM) [91] has also been suggested.

Note that this method neglects the dependence of C_f, C, and G on frequency and ignores the fact that aperture field distribution at the interface creates radiation on the outside. Radiation effect and the presence of significant higher mode effects will reduce the merit of this method at frequencies higher than a few GHz.

Stationary Solution of Aperture Admittance

To remedy some of the shortcomings of the previous method, a stationary solution of the aperture admittance as found in Marcuvitz [59] is suggested [58,78–80]. In this approach the admittance of the probe is represented by a stationary integral. The integral is solved by a series expansion technique and the final solutions are in the form

$$G = G_1 \varepsilon^{5/2} f^4 + G_2 \varepsilon^{7/2} f^6 + B_5 \varepsilon^{9/2} f^8 + \cdots \tag{8.54}$$

$$B = B_1 \varepsilon f + B_3 \varepsilon^2 f^3 + B_5 \varepsilon^3 f^5 + \cdots \tag{8.55}$$

where G_i and B_i are constants related to the coaxial line parameters. A good approximation is to consider only the first term of Eq. (8.54) and the first two terms of Eq. (8.55) [79]. Note that in this formulation, the impact of higher order modes inside the waveguide on the aperture admittance is ignored [78–80].

Full Wave Modeling

The full wave solution is possible by using a well-known numerical technique such as mode matching and MOM. Mosig et al. [73] present a formulation for the aperture using this method. They use a modal expansion (only TEM and TM_{0n} modes) of the field inside the coaxial probe. They even provided some charts for this purpose at different frequencies. The charts are for a SR7 coaxial cable (i.e., $a = 1.05\,\text{mm}$, $b = 3.675\,\text{mm}$, $\varepsilon_r = 2.3$) and at frequencies 1, 3, and 10 GHz. However, due to the computational cost and the limited computational power at the time, their method did not seem to be a practical one.

Stuchly et al. [82] also use MOM to find the aperture admittance for a wide range of inner conductor diameter and permittivity of a 50-Ω coaxial cable. Recognizing the need for introducing such data in a suitable form to be used for complex permittivity extraction, they fit the numerical results to a rational function that is

$$Y(j\omega a,\ \varepsilon) = \frac{\sum_{n=1}^{N}\sum_{p=1}^{P}\alpha_{np}\zeta^{p}(j\omega a)^{n}}{1+\sum_{m=1}^{M}\sum_{q=0}^{Q}\beta_{mq}\zeta^{q}(j\omega a)^{m}} \tag{8.56}$$

where $\zeta = \sqrt{\varepsilon}$ and α_{np} and β_{mq} are parameters of the function. For fitting they employ 56 dielectric constants in the range of $1 \leqslant \varepsilon' \leqslant 80$ and 20 normalized frequency in the range of $0.01 \leqslant ka \leqslant 0.19$, resulting in a total of 1120 data points. For fitting the simulated data to the above function they use a modified Levenberg-Marquardt algorithm [92] to minimize the total sum of absolute error between the admittance from Eq. (8.56) and admittance obtained by MOM at the 1120 points. This procedure yields fitting parameters α_{np} and β_{mq}. Truncation of the series is made by $M = N = 4$ and $P = Q = 8$. Therefore 68 parameters are given to yield a function that can represent the aperture admittance explicitly. If measured admittance is available, Eq. (8.56) can be solved for the complex permittivity of the medium.

Virtual Line Method

This method models the sample medium by a virtual transmission line with length L filled with the unknown dielectric of the specimen. If a coaxial line of length D is terminated to the sample, the following relation is valid [87]:

$$\varepsilon = \frac{-jc\sqrt{\varepsilon_c}}{2\pi fL}\,\frac{1-\Gamma_m e^{2jkD}}{1+\Gamma_m e^{2jkD}}\cot\frac{2\pi fL\sqrt{\varepsilon}}{c} \tag{8.57}$$

where ε_c is the dielectric constant of the probe, k is the wave number of the probe, c is the speed of light in free space, and Γ_m is the reflection coefficient measured at the input of the probe. In this technique D and L are measured from measurement of two known media, i.e., air and distilled water. Knowing these quantities, complex permittivity of an unknown medium can be evaluated by measuring the reflection coefficient through Eq. (8.57).

8.3.4. Free-Space (Quasioptical) Measurement Techniques

This method is based on placing a flat sample of thickness d between transmitting and reflecting horn antennas. This method is particularly suitable for W-band (70–120 GHz) as reported by Friedsam and Biebl [93] and Afsar et al. [94], at which quasioptical method based on a gaussian beam can be

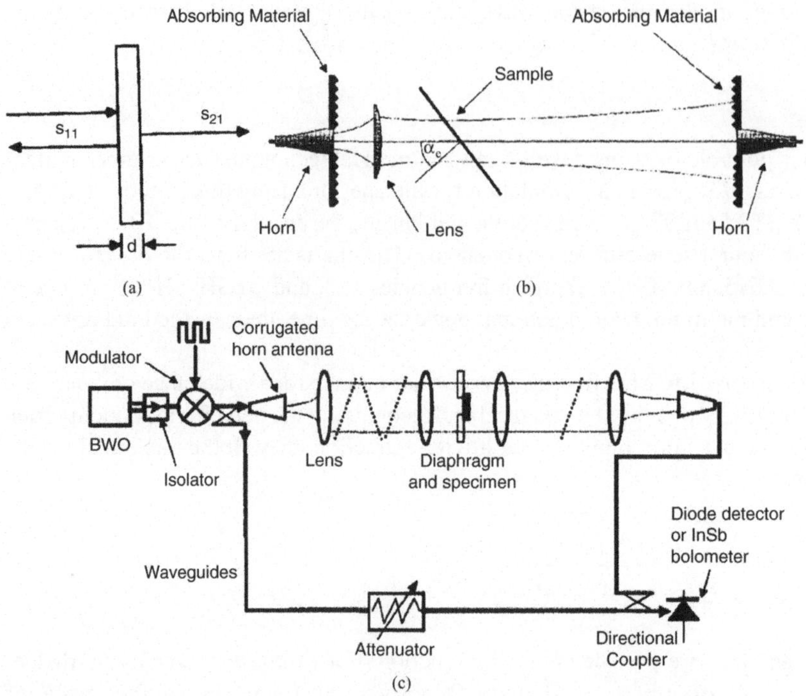

Figure 8.11 (a) Reflection and transmission of a dielectric sample, (b) Friedsam and Biebl [93] setup, and (c) Afsar et al. method [94] (with permission from IEEE).

applied. However, it may also be applied at lower frequencies as has been the case for a system reported by Ghodgaonkar et al. [95] at 8.6 to 13.4 GHz.

Figure 8.11 illustrates the set up for these three methods. In Figure 8.11a, from simple plane wave theory, knowing the transmission and reflection coefficient (S_{11} and S_{21}), we obtain the complex permittivity and permeability as [95]

$$\varepsilon = \frac{\gamma}{\gamma_0}\left(\frac{1-\Gamma}{1+\Gamma}\right) \tag{8.58}$$

$$\mu = \frac{\gamma}{\gamma_0}\left(\frac{1+\Gamma}{1-\Gamma}\right) \tag{8.59}$$

In the above relations, $\gamma_0(=j2\pi/\lambda_0)$ is the propagation constant of free space and

$$\gamma = \frac{\ln(1/T)}{d} \tag{8.60}$$

$$\Gamma = K \pm \sqrt{K^2 - 1} \tag{8.61}$$

where

$$K = \frac{S_{11}^{2} - S_{21}^{2} + 1}{2S_{11}} \tag{8.62}$$

$$T = \frac{S_{11} + S_{21} - \Gamma}{1 - (S_{11} + S_{21})\Gamma} \tag{8.63}$$

In their method they use a TRL calibration method [96–98] for the accurate measurement of S_{11} and S_{21} using a HP8510B network analyzer.

If the material is nonmagnetic, the transmission measurement is adequate. In the Friedsam and Biebl [93] set up (Fig. 8.11b), for 75 to 95 GHz, a sheet of sample is located between corrugated horn antennas. By using a collimating lens before the sample, a quasioptical gaussian beam illuminates the sample. Calibration procedure involved performing a reference measurement. This is performed by normalizing the transmission coefficient with sample in place to the case without sample. The mean value of the complex permittivity is obtained for various angle α_e and a nonlinear least square method to minimize an error function of difference between the theoretical and measured transmission coefficients. The reported uncertainties are 0.1% for dielectric constant and 2×10^{-4} for loss tangent.

In Afsar et al. method [94] (cf. Fig. 8.11c), an unbalanced bridge, with a waveguide as the reference arm, is used. The refractive and absorption indices are obtained from the distance between the maxima and the amplitude of transmittance spectra.

8.3.5. Two-Port Network Analyzer Measurement/Extraction

One-port measurement and extraction of dielectric properties of materials using a coaxial probe is generally adopted for measurement of biological tissues up to 20 GHz. There are some advantages offered by using two-port calibration and measurements. A powerful calibration technique known as TRL [96–98] is available for modern network analyzers. The TRL method potentially provides much better accuracy than the traditional calibration methods (which use imperfect match and open standards), specially at higher frequencies, and is inherently two port. In addition, a two-port measurement [99–102] gives more information (four S parameters) than a one-port measurement with only the reflection coefficient being measured.

Belhadj-Tahar et al. [99] presented a technique for the simultaneous measurement of the complex permittivity and permeability of a given material using ANA. A gap built in a coaxial line was filled with the material under test. Complex permittivity and permeability were computed from the two-port S parameters for several materials from 45 MHz to 18 GHz. They employed a mode matching method for the numerical solution of the structure, and the gradient method for the inverse problem solution. HP8510A ANA and APC7 coaxial standards were used. For avoiding the contact resistance and capacitance, the sample was metalized on the surfaces that were in contact with the coaxial connectors.

Measurement of teflon and alumina yielded accurate results for the dielectric constant; however, they concluded that an accurate measurement is not possible for the imaginary part of permittivity in the range of less than 0.1. They also commented that by using a 2.4-mm connector the method could be extended to 50 GHz.

Abdulnour et al. [100] developed a generic approach for permittivity measurement in microwave and millimeter wave frequencies. They first determined the scattering parameters of a discontinuity containing a material having a wide range of complex permittivity that was known a priori. The discontinuity was a tube containing a sample under test (i.e., a cylindrical shape dielectric). For the direct problem, they used a boundary integral equation method combined with a modal expansion approach. They developed some simple generic formulas from graphs for constant ε' and constant ε'' on S_{21} plane, which directly provide ε' and ε'' from measured S_{21}. An accuracy of better than 1% was claimed.

To cover a broad measurement range, they used a microstrip structure for 1 to 7 GHz (covering L-, S-, C-bands), WR90, WR62, and WR42 waveguides for 8 to 26 GHz (covering X-, Ku-, and K-bands). The frequency bandwidth was between 1.2 to 1.8 times the cutoff frequency of the dominant TE_{10} mode of the rectangular waveguide and up to half of the cutoff frequency for the microstrip

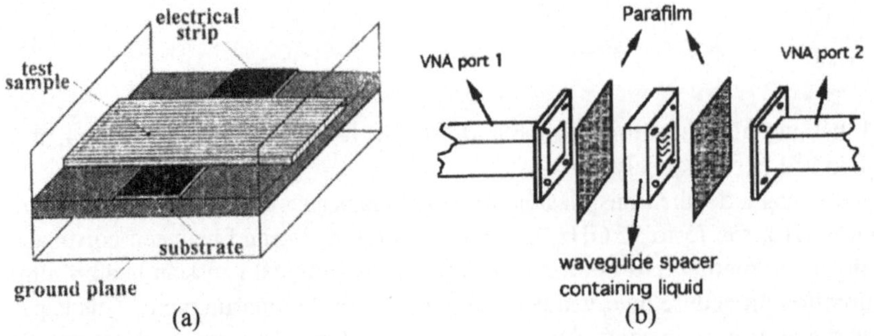

Figure 8.12 (a) Microstrip test structure used by Queffelec and Gelin [101], and (b) waveguide technique introduced by Vander Vorst and colleagues [103] (with permission from IEEE).

structure. They verified the complex permittivity of teflon, plexiglas, and some polymers. It should be mentioned that they used the HP8510B along with the TRL calibration method.

Queffelec and Gelin [101] also applied the microstrip line and two-port measurement by placing a dielectric sample with specified dimensions over a microstrip line, where the sample covers the whole width of the enclosure (cf. Fig. 8.12a). A 25 mils thick alumina substrate and an enclosure width of 25.4 mm were considered. They analyzed the structure based on the mode matching technique, over the frequency range of 45 MHz to 14 GHz. TRL method was used for calibration. The accuracy of the method was claimed to be better than 5% over this frequency range and for the dielectric constant less than 10. However, the measurement accuracy deteriorated for higher permittivity values.

It is worth mentioning that none of the above mentioned techniques was developed for the measurement of biological samples. A very simple technique is introduced by Vander Vorst and colleagues [103,104] for the measurement of blood, dioxane, and methanol up to 110 GHz. This method applies an HP network analyzer system with waveguide standards. They introduced LL calibration that is a reduced form of TRL calibration, and is designed for evaluating complex propagation constant γ. In their waveguide measurement system, a liquid sample holder is used, where its cross section exactly matches with the reference air-filled waveguide cross section.

A newer technique is the one introduced by Tofighi and Daryoush [25,105–109], based on a two-port microstrip test fixture. Fig. 8.13 depicts this two-port test fixture. Open-circuited microstrip transmission lines are coupled to tissue under test (TUT) through two small apertures. The sample is sandwiched between glass plates and then inserted between the microstrip ground planes. Planes 1–1′ and 2–2′ are the reference planes where the effects of embedding networks are removed. Microstrip lines and apertures are etched over the two sides of a fused silica substrate ($\varepsilon_r = 4.1$). Two coaxial to microstrip line launchers provide transitions from the network analyzer test set cables to the fixture.

A 100 mil (i.e., 2.54 mm) diameter circular aperture is considered (cf. Fig. 8.13b), as a compromise between the coupling factor and the spatial resolution. To have a realizable microstrip line test fixture with the aperture on the ground plane facing the sample, the width of the fused silica substrate should be slightly more than the width of air-filled space above. The substrate sits inside lips provided in the enclosure. Therefore, the microstrip cross section is not a complete rectangle (cf. Fig. 8.13c). This technique has been used for the extraction of the complex permittivity of brain grey and white matters, neurological cell solutions, and dielectric imaging of brain slices up to 50 GHz and will be explained in detail in later sections.

8.3.6. Characteristic Impedance Determination Method

If the propagation constant of a transmission line with the specimen as part of its structure is known, then the complex permittivity can be obtained [105,110,111]. The requirements for the applicability

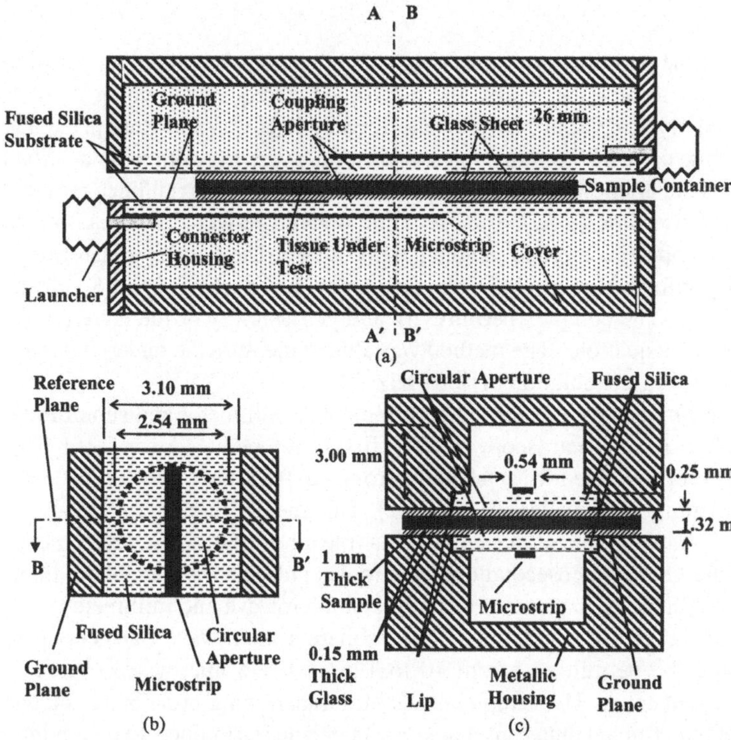

Figure 8.13 A two-port microstrip test fixture for complex permittivity measurement of biological materials; schematic of (a) side, (b) top, and (c) front views, where the two microstrip lines are coupled through two apertures, glass sheets, and TUT [108] (with permission from IEEE).

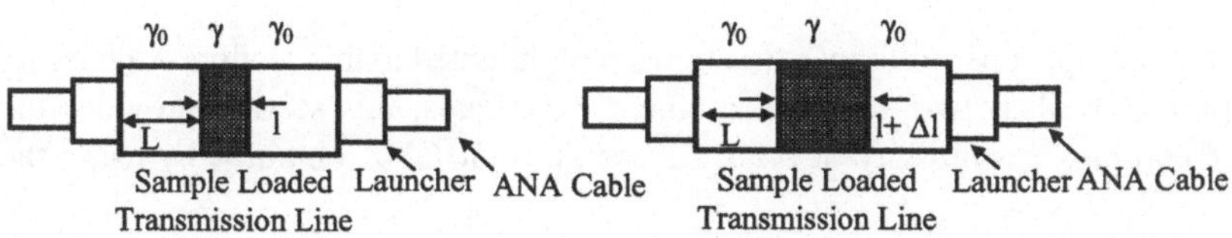

Figure 8.14 Conceptual diagram of method for propagation constant measurement with the transmission lines are loaded with two uniform specimen with difference in length of Δl.

of this method are that the specimen should be uniform across its length and the transmission line loaded with it operates only at its dominant mode.

Although it is conceptually a simple method, its general theory was formalized quite recently. The formulation for this technique is similar to the propagation constant determination as part of TRL [96–98,105] or multiline [112] calibrations and is omitted here. The propagation constant, γ, can be found from at least two successive two-port measurements of the transmission line with samples having two different lengths (cf. Fig. 8.14). The length difference of a quarter wavelength (wavelength of the transmission line loaded with the sample) at the middle of frequency band yields the most reliable results.

This method is used by Janezic and Jargon [110] for the measurement of polystyrene from 8 to 12 GHz using a coaxial line, by Wan et al. [111] for PVC in a rectangular waveguide at 8 to 18 GHz, and, by Tofighi [105] for fused silica as the substrate of a microstrip line of Fig. 8.12 from 5 to 50 GHz.

8.3.7. Other Methods

Methods for the measurement of the permittivity of materials are not restricted to those explained in the previous sections. Time-domain techniques are also alternatives to the frequency-domain methods [113–115]. One method, particularly popular in 1960s is described by Nicolson and Ross [113] and involves measurements in the time domain. They insert a sample with a certain width in an air-filled coaxial cable. The sample forms the unknown dielectric of a coaxial line. The coaxial line is excited by a pulse with subnanosecond rise time (e.g., $\leqslant$100 ps). Time-domain reflected and transmitted waveforms are taken by using a computer controlled sampling oscilloscope. The Fourier transforms of these signals provide necessary information to obtain the frequency dependent S_{11} and S_{21} of the dielectric field coaxial line section. The complex permittivity and permeability of the material can be retrieved using transmission line equations. The method was able to measure the material parameters from 0.4 to 10 GHz with a spectral resolution of 400 MHz.

One of the well-known techniques for dielectric measurement at millimeter and submillimeter range is dispersive fourier transform spectroscopy (DFTS) first developed by Afsar [116]. This method has been claimed to be very accurate for wavelengths from 3 to 0.25 mm (100 to 1200 GHz) and is extended at best up to 5 mm long wavelength (60 GHz). This method includes a two-beam interferometer with the sample located in one of the active arms (mirror arm) of the interferometer.

The recent advances in the field of picosecond optoelectronics pulses have made them interesting tools for being used in a reliable measurement technique in microwave and millimeter wave frequencies. Arjavalingam et al. [117] use a technique called coherent microwave transient spectroscopy (COMITS) to measure the permittivity from 10 to 125 GHz. No microwave source or detector is used in their experiment setup. The setup consists of a transmitting broadband antenna and an identical receiving antenna. The antennas are made from coplanar strip lines, exponentially tapered at one end (e.g., Vivaldi antenna).

8.4. CHARACTERIZATION OF BIOLOGICAL MATERIALS

Certain technical issues for complex permittivity extraction are highlighted in this section. Addressing these issues for one-port as well as two-port measurement techniques, this section provides the extracted results of the complex permittivity of brain tissues up to 50 GHz, obtained by using the two-port approach.

8.4.1. Open-Ended Coaxial Probe

As explained before, biological materials experience a large amount of dispersion at microwave and millimeter wave frequencies due to the water like dipolar absorption around 20 GHz. To appropriately characterize them, complex permittivity in a broad frequency range has to be obtained and fit to the Cole-Cole representation. Fortunately, such dispersion, manifested by power loss, has propelled the medical and industrial applications of microwave heating. Furthermore, as the signal attenuates into the medium, a finite *sensing volume* exists around the region of the source contact to the medium, which gives the justification to using measurement techniques such as open-ended coaxial line on finite samples.

As stated before, a lot of studies have been reported that provide various techniques for permittivity measurement at microwave frequencies, and attempts have been made to extend this knowledge beyond 20 GHz. As a result of these efforts the complex permittivity of tissues and liquids are generally well-known below 20 GHz [118–130].

Regarding to both measurement and characterization, there are several technical issues that are covered in this section. It is almost impossible to cover these issues for all the measurement techniques mentioned in the previous section. However, in what follows, we try to address issues related to new one-port and two-port approaches that employ network analyzers for reflection and/or transmission measurements.

One-Port Measurement System

One of the well-known studies with identifying practical details has been reported by Burdette et al. [71,72]. They were able to perform in vivo measurement, obtain continuous data from 0.1 GHz to 11 GHz, and process data in real time. They used probes with 0.085 in (2.16 mm) diameter semirigid coaxial cable as shown in Fig. 8.15.

The set up of their measurement system was based on an HP 8410B network analyzer. Short circuit, open circuit, and matched loads were employed for network analyzer calibration. Data collection was accomplished by using a semiautomated data acquisition and processing system, whose key component was an A/D converter. The system was utilized for the determination of the in vitro and in vivo dielectric properties of various material, which included saline, distilled water, methanol, ethylene glycol, canine and rat muscle, canine kidney, canine fat, rat brain, and rat blood. The standard error of mean of their results for these measurements was at most ± 3.25 for real part and ± 2.25 for imaginary part of ε. However, they found it hard to comment on the absolute accuracy of their measurement since the variability of data from the reference literature was greater.

Burdette's work is one of the earliest ones in terms of applying the automatic network analyzer for measurements. It also stands out because they did in vivo measurement and identified the sources of inaccuracy. Those sources are tissue dehydration, accumulation of dried tissue at the probe tip, variation of probe contact pressure, improper probe positioning, temperature change, and tissue inhomogeneity. However, the method used was based on the lumped element model.

Athey and Stuchly [74] used a similar system up to 1 GHz and introduced the uncertainty analysis of the resulting complex permittivity. The reported uncertainties were due to the error in the probe termination capacitance, line characteristic impedance, and measured reflection coefficient. They reported the measurement uncertainty for distilled water, NaCl solutions, and low and high water content tissues.

As another example, Fig. 8.15b illustrates an Agilent 85070D probe [45]. The specified operating ranges are -40 to 200°C for temperature, 200 MHz to 20 GHz for frequency, which requires greater than 20 mm for the sample diameter and $20/\sqrt{\varepsilon}$ for its thickness. The accuracy is claimed to be 5% for ε' and ± 0.05 for $\tan\delta$. The typical repeatability is specified as four time better than the

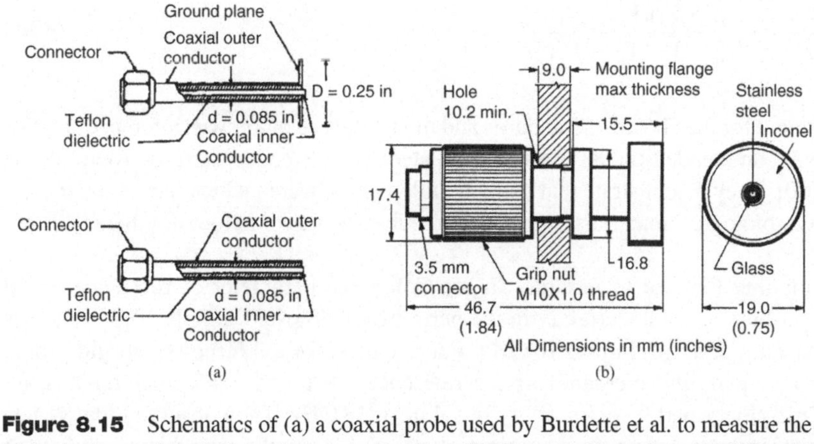

Figure 8.15 Schematics of (a) a coaxial probe used by Burdette et al. to measure the complex permittivity of biological materials [71] (with permission from IEEE), and (b) Agilent 85070D probe [45] (© 2002 Agilent Technologies, Inc. Reproduced with permission, Courtesy of Agilent Technologies, Inc.).

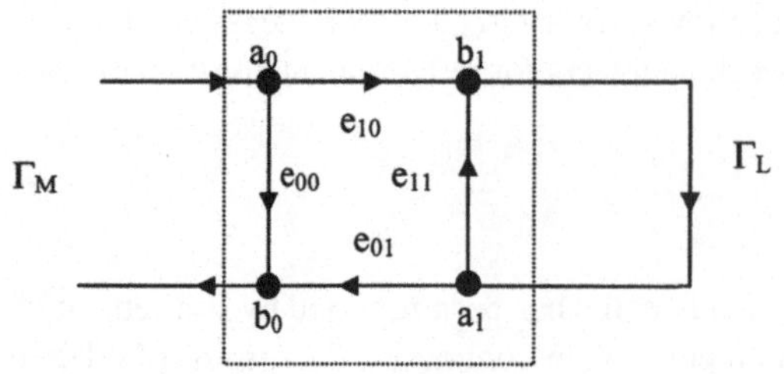

Figure 8.16 One-port signal flowchart.

accuracy. The recommended range of materials to be measured is $\varepsilon' < 100$ and $\tan\delta > 0.05$. Material should make a flat and air-gap free contact with the probe.

One-Port Calibration

An imperfect reflectrometer can be modeled by taking all the linear errors of the system and combining them into a two-port error adapter between the reflectrometer and the unknown one-port (Fig. 8.16) [97]. This linear error adaptor is associated with the network analyzer internal test set and transfer switches as well as the external cables and adaptors embedded up to measurement reference plane. In the case of noncoaxial media such as microstrip or waveguide, this adaptor includes coaxial to microstrip or waveguide launchers and the length of microstrip or waveguide from the launcher to the measurement reference plane.

The relationship between the actual and measured reflection coefficients (Γ_L and Γ_M) can easily be obtained using flowchart reduction techniques [97].

$$\Gamma_L = \frac{\Gamma_M - e_{00}}{e_{11}(\Gamma_M - e_{00}) + e_{01}e_{10}} \tag{8.64}$$

The error parameters are obtained from the measurement with three calibration standards usually open, short, and match termination [71,72,132,133]. Note that e_{01} and e_{10} cannot be separated that does not matter in practice. A direct formulation for an unknown load, knowing Y_i and Γ_i ($i = 1, 2, 3$), the aperture admittance and reflection coefficient of the standard, is [80]

$$\frac{(Y_L - Y_1)(Y_2 - Y_3)}{(Y_L - Y_2)(Y_3 - Y_1)} = \frac{(\Gamma_M - \Gamma_1)(\Gamma_2 - \Gamma_3)}{(\Gamma_M - \Gamma_2)(\Gamma_3 - \Gamma_1)} \tag{8.65}$$

There is no limitation for the choice of standards and in fact the use of known (reference) liquids for standards is highly recommended [80,81]. A thorough study of this concept can be found in the report by Misra et al. [80]. They demonstrate that using liquids such as water, which have similar electromagnetic properties as biological materials, significantly reduces the error in measurement that otherwise will be high.

It is worth mentioning that the uncertainty in the Cole-Cole parameter of reference liquid might significantly contribute to the uncertainty in the permittivity of tested sample. This issue has been studied by Nyshadham et al. [81] up to 18 GHz. For instance, for a 3.6-mm semirigid probe, and by employing short, open, and methanol as the reference, 10% and −4% uncertainties are observed for real and imaginary part of permittivity of saline at 18 GHz. They observed that generally the resulting uncertainty in the permittivity of the test material is smaller than the uncertainty in the reference liquid itself.

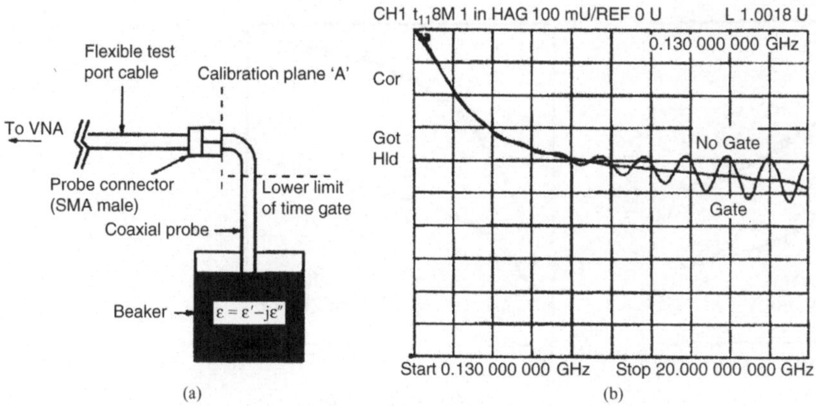

Figure 8.17 (a) Open-ended coaxial line probe system used by Anderson et al. [83], and (b) improvement by gating out the effect of connector in time domain (with permission from IEEE).

Besides reference calibration, ANA time gating feature [80,83], can also be helpful, in some applications, to remove the ripple artifact due to embedded adaptors between the ANA and the probe. Anderson et al. [83] employ this technique (cf. Fig. 8.17) in their study. They use this method in the absence of reference liquid calibration. In this method, the probe should be sufficiently long (>7 cm) to allow separation of the desired time domain components from the connector reflection.

Modeling Issues

Gabriel et al. [85] compared the calculated results using the static capacitance model, Marcuvitz model, and the MOM solution. It was concluded that the Marcuvitz formulation is accurate enough to predict the complex permittivity up to 20 GHz. They used this model for their later comprehensive measurement of various biological tissues up to 20 GHz [5].

A similar study was done by Misra et al. [80]. They used water, methanol, and dioxane as the standard liquid and measured various water–dioxane mixtures from 1 to 18 GHz. The study was performed for both lumped element and the Marcuvitz model for probe admittance. For a 80% water and 20% dioxane mixture, the deviation of measured permittivity from the literature values at 18 GHz was negligible for the Marcuvitz method but was 15% and 28% for real and imaginary parts for the lumped element model.

At higher frequencies and for better accuracy, resorting to numerical techniques seems inevitable [105,106]. In this scenario, it is useful to fit the modeling data to a function [82,107] by Eq. (8.56), a well-known example for open-ended probe. Numerical simulations such as FDTD [134,135] are particularly useful when the impact of deviation from ideal model such as probe flange [86] or sample container effects [89] are studied.

Sensitivity and Uncertainty

Qualitatively speaking, the higher is the change in the reflection coefficient of a probe (a more sensitive probe) the lower is the measurement uncertainty. For lumped-capacitance model, Stuchly and coworkers [43,66,67] quantifies this in terms of the optimum fringing capacitance in air (C_0) of a probe that yields the lowest uncertainty:

$$C_0 = \frac{1}{2\pi f\, Z_0} \frac{1}{\sqrt{\varepsilon'^2 + \varepsilon''^2}} \tag{8.66}$$

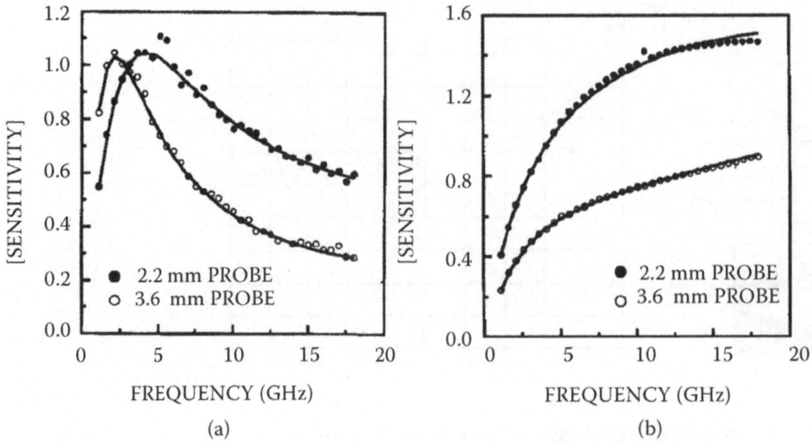

Figure 8.18 Sensitivity for (a) water and (b) methanol of 2.2-mm and 3.6-mm probes up to 18 GHz [83] (with permission from IEEE).

where Z_0 is the characteristic impedance of the probe. This quantity sets the probe dimension and apparently depends on the range of material to be measured. For instance, for water at 10 GHz the optimum capacitance is 0.005 pF [43].

A formal definition of the sensitivity is given by Stuchly et al. [82]. The sensitivity of a parameter Γ with respect to the change of another parameter ε is defined as

$$S_{\varepsilon}^{\Gamma} = \frac{\partial\Gamma/|\Gamma|}{\partial\varepsilon/|\varepsilon|} = \frac{|\varepsilon|}{|\Gamma|}\frac{\partial\Gamma}{\partial\varepsilon} \tag{8.67}$$

The magnitude of the sensitivity of a response to the change in the unknown parameter (which is to be extracted from the response) has some practical implications in error/uncertainty evaluation. For instance, a sensitivity of 2 dB implies that with a 10% error in reflection coefficient response, the error in the extracted parameter is 8%. Clearly, a negative sensitivity (in dB scale) is not very demanding, whereas a highly positive one can significantly reduce the impact of measurement error (systematic or random) on the extracted parameters. Knowing the relation between Γ and ε in closed form [see Eq. (8.56)], one can calculate this quantity. Figure 8.18 represents the sensitivity for water and methanol of 2.2 mm and 3.6 mm probes up to 18 GHz [83].

Higher Order Modes

For a fixed probe size, higher order modes' effect starts to appear at higher frequencies. This is the limiting factor of using the standard probe sizes in frequencies above 20 GHz. For a coaxial cable the cutoff frequency of first higher order (i.e., TE_{11}) mode is given by

$$f_c \approx \frac{v}{\pi(a+b)} \tag{8.68}$$

where v is the phase velocity of the dominant TEM mode in the coaxial probe. Even a nonpropagating higher order mode, if it reaches the adaptor connecting the ANA cable to the probe, after reflecting at the probe aperture, can affect the measurement results.

Cole-Cole Fitting

In a recent survey, Gabriel et al. [123] have represented a comprehensive collection of complex permittivity of various tissues. They also measured the tissue dielectric parameter [5] from 10 Hz to 20 GHz at 37°C, and provide parametric models for 17 types of tissues [123]. The parametric model they used is a four-term Cole-Cole relation:

$$\varepsilon(\omega) = \varepsilon' - j\varepsilon'' = \varepsilon_\infty + \sum_{n=1}^{4} \frac{\Delta\varepsilon_n}{1+(j\omega\tau_n)^{1-\alpha_n}} - \frac{j\sigma_I}{\omega\varepsilon_0} \tag{8.69}$$

This equation corresponds to four different dispersion regions and the corresponding parameters are obtained by fitting the measurement results to the above equation. These parameters are also tabulated by Gabriel et al. [123]. Furthermore, they provide graphs for complex permittivity of the 17 tissue types from dc up to 100 GHz obtained from Eq. (8.69). The complex permittivity is given in terms of the real part (i.e., ε') and the total conductivity of $\sigma = \omega\varepsilon_0\varepsilon''$.

In another study, Bao et al. [88] measure the complex permittivity of the rat brain grey and white matters up to 26.5 GHz at 25°C (24°C for white matter) and 37°C. Using a nonlinear least square algorithm, they fit the measured results to a two-term Cole-Cole relation covering 45 MHz to 26.5 GHz.

8.4.2. A Two-Port Microstrip Measurement System

This section provides an overview of a state-of-the art technique to extract complex permittivity of biological tissues up to 50 GHz, based on the microstrip test fixture shown in Fig. 8.13. This test fixture is employed as part of the two-port measurement system, where complex permittivity of dielectric and biological samples under test are extracted by comparing the measured scattering parameters with the simulated ones.

Test Fixture Setup

Figure 8.19 presents a close look photo of the setup developed by Tofighi and Daryoush [25, 105–109]. The test fixture microstrip cavity structure, ANA test port cables and connectors, input launchers, and TUT embedded in the test fixture are illustrated in the figure. TUT is placed between two microstrip cavities as shown in Fig. 8.19. A special sample container is also designed for containing biological tissues and liquids (cf. Fig. 8.19b). The sample container consists of a rectangular frame of an acrylic material made from a 0.04-in (1.02 mm) thick polycarbonate sheet. Two glass coverslips (Corning No.1 cover glass, $\varepsilon_r = 6.6$) with dimensions of $22 \times 50\,\text{mm}^2$ and 0.15 mm thick, are glued to the two sides of the frame using rubber cement glue. Liquids (e.g., water and saline) are injected inside the container. This test fixture has been used for the extraction of complex permittivity of white and grey brain matters of rat brain slices (cf. Fig. 8.19c) at 27°C at frequency range of 15 to 50 GHz as explained later in this section.

Two-Port Calibration

A 10-term model (cf. Fig. 8.20) is the simplified version of a well-known 12-term model [97], assuming that ANA transfer switches do not introduce different impedances in forward and reverse regimes. Using measurements of known standards (e.g., short, open, matched load, transmission), a efficient number of independent equations are obtained, where their simultaneous solutions will lead to the error terms [97,134] necessary to deembed the error networks. Error terms e_{30} and e_{03} are due to isolation and are usually negligible in low to moderate insertion loss measurements.

Thru reflect line (TRL) calibration is one of the most accurate two-port calibration techniques (cf. Fig. 8.21). Most of the today's ANA are capable of performing this type of calibration [131,132].

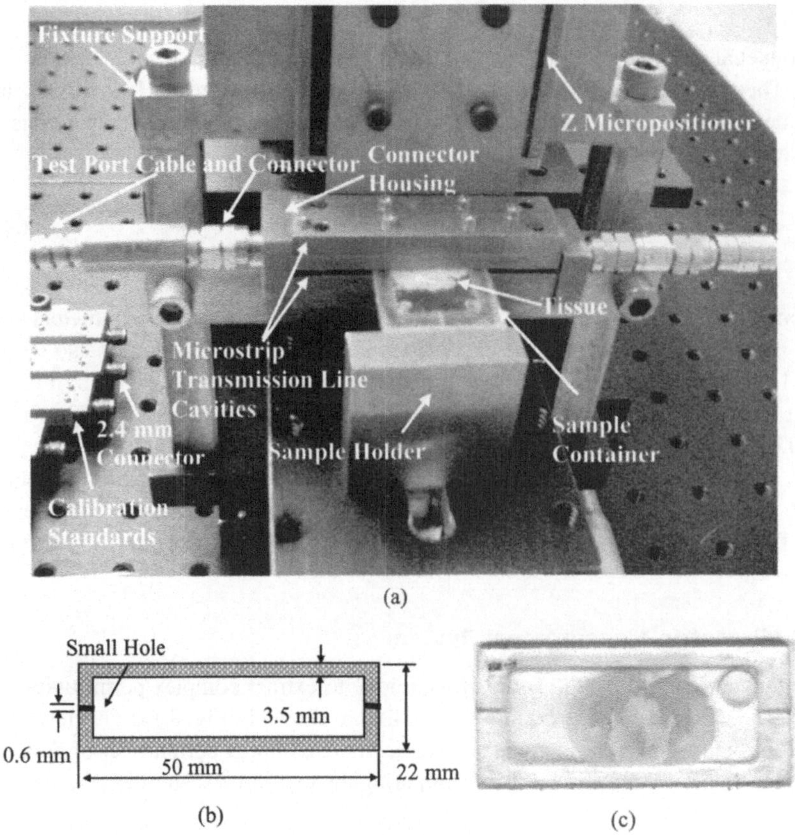

Figure 8.19 (a) Close-up photos of the experimental setup of the test fixture. This structure is designed for characterizing the complex permittivity of TUT with different thicknesses. (b) Sample container structure made from an acrylic sheet (1.02 mm thick). (c) Photos of a typical slice of rat brain inside a sample container filled with saline. (With permission from IEEE.)

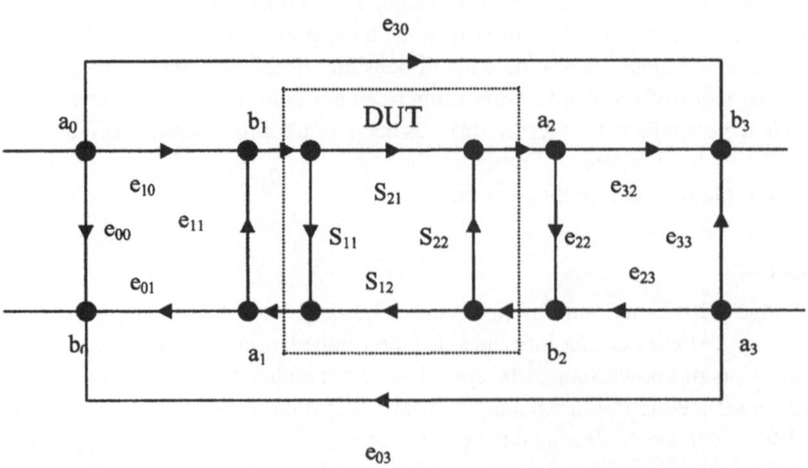

Figure 8.20 Ten-term error model of a two-port network used to identify the embedded network from the DUT.

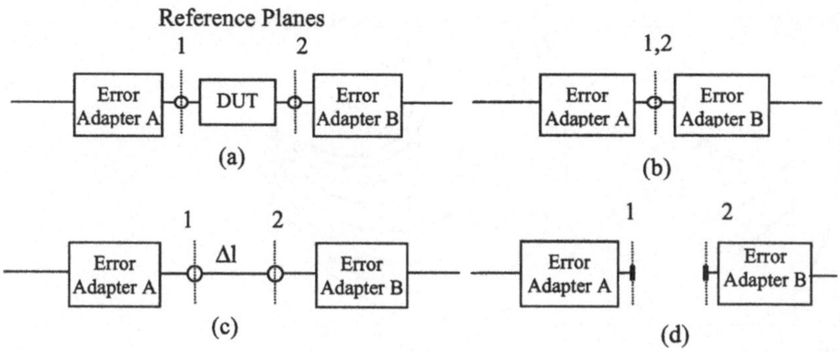

Figure 8.21 TRL calibration procedure to calibrate ANA at the reference planes 1 and 2: (a) DUT and error adapters, (b) through connection, (c) delay line with a length of L, and (d) reflect are placed between planes 1 and 2.

This technique requires the availability of two transmission lines, which are 90°($= \beta\Delta l$) different in their electrical length at the center frequency of the calibration (cf. Fig. 8.21c). A third standard is also needed, which should be a reflective one-port network either open or short (not necessarily perfect ones). Besides the input and output error networks, the propagation constant of the microstrip line and consequently the substrate permittivity can be obtained by TRL calibration routine. The theory behind TRL calibration is somewhat involved and can be found everywhere [96].

Higher Order Modes

To analyze the test fixture, a commercial finite element method (FEM) package from Agilent (HFSS 5.2) is used for full-wave characterization of the test fixture [105–107]. Based on the modal analysis, a microstrip line with the dimensions as shown in Fig. 8.13 on a fused silica substrate supports single mode operation up to 53.5 GHz. The second order mode has a 37 dB/cm attenuation at 50 GHz. In other words, the second order mode attenuates about 96.7 dB at 50 GHz before reaching the launchers for a launcher to launcher distance (see Fig. 8.13a) of 52 mm used in the design. The characteristic impedance of the line should be close to 50 Ω. Although having a 50 Ω characteristic impedance is not a requirement, it provides a lower return loss for coaxial to microstrip launcher than any other choice of the line characteristic impedance. As a result, the dynamic range of the system is expected to be maximized for this choice.

Modeling and Extraction Procedures

Performance of the test fixture including the TUT is evaluated using the FEM model in terms of its S parameters. The model includes the microstrip line, apertures, glass plates, TUT, and metallic housing. The absorbing boundary condition is defined to delimit the tissue boundary. A homogeneous distribution of tissue is considered in the modeling [105,106]. The modeling is performed for various frequencies from 5 to 50 GHz and for values of tan δ and ε', over the range of $0.15 \leqslant \tan\delta \leqslant 2.5$ and $2.5 \leqslant \varepsilon' \leqslant 75$. These values are within the expected range of biological tissues reported in the literature [122]. The simulation is performed once and the results are fitted to a complex function. Tofighi and Daryoush [107] employ a rational function relation similar to the one presented in Eq. (8.56) for their two-port test fixture. The S parameter at each frequency is fitted to a rational function of complex permittivity, as described in the following equation, as opposed to fitting the modeling S parameter for all frequencies to a single function:

$$S_{ij} = \frac{\sum_{p=0}^{P} A_p \varepsilon^P}{1 + \sum_{q=1}^{Q} B_q \varepsilon^q} \tag{8.70}$$

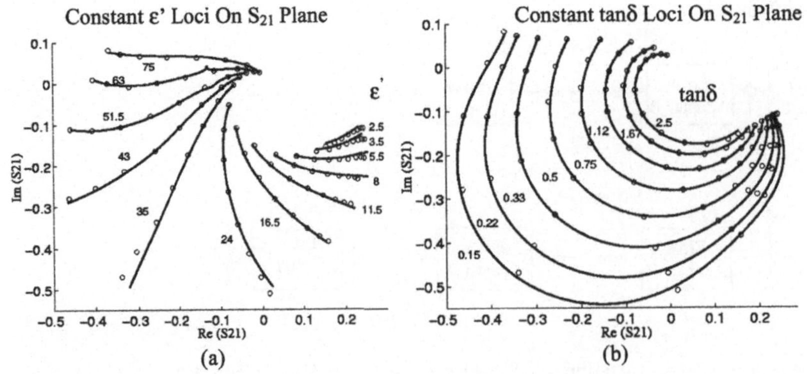

Figure 8.22 The loci of fixed (a) ε' and (b) $\tan\delta$ plotted in S_{21} plane at 15 GHz, obtained by the fitting method (—) to the data from the modeling (o) (with permission from IEEE).

A plot representing loci of S_{21} for fixed values of ε' and $\tan\delta$ is very helpful to visualize the success of this fitting procedure. Figure 8.22 represents a comparison between the simulated S_{21} and S_{21} fitted to Eq. (8.70) at 15 GHz. These loci are, in fact, perpendicular to one another at each modeling point (o).

Measurement and Extraction of the Complex Permittivity

Measurements are performed for distilled water and biological tissues (white and grey brain matters) at 27 ± 0.5°C. The extracted ε' and ε'' for distilled water are given at Fig. 8.23 for frequencies of 15 to 50 GHz. A single-term Debye relaxation is generally accepted for water dispersion at microwave frequencies [129]. The ripple-like behavior is a result of the lack of repeatability in performance of the coaxial to microstrip launchers, which were employed in TRL calibration standards and the test fixture. They can be removed by a correction technique, which employs S_{21} of water as a known reference material [107,108].

Biological tissue characterization can also be performed using this test fixture. S parameter measurements were performed for the cerebral cortex in the front brain and pons in the back of the brain as grey and white matters respectively. A relatively high proportion of nerve cell nuclei exist in grey matter, whereas white matter consists mainly of axons. Small ripple-like behaviors are observed for the extracted ε' and ε''. The results are presented in Fig. 8.24. The results are compared to the results of a four-term Cole-Cole dispersion relation provided by Gabriel et al. (the relation is obtained based on the measured data below 20 GHz) [123], and a newer two-term Cole-Cole dispersion relation for rat brain at 24°C for grey matter and 25°C for white matter given by Bao et al. (the relation is obtained from measured data below 26.5 GHz) [88]. These results are also compared to the measurement results for rabbit by Steel and Sheppard [47] at 35 GHz, by applying a linear interpolation for 27°C from their tabulated results given at two different temperatures (i.e., 20°C and 37°C). The results suggest that the extracted complex permittivity values for grey matter and white matter match better to the model provided by Bao et al. [88] and Gabriel et al. [123], respectively. Nonetheless, a further refinement of the published Cole-Cole models seems necessary as the extracted results are matched with the published ones at 35 GHz [47].

Fitting the Complex Permittivity to Cole-Cole Relation

The extracted results show Cole-Cole like dispersion characteristics, with a characteristic frequency (peak of absorption) around 22 GHz for both grey and white matters. Using a nonlinear least square

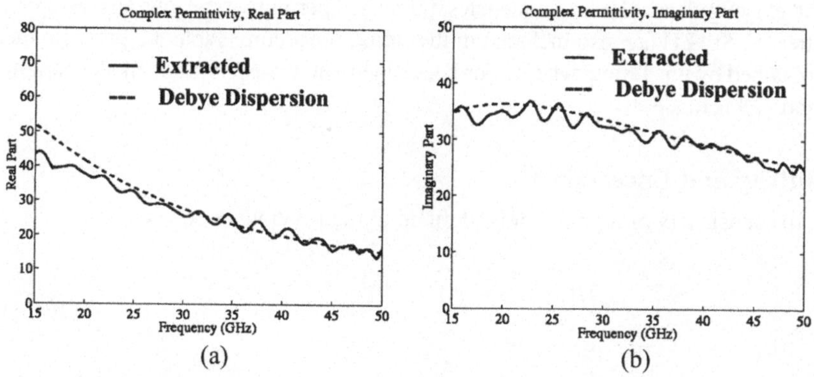

Figure 8.23 Extracted (—) and Debye model [129] (---) results of the complex permittivity ε ($\varepsilon = \varepsilon' - j\varepsilon''$) for distilled water at 27°C as a function of frequency: (a) real part (ε') and (b) imaginary part (ε'') [107] (with permission from IEEE).

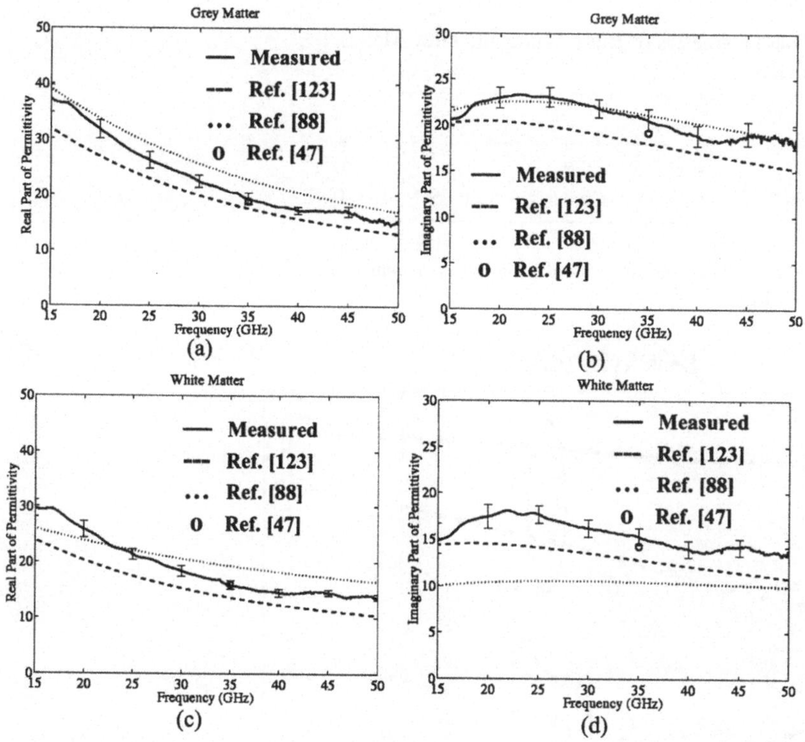

Figure 8.24 Extracted complex permittivity ε ($\varepsilon = \varepsilon' - j\varepsilon''$) for grey and white matters at 27°C as a function of frequency compared with the literature: (a) grey, real part (ε'), (b) grey, imaginary part (ε''), (c) white, real part (ε'), and (d) white, imaginary part (ε'') [107] (with permission from IEEE).

fitting method [92], the results of Fig. 8.24 are fitted to a single-term Cole-Cole relation for γ dispersion:

$$\varepsilon = \varepsilon_\infty + \frac{(\varepsilon_s - \varepsilon_\infty)}{1 + (j\omega\tau)^{1-\alpha}} + \frac{\sigma}{j\omega\varepsilon_0} \tag{8.71}$$

To account for the existing data at lower frequencies, published permittivity results above 1 GHz for white and grey matters [120,121] are also included in the fitting procedure. Table 8.2 provides the Cole-Cole parameters obtained by this fitting, where a further refinement to the published model parameters in references 88 and 123 is made.

Measurement Sensitivity and Uncertainty

Figure 8.25 illustrates the sensitivity of S parameters, defined in the previous sections

$$S_{\varepsilon}^{S_{ij}} = \frac{|\varepsilon|}{|S_{ij}|}\frac{\partial S_{ij}}{\partial \varepsilon}, \tag{8.72}$$

for water and brain grey and white matters (i.e., high water content tissues) and methanol (i.e., a low-loss liquid). Figure 8.25 shows that for water and high water content tissues (i.e., waterlike media) the sensitivity of S_{21} is better than S_{11} sensitivity by 15 dB. In this case S_{21} sensitivity is 3–7 dB. Recalling Fig. 8.18, we note that the sensitivity of a 2.2 mm is at best 1 (= 0 dB) at 5 GHz that reduces to −4 dB

Table 8.2 The Cole-Cole Parameters of Brain White and Grey Matters for γ-Dispersion above 1 GHz

	τ(ps)	ε_s	ε_∞	σ(S/m)	α
Grey matter[a]	6.75	49.5	5.8	0.96	0
White matter	6.34	37.3	5.3	0.79	0

[a]Measurement results of this study above 15 GHz at 27°C and the published results of the literature [120,121] above 1 GHz were included to obtain these parameters [107] (with permission from IEEE).

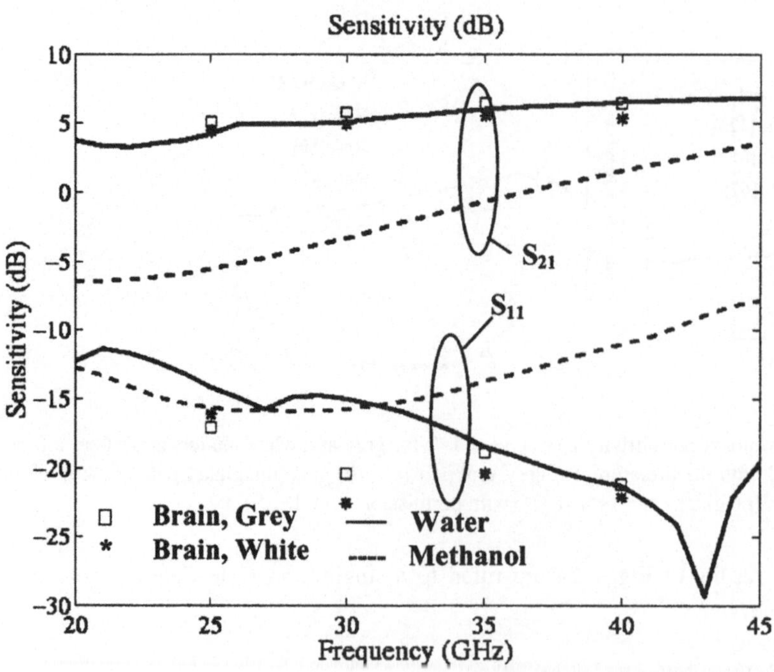

Figure 8.25 The sensitivity of S_{21} and S_{11} for water, brain (grey and white) matters, and methanol.

Table 8.3 Measurement Uncertainty, δ_i' and δ_i, from Various Sources, the Total Uncertainties, $\delta\varepsilon_{tot}'$ and $\delta\varepsilon_{tot}''$, and the Total Relative Uncertainties, $\delta_{tot}'/\varepsilon'$ and $\delta_{tot}''/\varepsilon''$

	$\delta\varepsilon_1'$	$\delta\varepsilon_1''$	$\delta\varepsilon_2'$	$\Delta\varepsilon_2''$	$\delta\varepsilon_3'$	$\delta\varepsilon_3''$	$\delta\varepsilon_4'$	$\delta\varepsilon_4''$	$\delta\varepsilon_{tot}'$	$\delta\varepsilon_{tot}''$	$\delta\varepsilon_{tot}'\varepsilon'$	$\delta\varepsilon_{tot}''\varepsilon''$
Grey	0.7	0.1	0.1	0.2	0.2	0.2	1.2	1.0	1.4	1.1	0.06	0.05
White	0.6	0.1	0.1	0.2	0.1	0.2	1.0	0.9	1.1	1.0	0.06	0.06

[a] $i = 1$; fitting error, 2; tissue thickness, 3; reference temperature, and 4; aperture placement on tissue.
[b] Obtained for white and grey matters at 30 GHz [108].
Source: With permission from IEEE.

at 18 GHz and expected to decrease even further at higher frequencies. This argument highlights the advantage of using transmission parameter (S_{21}) measurement as opposed to the reflection parameter (S_{11}) measurement as far as the extraction accuracy is concerned [108].

The higher sensitivity implies lower uncertainty in the extracted results [107]. The uncertainty in the extracted values of ε' and ε'' are due to S_{21} error and are associated with a number of error sources: (1) sample container thickness tolerance, (2) fitting error, and (3) error in the reference water complex permittivity due to its temperature uncertainty. The measurement's systematic errors are already reduced to a level sufficiently below the other errors by TRL calibration and water based correction methods that were explained before. Numerical modeling error is also reduced by careful study of the fixture response for known two-port networks formed by removing TUT (i.e., replacing it with air gap) and changing the separation between the apertures.

The uncertainties are referred to as δ_i' and δ_i'' (for real and imaginary parts respectively), where i is the index for identifying various independent contributors. Then, the total uncertainties are

$$\delta\varepsilon_{tot}' = \sqrt{\sum_i (\delta\varepsilon_i')^2} \qquad \delta\varepsilon_{tot}'' = \sqrt{\sum_i (\delta\varepsilon_i'')^2} \tag{8.73}$$

The uncertainties in S_{21}, i.e., $\delta S_{21,i}$ are estimated through modeling and simulation [107]. Once they are known the corresponding uncertainty in the complex permittivity can be estimated through relation

$$\delta\varepsilon_i = \delta\varepsilon_i' - j\delta\varepsilon_i'' = \frac{\delta S_{21,i}}{(\partial S_{21}/\partial\varepsilon)} \tag{8.74}$$

where the denominator can be easily obtained using Eq. (8.70).

On the other hand, there is some uncertainty due to the placement of aperture on top of a white or grey tissue region, which might not be exactly of the same texture, composition, or placement for all measurements (i.e., placement uncertainty, $\delta S_{21,4}$). Table 8.3 lists the result of various uncertainties and the total uncertainties obtained from Eq. (8.73) for measurement at 30 GHz [108]. The placement uncertainty ($\delta\varepsilon_4$) is obtained by repetitive tissue measurements and taking the standard deviation of the extracted results. This source of uncertainty is clearly the dominant factor. The same analysis is repeated for selected frequencies from 15 to 50 GHz. The results of this analysis have been already shown as error bars in Fig. 8.24. The uncertainty for both real and imaginary parts of the complex permittivity varies from 4% to 8% for the entire range, except the real part above 45 GHz, where error exceeds 10%.

8.4.3. Tissue Cole-Cole Parameters

A large body of knowledge has been accumulated on the dielectric properties of biological materials since the pioneering works by Cook [118] and Schwan [30] in the 1950s. These data extended to in vivo for some tissues, following the work by Burdette et al. and others [71,72,119].

Table 8.4 The Cole-Cole Parameters for a Variety of Biological Tissues [123]

Tissue type	ε_∞	$\Delta\varepsilon_1$	τ_1 (ps)	α_1	$\Delta\varepsilon_2$	τ_2 (ns)	α_2	$\Delta\varepsilon_3$	τ_3 (μs)	α_3	$\Delta\varepsilon_4$	τ_4 (ms)	α_4	σ_1
Blood	4.0	56.0	8.38	0.10	5200	132.63	0.10	0.0			0.0			0.7000
Bone (cancellous)	2.5	18.0	13.26	0.22	300	79.58	0.25	2.0×10^4	159.15	0.20	2.0×10^7	15.915	0.00	0.0700
Bone (cortical)	2.5	10.0	13.26	0.20	180	79.58	0.20	5.0×10^3	159.15	0.20	1.0×10^5	15.915	0.00	0.0200
Brain (grey matter)	4.0	45.0	7.96	0.10	400	15.92	0.15	2.0×10^5	106.1	0.22	4.5×10^7	5.305	0.00	0.0200
Brain (white matter)	4.0	32.0	7.96	0.10	100	7.96	0.10	4.0×10^4	53.05	0.30	3.5×10^7	7.958	0.02	0.0200
Fat (infiltrated)	2.5	9.0	7.96	0.20	35	15.92	0.10	3.3×10^4	159.15	0.05	1.0×10^7	15.915	0.01	0.0350
Fat (not infiltrated)	2.5	3.0	7.96	0.20	15	15.92	0.10	3.3×10^4	159.15	0.05	1.0×10^7	7.958	0.01	0.0100
Heart	4.0	50.0	7.96	0.10	1200	159.15	0.05	4.5×10^5	72.34	0.22	2.5×10^7	4.547	0.00	0.0500
Kidney	4.0	47.0	7.96	0.10	3500	198.94	0.22	2.5×10^5	79.58	0.22	3.0×10^7	4.547	0.00	0.0500
Lens Cortex	4.0	42.0	7.96	0.10	1500	79.58	0.10	2.0×10^5	159.15	0.10	4.0×10^7	15.915	0.00	0.3000
Liver	4.0	39.0	8.84	0.10	6000	530.52	0.20	5.0×10^4	22.74	0.20	3.0×10^7	15.915	0.05	0.0200
Lung (inflated)	2.5	18.0	7.96	0.10	500	63.66	0.10	2.5×10^5	159.15	0.20	4.0×10^7	7.958	0.00	0.0300
Muscle	4.0	50.0	7.23	0.10	7000	353.68	0.10	1.2×10^6	318.31	0.10	2.5×10^7	2.247	0.00	0.2000
Skin (dry)	4.0	32.0	7.23	0.00	1100	32.48	0.20	0.0			0.0			0.0002
Skin (wet)	4.0	39.0	7.96	0.10	280	79.58	0.00	3.0×10^4	1.59	0.16	3.0×10^4	1.592	0.20	0.0004
Spleen	4.0	48.0	7.96	0.10	2500	63.66	0.15	2.0×10^5	265.26	0.25	5.0×10^7	6.366	0.00	0.0300
Tendon	4.0	42.0	12.24	0.10	60	6.37	0.10	6.0×10^4	318.31	0.22	2.0×10^7	1.326	0.00	0.2500

Source: With permission from IOP Publishing Ltd.

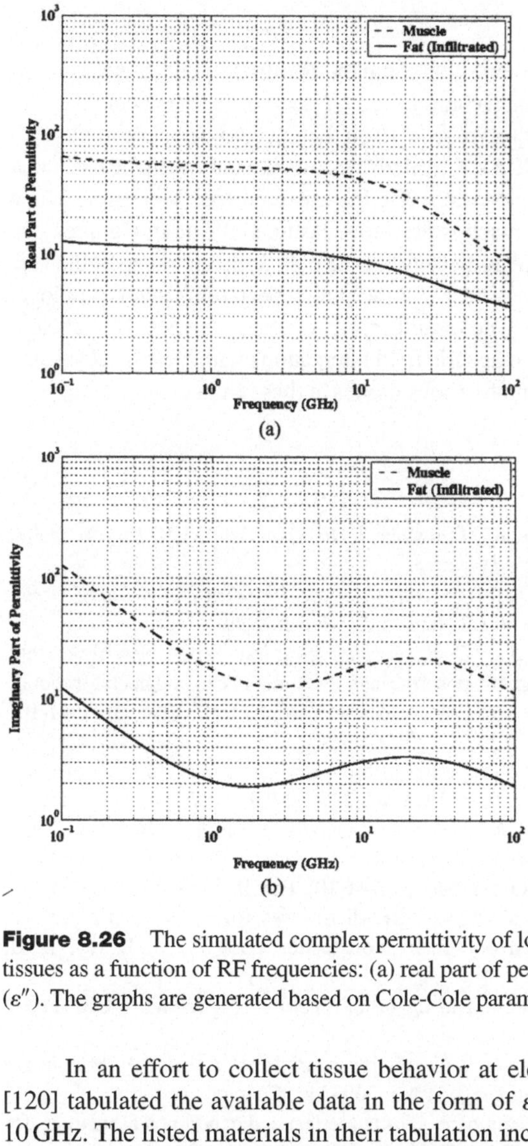

Figure 8.26 The simulated complex permittivity of low (fat) and high (muscle) water content biological tissues as a function of RF frequencies: (a) real part of permittivity (ε') and (b) imaginary part of permittivity (ε''). The graphs are generated based on Cole-Cole parameters from Table 8.4.

In an effort to collect tissue behavior at electromagnetic frequencies, Stuchly and Stuchly [120] tabulated the available data in the form of ε' and ε'' over the frequency range of 10 KHz to 10 GHz. The listed materials in their tabulation include a large body of biological tissues and phantoms for some of these tissues [120]. A complete set of information on the biological tissues including their electrical properties is also summarized in a book by Duck [121]. The previously mentioned parametric model of various tissues' complex permittivity using a four-term Cole-Cole relation of Eq. (8.69), given by Gabriel et al. [123], complements these studies. Table 8.4 lists parameters of Eq. (8.69) used to predict the dielectric properties of tissues.

To appreciate the impact of Cole-Cole parameters on the complex permittivity of biological tissues, two examples of high and low water contents such as muscle and fat are depicted over RF frequencies of interest (viz. 100 MHz to 100 GHz). Figure 8.26 depicts the real (a) and imaginary part (b) of complex permittivity. The overall range of variation from dc to 100 GHz is over seven orders of magnitude, however over frequencies of 100 MHz to 100 GHz this variation is as high as one order of magnitude. Note that above 400 MHz, where the dipolar relaxation of water is the dominant polarization mechanism, the single term relaxation relation of Eq. (8.71) can be used. Particularly, σ in that equation represents the lower frequency polarization mechanisms, for which $\omega\tau \gg 1$ at RF frequencies. In fact, the conductivity due to the ionic drift is negligible at these frequencies as well.

8.5. CONCLUSION

A number of techniques dealing with measurements of complex permittivity of dielectrics are reviewed, and strengths and weaknesses of each technique are presented. These techniques are extended for modeling and measurements of biological tissues. Even though complex permittivity data on biological tissues are primarily limited to 20 GHz today, narrowband measurements of biological tissues in millimeter wave frequencies of even close to 100 GHz are reported. Moreover, a new methodology based on two-port measurement approach is used to measure and extract complex permittivity of white and gray brain matters up to 50 GHz. Because of the high dynamic range of the variation for S_{21} compared to S_{11}, this technique provides great accuracy and repeatability and is employed to correct the Cole-Cole parameters for brain matter up to millimeter wave frequencies. Nonetheless, more studies are needed to characterize the electromagnetic field interactions with biological systems in millimeter wave frequencies, by characterization of various tissues at these frequencies.

REFERENCES

1. Javadi, H.S. Microwave Material, In: *Handbook of Microwave Technology*; Ishii, T.K. Ed.; Academic Press: San Diego, 1995; Vol. 2, 605–643, Chapter 19.
2. Baker-Jarvis, J.; Bill Riddle.; Janezic, M.D. *Dielectric and Magnetic Properties of Printing Wiring Boards and Other Substrate Material*; NIST Technical Note 1512, **March 1999**.
3. Rosen, A.; Rosen, H. (Eds.). *New Frontiers in Medical Device Technologies*; John Wiley: New York, 1995.
4. Gandhi, O.P.; Lazzi, G.; Furse, C.M. Electromagnetic absorption in the human head and neck for mobile telephones at 835 and 1900 MHz. IEEE Trans. Microwave Theory Techn. **Oct. 1996**, *MTT-44*(10), 1884–1897.
5. Gabriel, S.; Lau, R.W.; Gabriel, C. The dielectric properties of biological tissues: **II**. Measurements in the frequency range 10 Hz to 20 GHz. Phys. Med. Biol. **1996**, *41*, 2251–2269.
6. Larsen, L.E.; Jacobi, J.H. (Eds.). *Medical Application of Microwave Imaging*; IEEE Press: New York, 1986.
7. Special Issue on Medical Application and Biological Effects of RF/Microwaves, Rosen, A., Vander Vorst, A. Eds.; IEEE Trans. Microwave Theory Tech. **Oct. 1996**, *MTT-44*(10), Part II, 1753–1973.
8. Special Issue on Medical Application and Biological Effects of RF/Microwaves, Rosen, A., Vander Vorst, A., Kotsuka, Y. Eds.; IEEE Trans. Microwave Theory Tech. **Nov. 2000**, *MTT-48*(11), Parts I & II, 1781–2198.
9. Polk, C.; Postow, E. (Eds.). *CRC Handbook of Biological Effects of Electromagnetic Fields*; CRC Press: Boca Raton, Florida, 1986.
10. Battocletti, J.H. Biomedical applications of microwave engineering. In *Handbook of Microwave Technology*; Ishii, T.K. Ed.; Academic Press: San Diego, 1995; 309–345, Chapter 11.
11. Gandhi, O.P.; Lazzi, G.; Tinniswood, A.; Yu, Q. Comparison of numerical and experimental methods for determination of SAR and radiation patterns of handheld wireless telephones. Bioelectromagnetics **1999**, *20*, 93–101.
12. Okoniewski, M.; Stuchly, M. A study of the handset antenna and human body interaction. IEEE Trans. Microwave Theory Techn. **Oct. 1996**, *44*(10), 1855–1865.
13. Jensen, M.; Rahmat-Samii, Y. EM interaction of handset antennas and a human in personal communications. Proc. IEEE **Jan. 1995**, *83*(1), 7–17.
14. Dimblow P.J.; Mann, S.M. SAR calculations in an anatomically realistic model of the head for mobile communication transceivers at 900 MHz and 1.8 GHz. Phys. Med. Biol. **1994**, *39*, 1537–1533.
15. Bernardi, P.; Cavagnaro, M.; Pisa, S.; Piuzzi, E. Human exposure to radio base-station antennas in urban environment. IEEE Trans. Microwave Theory Tech. **Nov. 2000**, *MTT-48*(11), Part II, 1996–2002.
16. Schiavoni, A.; Bertotto, P.; Richiardi, G.; Bielli, P. SAR generated by commercial cellular phone modeling, head modeling, and measurements. IEEE Trans. Microwave Theory Tech. **Nov. 2000**, *MTT-48*(11), Part II, 2064–2071.
17. Rosen, A., Rosen, D.; Tuma, G.A.; Bucky, L.P. RF/Microwave-aided tumescent liposuction. IEEE Trans. Microwave Theory Tech. **Nov. 2000**, *MTT-48*(11), Part I, 1879–1884.

18. Sterzer, F.; Mendecki, J.; Mawhinney, D.D.; Friedenthal, E.; Melman, A. Microwave treatments for prostate disease. IEEE Trans. Microwave Theory Tech. **Nov. 2000**, *MTT-48*(11), Part I, 1885–1891.
19. Hiraoka, M.; Mitsumori, M.; Hiroi, N.; Ohno, S.; Tanaka, Y.; Kotsuka, Y.; Sugimachi, K. Development of RF and microwave heating equipment and clinical applications to cancer treatment in Japan. IEEE Trans. Microwave Theory Tech. **Nov. 2000**, *MTT-48*(11), Part I, 1789–1799.
20. Dunn, D.; Rappaport, M.; Terzuoli, A.J. Verification of deep-set brain tumor hyperthermia using a spherical microwave source distribution. IEEE Trans. Microwave Theory Tech. **Oct. 1996**, *MTT-44*(10), 1769–1776.
21. Camart, J.; Despretz, D.; Chive, M.; Pribetich, J. Modeling of various kinds of applicators used for microwave hyperthermia based on the FDTD method. IEEE Trans. Microwave Theory Tech. **Oct. 1996**, *MTT-44*(10), 1811–1818.
22. Labonte, S.; Blais, A.; Legault, S.R.; Ali, H.O.; Roy, L. Monopole antennas for micro-wave catheter ablation. IEEE Trans. Microwave Theory Techn. **Oct. 1996**, *MTT-44*(10), 1832–1839.
23. Jofre, L.; Hawley, M.S.; Broquetas, A.; Reyes, E.; Ferrando, M.; Elias-Fuste, A.R. Medical imaging with a microwave tomographic scanner. IEEE Trans. Biomed. Eng. **Mar, 1990**, *BME-37*(3), 303–310.
24. Hagness, S.C.; Taflove, A.; Bridges, J.E. Two-dimensional FDTD analysis of a pulsed microwave confocal system for breast cancer detection: fixed-focus and antenna-array sensors. IEEE Trans. Biomed. Eng. **Dec. 1998**, *BME-45*(12), 1470–1479.
25. Tofighi, M.R.; Daryoush, A.S. Near field microwave imaging of brain. Electron. Lett. **June 2001**, *37*(13), 807–808.
26. Frohlich, H. *Theory of Dielectrics, Dielectric Constants, and Dielectric Loss*; Oxford University Press: Amen House, London, 1958.
27. Daniel, V.V. *Dielectric Relaxation*; Academic Press: London, 1967.
28. Pethig, R. *Dielectric and Electronic Properties of Biological Materials*; Wiley: New York, 1979.
29. Pethig, R.; Kell, D.B. The passive electrical properties of biological systems: their significance in physiology, biophysics and biotechnology. Phys. Med. Biol. **1987**, *32*(8), 933–970.
30. Schwan, H.P. Electrical properties of tissues and cell suspensions. Advanced Phys. Med. Biol. **1957**, *5*, 147–209.
31. Schwan, H.P.; Foster, K.R. RF-field interactions with biological systems: electrical properties and biophysical mechanism. Proc. IEEE **1980**, *68*, 104–113.
32. Schwan, H.P. Dielectric properties of biological tissues and biophysical mechanisms of electromagnetic field interaction. In *Biological Effects of Nonionizing Radiation*; Illinger, K.H. Ed.; ACS Symposium Series: Washington D.C., 1981, 109–131.
33. Cole, K.S.; Cole, R.H. Dispersion in dielectrics; I. Alternating current characteristics. J. Chem. Phys. **Apr. 1941**, *9*, 341–351.
34. Grant, E.H.; Keefe, S.E.; Takashima, S. The dielectric behavior of aqueous solutions of bovine serum albumin from radiowave to microwave frequencies. J. Phys. Chem. **1968**, *72*, 4373–4380.
35. Foster, K.R.; Schepps, J.L.; Schwan, H.P. Microwave dielectric relaxation in muscle: A second look. Biophys. J. **1980**, *29*, 271–281.
36. Afsar, M.N.; Hasted, J.B. Measurements of optical constants of liquid H_2O and D_2O. J. Opt. Soc. Amer. **July 1977**, *67*, 902–904.
37. Grant, E.H.; Nightingale, R.V.; Sheppard, R.J. Dielectric properties of water in myoglobin solution. In *Biological Effects of Nonionizing Radiation*; Illinger, K.H. Ed.; ACS Symposium Series: Washington D.C., 1981, 57–62.
38. Grant, E.H.; Szwarnowski, S.; Sheppard, R.J. Dielectric properties of water in microwave and far-infrared regions. In: *Biological Effects of Nonionizing Radiation*; Illinger, K.H. Ed.; ACS Symposium Series: Washington D.C., 1981, 47–56.
39. Westphal, W.B. Dielectric measuring techniques. In *Dielectric Material and Applications*; Von Hippel, A.R. Ed.; Wiley: New York, 1954, 63–122.
40. Fox, J.; Sucher, M. In: *Handbook of Microwave Measurements*; Polytechnique Institute of Brooklyn: New York, 1954.
41. Bussey, H.E. Measurement of RF Properties of Materials, A Survey. Proc. IEEE **June 1967**, *55*(6), 1046–1053.
42. Grant, E.H.; Sheppard, R.J.; South, G.P. *Dielectric Behaviour of Biological Molecules in Solution*; Oxford University Press: Oxford, 1978.

43. Stuchly, M.M.; Stuchly, S.S. Coaxial line reflection methods for measuring dielectric properties of biological substances at radio and microwave frequencies— A review. IEEE Trans. Instrum. Meas. **Sept. 1980**, *IM-29*(3), 176–183.
44. Afsar, M.N.; Birch, J.R.; Clarke, R.N. The measurement of the properties of materials. Proc. IEEE **Jan. 1986**, *74*(1), 183–199.
45. Agilent 85070D Dielectric Probe Kit, Product Overview, Agilent Technology, www.agilent.com.
46. Steel, M.C.; Sheppard, R.J. The dielectric properties of rabbit tissue, pure water and various liquids suitable for tissue phantoms at 35 GHz. Phys. Med. Biol. **1988**, *33*, 467–472.
47. Steel, M.C.; Sheppard, R.J.; Collin, R. Precision waveguide cells for the measurement of complex permittivity of lossy liquids and biological tissue at 35 GHz. J. Phys. E. **1987**, *20*, 872–877.
48. Harrington, R.F. *Time-Harmonic Electromagnetic Fields*; McGraw-Hill: New York, 1961.
49. Waldron, R.A. Theory of strip-line cavity for measurement of dielectric constants and gyromagnetic-resonance line-width. IEEE Trans. Microwave Theory Tech. **Jan. 1964**, *MTT-12*(1), 123–131.
50. Lakshminarayana, M.R.; Partain, L.D.; Cook, W.A. Simple microwave technique for independent measurement of sample size and dielectric constant with results for a gunn oscillator system. IEEE Trans. Microwave Theory Tech. **July 1979**, *MTT-27*(7), 661–665.
51. Parkash, A.; Vaid, J.K.; Mansingh, A. Measurement of dielectric parameters at microwave frequencies by cavity perturbation technique. IEEE Trans. Microwave Theory Tech. **Sept. 1979**, *MTT-27*(9), 791–795.
52. Jones, C.A.; Kantor, Y.; Grosvenor, J.H.; Janezic, M.D. *Striple Resonator for Electromagnetic Measurements of Materials*; NIST Technical Note 1505, National Institute of Standard and Technology: Boulder, Colorado, July 1998.
53. Jones, C.A. Permittivity and permeability measurements using strip-line resonator cavities—A comparison. IEEE Trans. Instrum. Meas. **Aug. 1999**, *IM-40*(4), 843–848.
54. Fenske K.; Misra, D. Dielectric materials at microwave frequencies. Applied Microwave & Wireless **Oct. 2000**, *12*(10), 92–100.
55. Land, D.V.; Campbell, A.M. A quick accurate method for measuring the microwave dielectric properties of small tissue samples. Phys. Med. Biol. **1992**, *37*(1), 183–192.
56. Carter, R.G. Accuracy of microwave cavity perturbation measurements. IEEE Trans. Microwave Theory Tech. **May 2001**, *MTT-49*(5), 918–923.
57. Tanabe, E.; Joines, W.T. A nondestructive method for measuring the complex permittivity of dielectric materials at microwave frequencies using an open transmission line resonator. IEEE Trans. Instrum. Meas. **Sept. 1976**, *IM-25*(3), 222–226.
58. Xu, D.; Liu, L.; Jiang, Z. Measurement of the dielectric properties of biological substances using and improved open-ended coaxial line resonator method. IEEE Trans. Microwave Theory Tech. **Dec. 1987**, *MTT-35*(12), 1424–1428.
59. Marcuvitz, N. *Waveguide Handbook*; McGraw-Hill: New York, 1951.
60. Jones, R.G. Precise dielectric measurements at 35 GHz using an open microwave resonator. Proc. IEE **Apr. 1976**, *123*(4), 285–290.
61. Clarke, R.N.; Rosenberg, C.B. Fabry-Perot and open resonators at microwave and millimeter wave frequencies, 2-300 GHz. J. Phys. E.: Sci. Instrum. **1982**, *15*, 9–24.
62. Hirvonen, T.M.; Vainikainen, P.; Lozowski, A.; Raisanen, A. Measurement of dielectrics at 100 GHz with an open resonator connected to a network analyzer. IEEE Trans. Microwave Theory Tech. **Aug. 1996**, *MTT-45*(4), 780–786.
63. Afsar, M.N.; Huachi, X. An automated 60-GHz open resonator system for precision dielectric measurement. IEEE Trans. Microwave Theory Tech. **Dec. 1990**, *MTT-38*(12), 1845–1853.
64. Afsar, M.N.; Ding, H.; Tourshan, K. A new 60 GHz open-resonator technique for precision permittivity and loss-tangent measurement. IEEE Trans. Instrum. Meas. **Apr. 1999**, *IM-48*(2), 626–630.
65. Afsar, M.N.; Ding, H. A novel open-resonator system for precise measurement of permittivity and loss tangent. IEEE Trans. Instrum. Meas. **Apr. 2001**, *IM-50*(2), 402–405.
66. Stuchly, S.S.; Rzepecka, M.A.; Iskander, M.F. Permittivity measurements at microwave frequencies using lumped elements. IEEE Trans. Instrum. Meas. **Mar. 1975**, *IM-23*(1), 57–62.
67. Rzepecka, M.A.; Stuchly, S.S. A lumped element capacitance method for the measurement of the permittivity and conductivity in the frequency and time domain—A further analysis. IEEE Trans. Instrum. Meas. **Mar. 1974**, *IM-24*(1), 27–32.

68. Iskander, M.F.; Stuchly, S.S. Fringing field effect in lumped-capacitance method for permittivity measurement. IEEE Trans. Instrum. Meas. **Mar. 1978**, *IM-27*, 107–109.
69. Bianco, B.; Corana, L.; Gogioso, L.; Ridella, S.; Parodi, M. Open-circuited coaxial lines as standards for microwave measurements. Electron. Lett. **1980**, *16*(10), 373–374.
70. Kraszewski, A.; Stuchly, S.S.; Stuchly, M.A.; Symons, S. On measurement accuracy of the tissue permittivity in-vivo. IEEE Trans. Instrum. Meas. **Mar. 1983**, *IM-32*(1), 37–42.
71. Burdette, E.C.; Cain, F.L.; Seals, J. In vivo probe measurement technique for determining dielectric properties at VHF through microwave frequencies. IEEE Trans. Microwave Theory Tech. **Apr. 1980**, *MTT-28*(4), 414–424.
72. Burdette, E.C.; Cain, F.L.; Seals, J. In-situ tissue permittivity at microwave frequencies: perspective, techniques, results. In: *Medical Application of Microwave Imaging*; Larsen, L.E., Jacobi, J.H. Eds.; IEEE Press: New York, 1986, 13–40.
73. Mosig, J.R.; Besson, J.E.; Gex-Fabry, M.; Gardiol, F.E. Reflection of an open-ended coaxial line and application to nondestructive measurement of materials. IEEE Trans. Instrum. Meas. **March 1981**, *IM-30*(1), 46–51.
74. Athey, T.W.; Stuchly, M.A.; Stuchly, S.S. Measurement of radio frequency permittivity of biological tissues with an open-ended coaxial line: Part I. IEEE Trans. Microwave Theory Tech. **Jan. 1982**, *MTT-30*(1), 82–86.
75. Kraszewski, A.; Stuchly, M.A.; Stuchly, S.S. ANA calibration methods for measurement of dielectric properties. IEEE Trans. Instrum. Meas. **June 1983**, *IM-32*(2), 385–386.
76. Kraszewski, A.; Stuchly, S.S. Capacitance of open-ended dielectric field coaxial lines—experimental results. IEEE Trans. Instrum. Meas. **Dec. 1983**, *IM-32*(4), 517–519.
77. Gajda, G.; Stuchly, S.S. An equivalent circuit of an open-ended coaxial line. IEEE Trans. Microwave Theory Tech. **May 1983**, *MTT-31*(5), 380–384.
78. Misra, D.K. A quasi-static analysis of open coaxial lines. IEEE Trans. Microwave Theory Tech. **1987**, *MTT-35*, 925–938.
79. Staebell, K.F.; Misra, D. An experimental technique for in vivo permittivity measurement of materials at microwave frequencies. IEEE Trans. Microwave Theory Tech. **March 1990**, *MTT-38*(3), 337–339.
80. Misra, D.M.; Chabbra, M.; Epstein, B.R.; Mirotznik, M.; Foster, K.R. Noninvasive electrical characterization of materials at microwave frequencies using an open-ended coaxial line: Test of an improved calibration technique. IEEE Trans. Microwave Theory Tech. **Jan. 1990**, *MTT-38*(1), 8–13.
81. Nyshadham, A.; Sibbald, C.L.; Stuchly, S.S. Permittivity measurements using open-ended sensors and reference liquid calibration—An uncertainty analysis. IEEE Trans. Microwave Theory Tech. **Feb. 1992**, *MTT-40*(2), 305–314.
82. Stuchly, S.S.; Sibbald, C.L.; Anderson, J.M. A new aperture admittance model for open-ended waveguides. IEEE Trans. Microwave Theory Techn. **Feb. 1994**, *MTT-42*(2), 192–198.
83. Anderson, J.M.; Sibbald, C.L.; Stuchly, S.S. Dielectric measurements using a rational function model. IEEE Trans. Microwave Theory Techn. **Feb. 1994**, *MTT-42*(2), 199–204.
84. Baker-Jarvis, J.; Janezic, M.D.; Domich, P.D.; Geyer, R.G. Analysis of an open-ended coaxial probe with lift off for nondestructive testing. IEEE Trans. Instrum. Meas. **Oct. 1994**, *IM-43*(5), 711–718.
85. Gabriel, C.; Chan, T.Y.A.; Grant, E.H. Admittance models for open-ended coaxial probes and their place in dielectric spectroscopy. Phys. Med. Biol. **1994**, *39*, 2183–2199.
86. Okoniewski, O.; Anderson, J.A.; Okoniewska, E.; Gupta, K.; Stuchly, S.S. Further analysis of open-ended sensors. IEEE Trans. Microwave Theory Techn. **Aug. 1995**, *MTT-43*(8), 1986–1989.
87. Berube, D.; Ghannouchi, F.M.; Savard, P. A comparative study of four open-ended coaxial probe models for permittivity measurements of lossy dielectric/biological material at microwave frequencies. IEEE Trans. Microwave Theory Tech. **Oct. 1996**, *MTT-44*(10), 1928–1934.
88. Bao, J.; Lu, S.; Hurt, W.D. Complex dielectric measurements and analysis of brain tissues in the radio and microwave frequencies. IEEE Trans. Microwave Theory Tech. **Oct. 1997**, *MTT-45*(10), 1730–1740.
89. Hoshina, S.; Kanai, Y.; Miakawa, M. A numerical study of the measurement region of an open-ended coaxial probe used for complex permittivity measurement. IEEE Trans. Magnetics. **Sep. 2001**, *MTT-37*(5), 3311–3314.
90. Itoh, T. (Ed.). *Numerical Techniques for Microwave and Millimeter-Wave Passive Structures*, Wiley: New York, 1989.
91. Jin, J. *The Finite Element Method in Electromagnetics*; Wiley: New York, 1993.

92. Press, W.H. *Numerical Recipes in C: The Art of Scientific Computing*; Cambridge University Press: New York, 1992.
93. Friedsam, G.L.; Biebl, E.M. Precision free-space measurements of complex permittivity of polymers in the W-band. 1997 IEEE MTT-S International Microwave Symposium Digest, Denver, CO, **June 1997**, *3*, 1351–1354.
94. Afsar, M.N.; Tkachov, I.I.; Kocharyan, K.N. A novel W-band spectrometer for dielectric measurements. IEEE Trans. Microwave Theory Tech. **Dec. 2000**, *MTT-48*(12), 2637–2643.
95. Ghodgaonkar, D.K.; Varadan, V.V.; Varadan, V.K. Free-space measurement of complex permittivity and complex permeability of magnetic materials at microwave frequencies. IEEE Trans. Instrum. Meas. **Apr. 1990**, *IM-39*(2), 387–394.
96. Eagen, G.F.; Hoer, C.A. Thru-reflect-line: An improved technique for calibrating the dual six-port automatic network analyzer. IEEE Trans. Microwave Theory Tech. **Dec. 1979**, *MTT-27*(12), 987–993.
97. Parisot, M.; Soares, R. S parameter measurements and their use in circuit design. In *GaAs MESFET Circuit Design*; Soares, R. Ed.; Artech House: Norwood, MA, 1988, Chapter 3.
98. Soares, R.A.; Gouzien, P.; Legaud, P.; Follot, G. A unified approach to two-port calibration techniques and some applications. IEEE Trans. Microwave Theory Tech. **Nov. 1989**, *MTT-37*(11), 1669–1673.
99. Belhadj-Tahar, N.; Fourrier-Lamer, A.; Chanterac, H. Broadband simultaneous measurement of complex permittivity and permeability using a coaxial discontinuity. IEEE Trans. Microwave Theory Tech. **Jan. 1990**, *MTT-38*(1), 1–7.
100. Abdulnour, J.; Akyel, C.; Wu, K. A Generic approach for permittivity of dielectric materials using a discontinuity in a rectangular waveguide or a microstrip line. IEEE Trans. Microwave Theory Tech. **May 1995**, *MTT-43*(5), 1060–1066.
101. Queffelec, P.; Gelin, P. Influence of higher order modes on the measurements of complex permittivity and permeability of materials using a microstrip discontinuity. IEEE Trans. Microwave Theory Tech. **June 1996**, *MTT-44*(6), 814–824.
102. Abbas, Z.; Pollard, R.D.; Kelsall, R.W. Complex permittivity measurements at Ka-band using rectangular dielectric waveguide. IEEE Trans. Instrum. Meas. **Oct. 2001**, *IM-50*(5), 1334–1342.
103. Duhamel, F.; Huynen, I.; Vander Vorst, A. Measurements of complex permittivity of biological and organic liquids up to 110 GHz. 1997 IEEE MTT-S International Microwave Symposium Digest, Denver, CO, **June 1997**, *1*, 107–110.
104. Fossion, M.; Huynen, I.; Vanhoenacker, D.; Vander Vorst, A. A new and simple calibration method for measuring planar lines parameter up to 40 GHz. Proc. 22nd European Microwave Conference: Espoo, Finland, **Aug. 1992**, 180–185.
105. Tofighi, M.R. Design and Implementation of a Two-Port Microstrip Test Fixture for Complex Permittivity Characterization and Near-Field Imaging of Biological Materials up to 50 GHz, PhD Thesis, Drexel University: Philadelphia, PA, 2001.
106. Tofighi, M.R.; Daryoush, A.S. Characterization of biological tissues up to millimeter wave: Test fixture design, 2000 IEEE MTT-S International Microwave Symposium Digest, Boston, MA, **June 2000**, *2*, 1041–1044.
107. Tofighi, M.R.; Daryoush, A.S. Characterization of the complex permittivity of brain tissues up to 50 GHz utilizing a two-port microstrip test fixture. IEEE Trans. Microwave Theory Tech. **Oct. 2002**, *MTT-50*(10), 2217–2225.
108. Tofighi, M.R.; Daryoush, A.S. Comparison of two post-calibration correction methods for complex permittivity measurement of biological tissues up to 50 GHz. IEEE Trans. Instrum. Meas. **Dec. 2002**, *51*(6), 1170–1176.
109. Tofighi, M.R.; Daryoush, A.S. Study of the activity of neurological cell solutions using complex permittivity measurement, 2002 IEEE MTT-S International Microwave Symposium Digest, Seattle, WA, **June 2002**, *2*, 1763–1766.
110. Janezic, M.D.; Jargon, J.A. Complex permittivity determination from propagation constant measurements. IEEE Microwave Guided Wave Lett. **Feb. 1999**, *9*(2), 76–78.
111. Wan, C.; Nauwelaers, B.; De Raedt, W.; Van Rossum, M. Two new measurements methods for explicit determination of complex permittivity. IEEE Trans. Microwave Theory Tech. **Nov. 1998**, *MTT-46*(11), 1614–1619.

112. Marks, R.B. A Multiline method of network analyzer calibration. IEEE Trans. Microwave Theory Tech. **July 1991**, *MTT-39*(7), 1205–1215.
113. Nicolson, A.M.; Ross, G.F. Measurement of the intrinsic properties of materials by time-domain techniques. IEEE Trans. Instrum. Meas. **Nov. 1970**, *IM-19*(4), 377–382.
114. Courtney, C.C. Time-Domain measurement of the electromagnetic properties of materials. IEEE Trans. Microwave Theory Tech. **May 1998**, *MTT-46*(5), 517–522.
115. Jargon, J.; Janezic, M.D. Measuring complex permittivity and permeability using time-domain network analysis, 1996 IEEE MTT-S International Microwave Symposium Digest, San Francisco, CA, **June 2002**, *2*, 1407–1409.
116. Afsar, M.N. Dielectric measurements of millimeter-wave materials. IEEE Trans. Microwave Theory Tech. **Dec. 1984**, *MTT-32*(12), 1598–1609.
117. Arjavalingam, G.; Pastol, Y.; Halbout, J.; Kopcsay, G.V. Broad band microwave measurements with transient radiation from optoelectronically pulsed antennas. IEEE Trans. Microwave Theory Tech. **May 1990**, *MTT-38*(5), 615–621.
118. Cook, H.F. The dielectric behavior of some types of human tissues at microwave frequencies. Brit. J. Appl. Phys. **1951**, *2*, 295–300.
119. Kraszewski, A.; Stuchly, M.A.; Stuchly, S.S.; Smith, M. In vivo and in vitro dielectric properties of animal tissues at radio frequencies. Bioelectromagnetics **1982**, *3*, 421–432.
120. Stuchly, M.A.; Stuchly, S.S. Dielectric properties of biological substances—tabulated. J. Microwave Power **1980**, *1*(15), 19–26.
121. Duck, F.A. *Physical Properties of Tissue: A Comprehensive Reference Book*; Academic Press: London, 1990.
122. Gabriel, C.; Gabriel, S.; Corthout, E. The dielectric properties of biological tissues: I. Literature survey. Phys. Med. Biol. **1996**, *41*, 2231–2249.
123. Gabriel, S.; Lau, R.W.; Gabriel, C. The dielectric properties of biological tissues: III. Parametric models for the dielectric spectrum of tissues. Phys. Med. Biol. **1996**, *41*, 2271–2293.
124. Malmberg, C.G.; Maryott, A.A. Dielectric constant of water from 0 to 100°C. J. Res. Nat. Bureau Stand. **Jan. 1956**, *56*(1), 1–7.
125. Tables of Dielectric Dispersion Data for Pure Liquids and Dilute Solutions, National Bureau of Standards Circular 589, Nov. 1958.
126. Grant, E.H.; Buchanan, T.J.; Cook, H.F. Dielectric behavior of water at microwave frequencies. J. Chem. Phys. **1957**, *26*, 156–161.
127. Von Hippel, A. The Dielectric relaxation spectra of water, ice, and aqueous solutions, and their interpretation: 1. Critical survey of the status quo for water. IEEE Trans. Electrical Insulation, **Oct. 1988**, *23*(5), 801–816.
128. Von Hippel, A. The Dielectric relaxation spectra of water, ice, and aqueous solutions, and their interpretation: 2. Tentative interpretation of the relaxation spectrum of water in the time and frequency domain. IEEE Trans. Electric. Insulation **Oct. 1988**, *23*(5), 817–823.
129. Stogryn, A. Equations for calculating the dielectric constant of saline water. IEEE Trans. Microwave Theory Tech. **Aug. 1971**, *MTT-19*(8), 733–736.
130. Jordan, B.P.; Sheppard, R.J.; Szwarnowski, S. The Dielectric properties of formamide, ethanol, and methanol. J. Phy. D: Appl. Phys. **1978**, *11*, 695–701.
131. Model 37XXXC Vector Network Analyzer Operational Manual, ANRITSU P/N: 10410–00226, June 2000.
132. 37100C/37200C/37300C Vector Network Analyzers Technical Data Sheet, ANRITSU P/N: 11410–00247, June 2000.
133. Rehnmark, S. On the calibration process of the automatic network analyzer. IEEE Trans. Microwave Theory Tech. **Apr. 1974**, *MTT-22*, 457–458.
134. Taflove, A. *Computational Electrodynamics: The Finite Difference Time-Domain Method*, Artech House: Dedham, MA, **1995**.
135. Yee, K.S. Numerical Solution of initial boundary value problems involving Maxwell's equations in isotropic media. IEEE Trans. Antennas Propagat. **May 1966**, *AP-14*, 302–307.

Appendix A
Some Useful Constants

Permittivity of free space (ε_0) = 8.854×10^{-12} F/m
Permeability of free space (μ_0) = $4\pi \times 10^{-7}$ H/m
Speed of electromagnetic waves in free space (c) = 3×10^8 m/s
Impedance of free space (Z_0 or η_0) = $376.7\ \Omega$
Boltzmann's constant (k) = 1.38×10^{-23} J/K
Charge of electron (e or q_e) = -1.602×10^{-19} C

Appendix B
Some Units and Conversions

Quantity	SI[a] unit	Conversion factor
Length	meter (m)	= 39.37 in
Mass	kilogram (kg)	= 2.21 pound-mass (lb_m)
Time	second (s)	
Frequency	hertz (Hz)	= 1 cycle/s
Force	newton (N)	= 0.2248 pound-force (lb_f)
Charge	coulomb (C)	
Charge density	coulomb/meter^3 (C/m^3)	
Current	ampere (A)	
Current density	ampere/meter^2 (A/m^2)	
Electric field	volt/meter (V/m)	
Electric flux density	coulomb/meter^2 (C/m^2)	
Magnetic field	ampere/meter (A/m)	
Magnetic flux density	tesla (T)	= 10,000 G
	or weber/meter^2 (Wb/m^2)	= 10,000 G
Resistance	ohm (Ω)	
Conductivity	siemens/meter (S/m)	
	or mho/m	
Capacitance	farad (F)	
Permittivity	farad/meter (F/m)	
Inductance	henry (H)	
Permeability	henry/meter (H/m)	

[a]SI = International System of Units.

Appendix C
Review of Vector Analysis and Coordinate Systems

Since the formulation and application of various electromagnetic laws is greatly facilitated by the use of vector analysis, this appendix presents a concise review of vector analysis and the principal coordinate systems.

C.1. SCALARS AND VECTORS

A *scalar* quantity can be expressed as a single real number. (It can be positive, negative, or zero.) For example, voltage and current are scalar quantities. In *ac* analysis it is mathematically convenient to use *phasors* to represent sinusoidally varying voltages and currents. Phasors are referred to as *complex scalars*, since they require complex numbers (either magnitude and phase or real and imaginary parts) for their specification.

A *vector* quantity (e.g., the electric field) requires both a magnitude and a direction for its specification. The magnitude is always positive (it may be zero).

C.2. THE RECTANGULAR COORDINATE SYSTEM

The rectangular coordinate system (Fig. C.1a) locates a point P in three-dimensional space by assigning to it the coordinates (x_1, y_1, z_1) within a frame of reference defined by three mutually orthogonal (perpendicular) axes: the x axis, the y axis, and the z axis. It is conventional to choose a *right-handed* coordinate system (and we will do so throughout this handbook). This choice simply means that if we first point the fingers of the right hand along the x axis and then curl them to point along the y axis, the extended thumb will align with the z axis.

To deal with vectors, we define a set of three unit vectors $\mathbf{a}_x$, $\mathbf{a}_y$, and $\mathbf{a}_z$ (each with a magnitude equal to one) aligned with (parallel to) the three axes. An arbitrary vector $\mathbf{A}$ may now be expressed as $\mathbf{A} = A_x\,\mathbf{a}_x + A_y\,\mathbf{a}_y + A_z\,\mathbf{a}_z$, where A_x, A_y, A_z are said to be its scalar components along the three axes. The vector $\mathbf{A}$ has a magnitude $A = [A_x^2 + A_y^2 + A_z^2]^{1/2}$. Figure C.1b shows a differential volume $dV = dxdydz$. The surfaces have differential areas, ds, of $dx\,dy$, $dy\,dz$, and $dz\,dx$.

C.3. SCALAR AND VECTOR FIELDS

The concepts of scalars and vectors introduced in Sec. C.1 can be extended to define scalar and vector fields. A scalar field associates a scalar quantity with every point in a region of space. If we set up a rectangular coordinate system to identify various points in the 3D space in a room, we may describe the temperature distribution (scalar field) as some function $T = f(x, y, z)$ so that at the point (x_1, y_1, z_1) the temperature $T(x_1, y_1, z_1)$ is given by the value of the function $f(x_1, y_1, z_1)$. In a similar fashion, if we associate a vector with every point in a region, we will have a vector field. In the

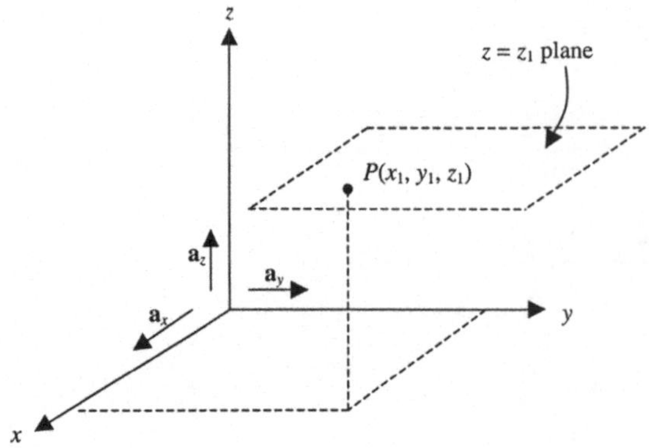

(a)

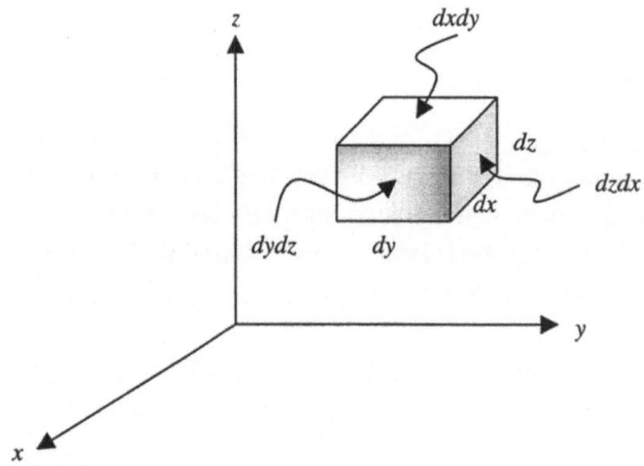

(b)

Figure C.1 The rectangular coordinate system: (a) the coordinates of a point and the unit vectors and (b) differential elements.

rectangular coordinate system, we can write a vector field in terms of its three components, each of which is a scalar field. For example, the velocity distribution in a river may be expressed as $\mathrm{v} = v_x(x, y, z)\mathbf{a}_x + v_y(x, y, z)\mathbf{a}_y + v_z(x, y, z)\mathbf{a}_z$.

C.4. VECTOR ADDITION AND SUBTRACTION

Two vectors **A** and **B** may be added together graphically by the familiar *parallelogram rule* shown in Fig. C.2. The addition can also be performed by adding the corresponding components of the two vectors.

If $\mathbf{A} = A_x\mathbf{a}_x + A_y\mathbf{a}_y + A_z\mathbf{a}_z$ and $\mathbf{B} = B_x\mathbf{a}_x + B_y\mathbf{a}_y + B_z\mathbf{a}_z$, their sum is a vector **C**, given as

$$\mathbf{C} = \mathbf{A} + \mathbf{B} = (A_x + B_x)\mathbf{a}_x + (A_y + B_y)\mathbf{a}_y + (A_z + B_z)\mathbf{a}_z$$

Vector addition always obeys the following laws:

$$\mathbf{A} + \mathbf{B} = \mathbf{B} + \mathbf{A} \qquad \text{(commutative)}$$

$$\mathbf{A} + (\mathbf{B} + \mathbf{C}) = (\mathbf{A} + \mathbf{B}) + \mathbf{C} \qquad \text{(associative)}$$

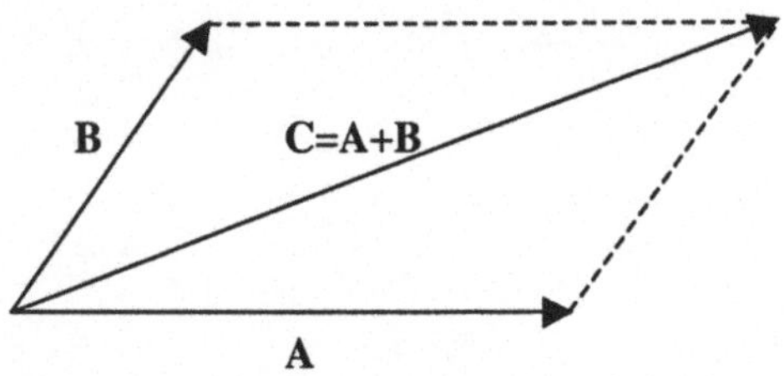

Figure C.2 Vector addition via the parallelogram rule.

Vector subtraction, $\mathbf{A} - \mathbf{B}$, is accomplished by reversing the direction of **B** to obtain another vector $-\mathbf{B}$ and then adding it to the vector **A**. Thus we have

$$\mathbf{D} = \mathbf{A} - \mathbf{B} = \mathbf{A} + (-\mathbf{B})$$

or

$$\mathbf{D} = (A_x - B_x)\mathbf{a}_x + (A_y - B_y)\mathbf{a}_y + (A_z - B_z)\mathbf{a}_z$$

where **A** and **B** have been expressed in terms of their rectangular components.

In dealing with vector fields, it is important to realize that we should be adding and subtracting only those vectors that are defined at the same point in space.

C.5. POSITION AND DISTANCE VECTORS

The position vector associated with a point P, which has the rectangular coordinates (x_1, y_1, z_1), is the vector extending from the origin $O(0, 0, 0)$ to the point P. It may be expressed as (Fig. C.3)

$$\mathbf{OP} = x_1\mathbf{a}_x + y_1\mathbf{a}_y + z_1\mathbf{a}_z \tag{C.1}$$

The distance vector **PQ** extends from the point $P(x_1, y_1, z_1)$ to the point $Q(x_2, y_2, z_2)$ and can be expressed as

$$\begin{aligned}\mathbf{PQ} &= \mathbf{OQ} - \mathbf{OP}\\ &= (x_2\mathbf{a}_x + y_2\mathbf{a}_y + z_2\mathbf{a}_z) - (x_1\mathbf{a}_x + y_1\mathbf{a}_y + z_1\mathbf{a}_z)\\ &= (x_2 - x_1)\mathbf{a}_x + (y_2 - y_1)\mathbf{a}_y + (z_2 - z_1)\mathbf{a}_z\end{aligned} \tag{C.2}$$

The scalar distance PQ is given by the magnitude of the vector **PQ**. Thus,

$$PQ = |\mathbf{PQ}| = \sqrt{(x_2 - x_1)^2 + (y_2 - y_1)^2 + (z_2 - z_1)^2} \tag{C.3}$$

C.6. VECTOR DIVISION AND MULTIPLICATION

The operation **A/B** is *not* defined. However, a vector can be divided by a scalar.

Two forms of vector to vector multiplication are useful in our work.

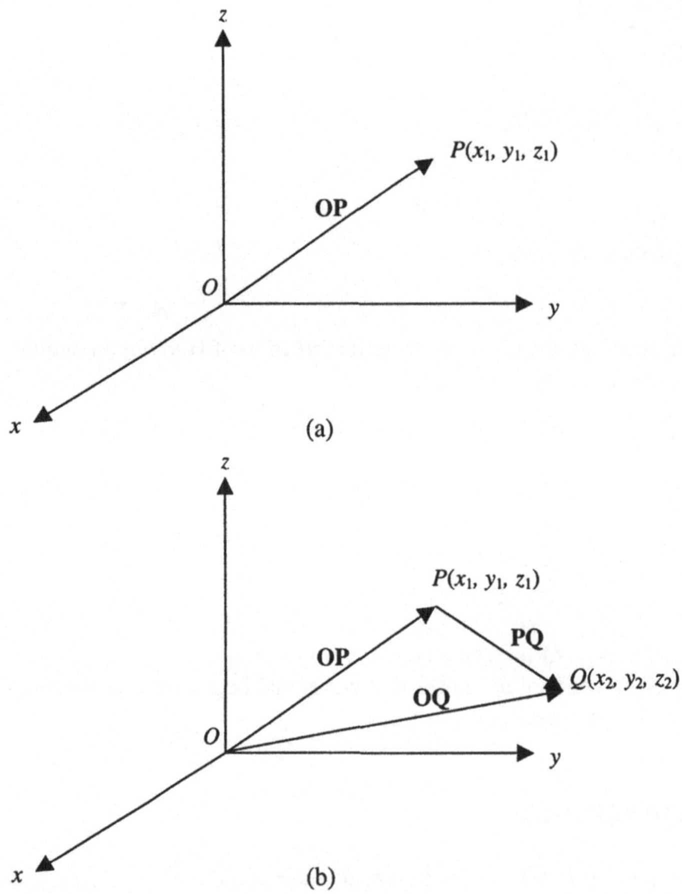

Figure C.3 (a) The position vector **OP** extends from the origin O to the point P. (b) The distance vector **PQ** extends from P to Q.

C.6.1. Scalar (Dot) Product

The scalar product of the two vectors **A** and **B** is represented symbolically as $\mathbf{A} \bullet \mathbf{B}$ (hence the alternate name, the *dot product*).

$$\mathbf{A} \bullet \mathbf{B} = |\mathbf{A}||\mathbf{B}| \cos \theta_{AB} \tag{C.4}$$

where θ_{AB} is the smaller angle between **A** and **B**. Also,

$$\mathbf{A} \bullet \mathbf{B} = A_x B_x + A_y B_y + A_z B_z \tag{C.5}$$

The scalar product is *commutative*, i.e., $\mathbf{A} \bullet \mathbf{B} = \mathbf{B} \bullet \mathbf{A}$. Also note that

$$\mathbf{A} \bullet \mathbf{a}_x = A_x$$
$$\mathbf{A} \bullet \mathbf{a}_y = A_y$$
$$\mathbf{A} \bullet \mathbf{a}_z = A_z$$

C.6.2. Vector (Cross) Product

The vector product between **A** and **B** is a vector represented as **A** × **B** and is given by

$$\mathbf{A}\times\mathbf{B} = |\mathbf{A}||\mathbf{B}||\sin\theta_{AB}|\mathbf{a}_n \tag{C.6}$$

where θ_{AB} is the smaller angle between **A** and **B**, and $\mathbf{a}_n$ is a unit vector normal to the plane containing **A** and **B**. (Since each plane has two normal vectors, it is important to note that $\mathbf{a}_n$ is the one obtained by the right-hand rule. If the fingers of the right hand are extended in the direction of **A** and then curled towards vector **B**, the direction of the outstretched thumb is the direction of $\mathbf{a}_n$.)

Also,

$$\mathbf{A}\times\mathbf{B} = (A_yB_z - A_zB_y)\mathbf{a}_x + (A_zB_x - A_xB_z)\mathbf{a}_y + (A_xB_y - A_yB_x)\mathbf{a}_z \tag{C.7}$$

C.7. THE CYLINDRICAL COORDINATE SYSTEM

While much of our work is carried out conveniently in the familiar rectangular coordinate system (introduced in Sec. C.2), some physical situations have a natural symmetry which makes the cylindrical coordinate system easier to use. Examples include a coaxial cable and an optical fiber.

The cylindrical coordinate system we will use is a natural extension of the two dimensional (xy plane) polar coordinates (ρ,ϕ), to three dimensions (ρ,ϕ,z). Figure C.4 shows the geometric relationship between the rectangular coordinates (x, y, z) of a point P and its cylindrical coordinates (ρ,ϕ,z). You will notice that the z coordinate is common to both systems, while ρ and ϕ are related to x and y as follows

$$\rho = +\sqrt{x^2 + y^2} \qquad \phi = \tan^{-1}\left(\frac{y}{x}\right) \tag{C.8}$$

$$x = \rho\cos\phi \qquad y = \rho\sin\phi, \tag{C.9}$$

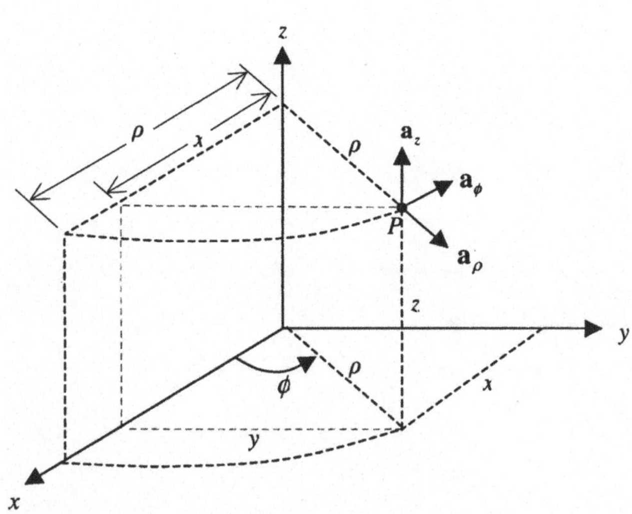

Figure C.4 The cylindrical coordinate system.

where ρ may be thought of as the "*horizontal*" radial distance from the origin to the point P, while ϕ measures the "*azimuthal*" angle from the x axis in a counterclockwise (toward the y axis) direction.

Just as, in the rectangular coordinate system, a point $P(x_1, y_1, z_1)$ corresponds to the intersection of the three mutually orthogonal *planar* surfaces: $x = x_1$, $y = y_1$, $z = z_1$, in the cylindrical system, $P(\rho_1, \phi_1, z_1)$ is located at the intersection of the three orthogonal surfaces:

$$\begin{aligned} \rho &= \rho_1 \quad \text{(cylinder)} \\ \phi &= \phi_1 \quad \text{(plane)} \\ z &= z_1 \quad \text{(plane)} \end{aligned}$$

A vector **A** may be expressed in the cylindrical system as

$$\mathbf{A} = A_\rho \mathbf{a}_\rho + A_\phi \mathbf{a}_\phi + A_z \mathbf{a}_z \tag{C.10}$$

with $|\mathbf{A}| = (A_\rho^2 + A_\phi^2 + A_z^2)^{1/2}$, where $\mathbf{a}_\rho$, $\mathbf{a}_\phi$, and $\mathbf{a}_z$, are mutually orthogonal unit vectors as shown in Fig. C.5. $\mathbf{a}_\rho$ points in the direction of increasing "horizontal" radial distance ρ, $\mathbf{a}_\phi$ also lies in a "horizontal" plane (parallel to the xy plane) and points in the direction of increasing ϕ, and finally $\mathbf{a}_z$ is parallel to the positive z axis (as before). Also note the right-hand rule relationship among $\mathbf{a}_\rho$, $\mathbf{a}_\phi$, and $\mathbf{a}_z$, i.e.,

$$\begin{aligned} \mathbf{a}_\rho \times \mathbf{a}_\phi &= \mathbf{a}_z \\ \mathbf{a}_\phi \times \mathbf{a}_z &= \mathbf{a}_\rho \\ \mathbf{a}_z \times \mathbf{a}_\rho &= \mathbf{a}_\phi \end{aligned} \tag{C.11}$$

A differential volume element dV in the cylindrical coordinate system is

$$dV = (d\rho)(\rho d\phi)(dz) \tag{C.12}$$

Note that $\rho d\phi$ (and not $d\phi$ by itself) represents an incremental distance in the direction of $\mathbf{a}_\phi$, since $d\phi$ is a dimensionless angular measure.

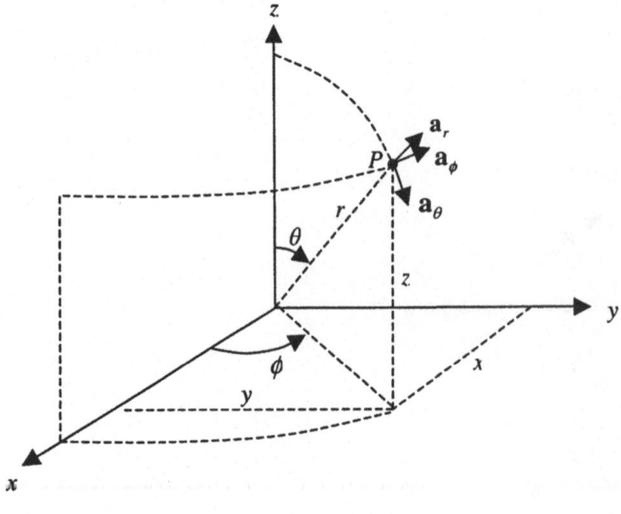

Figure C.5 The spherical coordinate system.

C.8. THE SPHERICAL COORDINATE SYSTEM

If one wishes to analyze the scattering of microwave radar signals from raindrops or the electromagnetic interaction between a cell phone and the human head, the spherical coordinate system may facilitate setting up the mathematical problem. In the spherical coordinate system (Fig. C.5), a point $P(r_1,\theta_1,\phi_1)$ lies at the intersection of the three mutually orthogonal surfaces

$$\begin{aligned} r &= r_1 \quad \text{(sphere)} \\ \theta &= \theta_1 \quad \text{(cone)} \\ \phi &= \phi_1 \quad \text{(plane)} \end{aligned}$$

r represents the three-dimensional distance between the origin and the point, θ ($0 \leqslant \theta < \pi$) is the "*elevation*" angle measured from the positive z axis, and ϕ ($0 \leqslant \phi < 2\pi$) is the "*azimuthal*" angle measured from the positive x axis (as in the cylindrical coordinate system). They are related to (x, y, z) as follows:

$$r = [x^2 + y^2 + z^2)^{1/2} \qquad \theta = \cos^{-1}\left(\frac{z}{\sqrt{x^2 + y^2 + z^2}}\right) \qquad \phi = \tan^{-1}\left(\frac{y}{x}\right) \tag{C.13}$$

$$x = r \sin\theta \cos\phi \qquad y = r \sin\theta \sin\phi \qquad z = r \sin\theta \tag{C.14}$$

A vector $\mathbf{A}$ in the spherical coordinate system is expressed as

$$\mathbf{A} = A_r\mathbf{a}_r + A_\theta\mathbf{a}_\theta + A_\phi\mathbf{a}_\phi \tag{C.15}$$

with $|\mathbf{A}| = (A_r^2 + A_\theta^2 + A_\phi^2)^{1/2}$.

The unit vectors $\mathbf{a}_r$, $\mathbf{a}_\theta$, and $\mathbf{a}_\phi$, are mutually orthogonal and follow the right-hand rule relationship embedded in

$$\begin{aligned} \mathbf{a}_r \times \mathbf{a}_\theta &= \mathbf{a}_\phi \\ \mathbf{a}_\theta \times \mathbf{a}_\phi &= \mathbf{a}_r \\ \mathbf{a}_\phi \times \mathbf{a}_r &= \mathbf{a}_\theta \end{aligned} \tag{C.16}$$

A differential volume element dV is written as

$$dV = (dr)(r\,d\theta)(r \sin\theta\,d\phi) = r^2 \sin\theta\,dr\,d\theta\,d\phi \tag{C.17}$$

C.9. COORDINATE AND VECTOR TRANSFORMATION

In working with the various coordinate systems, we may need to convert parameters given in one coordinate system into parameters in another coordinate system.

Vector Interconversion Strategy

The most common conversions in practice are those between rectangular and cylindrical coordinate systems and those between rectangular and spherical coordinate systems. Both types can be accomplished easily with the help of Table C.1.

Table C.1 Unit Vector Transformation

(a) Dot products of unit vectors (rectangular/cylindrical)

	$\mathbf{a}_\rho$	$\mathbf{a}_\phi$	$\mathbf{a}_z$
$\mathbf{a}_x$	$\cos\phi$	$-\sin\phi$	0
$\mathbf{a}_y$	$\sin\phi$	$\cos\phi$	0
$\mathbf{a}_z$	0	0	1

(b) Dot products of unit vectors (rectangular/spherical)

	$\mathbf{a}_r$	$\mathbf{a}_\theta$	$\mathbf{a}_\phi$
$\mathbf{a}_x$	$\sin\theta\cos\phi$	$\cos\theta\cos\phi$	$-\sin\phi$
$\mathbf{a}_y$	$\sin\theta\sin\phi$	$\cos\theta\sin\phi$	$\cos\phi$
$\mathbf{a}_z$	$\cos\theta$	$-\sin\theta$	0

Example C.1. Vector transformation (rectangular to cylindrical): Convert $\mathbf{A} = z\mathbf{a}_x + x\mathbf{a}_y$ to cylindrical coordinates.

We start by writing $\mathbf{A} = A_\rho\mathbf{a}_\rho + A_\phi\mathbf{a}_\phi + A_z\mathbf{a}_z$. Then

$$\begin{aligned} A_\rho &= \mathbf{A}\bullet\mathbf{a}_\rho \\ &= (z\mathbf{a}_x + x\mathbf{a}_y)\bullet\mathbf{a}_\rho \\ &= z(\mathbf{a}_x\bullet\mathbf{a}_\rho) + x(\mathbf{a}_y\bullet\mathbf{a}_\rho) \\ &= z\cos\phi + (\rho\cos\phi)(\sin\phi) \end{aligned}$$

$$\begin{aligned} A_\phi &= \mathbf{A}\bullet\mathbf{a}_\phi \\ &= (z\mathbf{a}_x + x\mathbf{a}_y)\bullet\mathbf{a}_\phi \\ &= z(\mathbf{a}_x\bullet\mathbf{a}_\phi) + x(\mathbf{a}_y\bullet\mathbf{a}_\phi) \\ &= z(-\sin\phi) + (\rho\cos\phi)(\cos\phi) \end{aligned}$$

$$\begin{aligned} A_z &= \mathbf{A}\bullet\mathbf{a}_z \\ &= (z\mathbf{a}_x + x\mathbf{a}_y)\bullet\mathbf{a}_z \\ &= z(\mathbf{a}_x\bullet\mathbf{a}_z) + x(\mathbf{a}_y\bullet\mathbf{a}_z) \\ &= z\cdot 0 + x\cdot 0 \end{aligned}$$

Therefore,

$$\mathbf{A} = (z\cos\phi + \rho\cos\phi\sin\phi)\mathbf{a}_\rho + (-z\sin\phi + \rho\cos^2\phi)\mathbf{a}_\phi$$

Example C.2. Vector transformation (spherical to rectangular): Convert $\mathbf{E} = E_o/(r^2)\ \mathbf{a}_r$ into rectangular coordinates.

We start by writing $\mathbf{E} = E_x\mathbf{a}_x + E_y\mathbf{a}_y + E_z\mathbf{a}_z$. Then,

$$\begin{aligned} E_x &= \mathbf{E}\bullet\mathbf{a}_x = \left(\frac{E_o}{r^2}\right)(\mathbf{a}_r\bullet\mathbf{a}_x) = \frac{E_o}{r^2}\sin\theta\cos\phi = \frac{E_o}{r^2}\,\frac{r\sin\theta\cos\phi}{r} \\ &= \frac{E_o}{r^3}r\sin\theta\cos\phi = \frac{E_o x}{(x^2+y^2+z^2)^{3/2}} \end{aligned}$$

Similarly,

$$E_y = \frac{E_o y}{(x^2 + y^2 + z^2)^{3/2}} \quad \text{and} \quad E_z = \frac{E_o z}{(x^2 + y^2 + z^2)^{3/2}}$$

C.10. VECTOR DIFFERENTIAL OPERATORS

Maxwell equations and the associated relationships are expressed in terms of vector differential operators. Therefore, we have tabulated the expressions for the various differential operators in all the coordinate systems below:

C.10.1. Divergence

Rectangular

$$\nabla \bullet \mathbf{D} = \frac{\partial D_x}{\partial x} + \frac{\partial D_y}{\partial y} + \frac{\partial D_z}{\partial z}$$

Cylindrical

$$\nabla \bullet \mathbf{D} = \frac{1}{\rho}\frac{\partial}{\partial \rho}(\rho D\rho) + \frac{1}{\rho}\frac{\partial D_\phi}{\partial \phi} + \frac{\partial D_z}{\partial z}$$

Spherical

$$\nabla \bullet \mathbf{D} = \frac{1}{r^2}\frac{\partial}{\partial r}(r^2 D_r) + \frac{1}{r \sin\theta}\frac{\partial}{\partial \theta}(D_\theta \sin\theta) + \frac{1}{r \sin\theta}\frac{\partial D_\phi}{\partial \phi}$$

C.10.2. Gradient

Rectangular

$$\nabla V = \frac{\partial V}{\partial x}\mathbf{a}_x + \frac{\partial V}{\partial y}\mathbf{a}_y + \frac{\partial V}{\partial z}\mathbf{a}_z$$

Cylindrical

$$\nabla V = \frac{\partial V}{\partial \rho}\mathbf{a}_\rho + \frac{1}{\rho}\frac{\partial V}{\partial \phi}\mathbf{a}_\phi + \frac{\partial V}{\partial z}\mathbf{a}_z$$

Spherical

$$\nabla V = \frac{\partial V}{\partial r}\mathbf{a}_r + \frac{1}{r}\frac{\partial V}{\partial \theta}\mathbf{a}_\theta + \frac{1}{r \sin\theta}\frac{\partial V}{\partial \phi}\mathbf{a}_\phi$$

C.10.3. Curl

Rectangular

$$\nabla \times \mathbf{E} = \left(\frac{\partial E_z}{\partial y} - \frac{\partial E_y}{\partial z} \right) \mathbf{a}_x + \left(\frac{\partial E_x}{\partial z} - \frac{\partial E_z}{\partial x} \right) \mathbf{a}_y + \left(\frac{\partial E_y}{\partial x} - \frac{\partial E_x}{\partial y} \right) \mathbf{a}_z$$

Cylindrical

$$\nabla \times \mathbf{E} = \left(\frac{1}{\rho} \frac{\partial E_z}{\partial \phi} - \frac{\partial E_\phi}{\partial z} \right) \mathbf{a}_\rho + \left(\frac{\partial E_\rho}{\partial z} - \frac{\partial E_z}{\partial \rho} \right) \mathbf{a}_\phi + \frac{1}{\rho} \left(\frac{\partial}{\partial \rho} \rho E_\phi - \frac{\partial E_\rho}{\partial \phi} \right) \mathbf{a}_z$$

Spherical

$$\nabla \times \mathbf{E} = \frac{1}{r \sin\theta} \left[\frac{\partial}{\partial \theta} (E_\phi \sin\theta) - \frac{\partial E_\theta}{\partial \phi} \right] \mathbf{a}_r + \frac{1}{r} \left(\frac{1}{\sin\theta} \frac{\partial E_r}{\partial \phi} - \frac{\partial}{\partial r} r E_\phi \right) \mathbf{a}_\theta + \frac{1}{r} \left(\frac{\partial}{\partial r} r E_\theta - \frac{\partial E_r}{\partial \theta} \right) \mathbf{a}_\phi$$

Index